Lecture Notes in Computer Science 6690

Commenced Publication in 1973
Founding and Former Series Editors:
Gerhard Goos, Juris Hartmanis, and Jan van Leeuwen

Advanced Research in Computing and Software Science

Subline of Lectures Notes in Computer Science

W0192950

Luke Ong (Ed.)

Typed Lambda Calculi and Applications

10th International Conference, TLCA 2011
Novi Sad, Serbia, June 1-3, 2011
Proceedings

 Springer

Volume Editor

Luke Ong
University of Oxford
Department of Computer Science
Wolfson Building, Parks Road
Oxford OX1 3QD, UK
E-mail: lo@comlab.ox.ac.uk

ISSN 0302-9743   e-ISSN 1611-3349
ISBN 978-3-642-21690-9   ISBN 978-3-642-21691-6   (eBook)
DOI 10.1007/978-3-642-21691-6
Springer Heidelberg Dordrecht London New York

Library of Congress Control Number: 2011928922

CR Subject Classification (1998): D.1.6, D.3.2, F.3, F.4, I.2.3

LNCS Sublibrary: SL 1 – Theoretical Computer Science and General Issues

Typesetting: Camera-ready by author, data conversion by Scientific Publishing Services, Chennai, India

Printed on acid-free paper

Springer is part of Springer Science+Business Media (www.springer.com)

Preface

This volume contains the papers of the 10th International Conference on Typed Lambda Calculi and Applications (TLCA 2011), which was held during 1-3 June 2011, in Novi Sad, Serbia. The Conference was part of the 6th International Conference on Rewriting, Deduction, and Programming (RDP 2011), together with the 22nd International Conference on Rewriting Techniques and Applications (RTA 2011), the 10th International Workshop on Reduction Strategies in Rewriting and Programming (WRS 2011), the Workshop on Compilers by Rewriting, Automated (COBRA 2011), the Workshop on Theory and Practice of Delimited Continuations (TPDC 2011), the Workshop on the Two Faces of Complexity (2FC 2011), and the annual meeting of the IFIP Working Group 1.6 on Term Rewriting.

The TLCA series of conferences serves as a forum for presenting original research results that are broadly relevant to the theory and applications of lambda calculus. Previous TLCA conferences were held in Utrecht (1993), Edinburgh (1995), Nancy (1997), L'Aquila (1999), Kraków (2001), Valencia (2003), Nara (2005), Paris (2007) and Brasília (2009).

A total of 15 papers were accepted out of 44 submissions for presentation at TLCA 2011 and for inclusion in the proceedings. Each submitted paper was reviewed by at least three members of the Programme Committee, who were assisted in their work by 67 external reviewers. I would like to thank the members of the Programme Committee and the external reviewers for their review work, as well as Andrei Voronkov for providing the EasyChair system which proved invaluable throughout the review process and the preparation of this volume. In addition to the contributed papers, the TLCA programme contained invited lectures by Alexandre Miquel, Vladimir Voevodsky and Stephanie Weirich. The RDP 2011 programme also included invited lectures by Sophie Tison and Ashish Tiwari.

Many people helped to make TLCA 2011 a success. I would like to thank the RDP'11 Conference Chair Silvia Ghilezan, the Organising Committee, the local organising team, TLCA Publicity Chair Luca Paolini, the TLCA Steering Committee and our sponsors: Ministry of Education and Science, Republic of Serbia; Provincial Secretariat of Science and Technological Development, Province of Vojvodina; City of Novi Sad; RT-RK Computer Based Systems; FIMEK, Faculty of Economy and Engineering Management; Telvent DMS LLC Novi Sad; Embassy of France in Serbia; and Embassy of the USA in Serbia.

March 2011   Luke Ong

TLCA 2011 Organisation

Programme Chair

Luke Ong — University of Oxford, UK

Programme Committee

Thorsten Altenkirch	University Nottingham, UK
Stefano Beradi	University of Turin, Italy
Adriana Compagnoni	Stevens Institute of Technology, New Jersey, USA
Giles Dowek	INRIA, Paris, France
Silvia Ghilezan	University of Novi Sad, Serbia
Hugo Herbelin	INRIA, Paris, France
Atsushi Igarashi	Kyoto University, Japan
Ranjit Jhala	University of California, San Diego, USA
Ugo dal Lago	University of Bologna, Italy
Ralph Matthes	CNRS, IRIT, France
Luke Ong	University of Oxford, UK
Rick Statman	Carnegie-Mellon University, Pittsburgh, USA
Tachio Terauchi	Tohoku University, Sendai, Japan
Nobuko Yoshida	Imperial College, London, UK

Conference Chair

Silvia Ghilezan — University of Novi Sad, Serbia

Organising Committee

Siniša Crvenković	University of Novi Sad, Serbia
Ilija Ćosić	University of Novi Sad, Serbia
Kosta Došen	Mathematical Institute, Belgrade, Serbia
Silvia Ghilezan	University of Novi Sad, Serbia
Predrag Janičić	University of Belgrade, Serbia
Zoran Marković	Mathematical Institute, Belgrade, Serbia
Zoran Ognjanović	Mathematical Institute, Belgrade, Serbia
Jovanka Pantović	University of Novi Sad, Serbia
Zoran Petrić	Mathematical Institute, Belgrade, Serbia
Miroslav Popović	University of Novi Sad, Serbia
Nataša Sladoje	University of Novi Sad, Serbia
Miroslav Vesković	University of Novi Sad, Serbia

TLCA Steering Committee Chair

Mariangiola Dezani-Ciancaglini   University of Turin, Italy

TLCA Steering Committee

Samson Abramsky   University of Oxford, UK
Henk Barendregt   Radboud University, Nijmegen,
                               The Netherlands
Pierre-Louis Curien   CNRS – Université Paris 7, France
Mariangiola Dezani-Ciancaglini   University of Turin, Italy
Roger Hindley   Swansea University, UK
Martin Hofmann   Ludwig-Maximilians-Universität, Munich
                               Germany
Simona Ronchi Della Rocca   University of Turin, Italy
Paweł Urzyczyn   University of Warsaw, Poland

TLCA Publicity Chair

Luca Paolini   University of Turin, Italy

External Reviewers

Federico Aschieri   Healfdene Goguen
Franco Barbanera   Jonathan Grattage
Bruno Barras   Ryu Hasegawa
Alexis Bernadet   Furio Honsell
Malgorzata Biernacka   David Hopkins
Paola Bruscoli   Mauro Jaskelioff
James Chapman   Yukiyoshi Kameyama
Pierre Clairambault   Delia Kesner
Robin Cockett   Pierre Lescanne
Tristan Crolard   Silvia Likavec
Pierre-Louis Curien   Peter Lefanu Lumsdaine
Philippe De Groote   Harry Mairson
Romain Demangeon   Simone Martini
Pierre-Malo Denilou   Damiano Mazza
Pietro Di Gianantonio   Conor Mcbride
Alessandra Di Pierro   Yasuhiko Minamide
Claudia Faggian   Aybek Mukhamedov
Andrzej Filinski   Guillaume Munch-Maccagnoni
Luca Fossati   Hiroshi Nakano
Marco Gaboardi   Keiko Nakata
Jacques Garrigue   Koji Nakazawa

Aleksandar Nanevski
Robin Neatherway
Paulo Oliva
Michele Pagani
Christophe Raffalli
Laurent Regnier
Claudio Sacerdoti Coen
Antonino Salibra
Carsten Schuermann
Matthieu Sozeau
Yiorgos Stavrinos
Lutz Strassburger
Thomas Streicher

Izumi Takeuti
Ross Tate
Kazushige Terui
Takeshi Tsukada
Nikos Tzevelekos
PawełUrzyczyn
Steffen Van Bakel
Femke Van Raamsdonk
Marek Zaionc
Noam Zeilberger
Margherita Zorzi
Dragisa Zunić.

Table of Contents

Abstracts of Invited Lectures

Contributed Papers

A Survey of Classical Realizability

Alexandre Miquel

ENS de Lyon, France
alexandre.miquel@ens-lyon.fr

The theory of classical realizability was introduced by Krivine [Kri09] in the middle of the 90's in order to analyze the computational contents of classical proofs, following the connection between classical reasoning and control operators discovered by Griffin [Gri90]. More than an extension of the theory of intuitionistic realizability, classical realizability is a complete reformulation of the very principles of realizability based on a combination [OS08, Miq10] of Kleene's realizability [Kle45] with Friedman's A-translation [Fri78].

One of the most interesting aspects of the theory is that it is highly modular: although it was originally developed for second order arithmetic, classical realizability naturally extends to more expressive frameworks such as Zermelo-Fraenkel set theory [Kri01] or the calculus of constructions with universes [Miq07]. Moreover, the underlying language of realizers can be freely enriched with new instructions in order to realize extra reasoning principles, such as the axiom of dependent choices [Kri03]. More recently, Krivine [Kri10] proposed an ambitious research programme, whose aim is to unify the methods of classical realizability with Cohen's forcing [Coh63, Coh64] in order to extract programs from proofs obtained by the forcing technique.

In this talk, I shall survey the methods and the main results of classical realizability, while presenting some of its specific problems, such as the specification problem. I shall also discuss the perspectives of combining classical realizability with the technique of forcing, showing that this combination is not only interesting from a pure model theoretic point of view, but that it can also bring new insights about a possible widening of the proofs-as-programs correspondence beyond the limits of pure functional programming.

References

[Coh63]   Cohen, P.J.: The independence of the continuum hypothesis. Proceedings of the National Academy of Sciences of the United States of America 50(6), 1143–1148 (1963)

[Coh64]   Cohen, P.J.: The independence of the continuum hypothesis II. Proceedings of the National Academy of Sciences of the United States of America 51(1), 105–110 (1964)

[Fri78]   Friedman, H.: Classically and intuitionistically provably recursive functions. Higher Set Theory 699, 21–28 (1978)

[Gri90]   Griffin, T.: A formulae-as-types notion of control. In: Principles of Programming Languages (POPL 1990), pp. 47–58 (1990)

[Kle45]   Kleene, S.C.: On the interpretation of intuitionistic number theory. Journal of Symbolic Logic 10, 109–124 (1945)

L. Ong (Ed.): TLCA 2011, LNCS 6690, pp. 1–2, 2011.
© Springer-Verlag Berlin Heidelberg 2011

[Kri01] Krivine, J.-L.: Typed lambda-calculus in classical Zermelo-Fraenkel set theory. Arch. Math. Log. 40(3), 189–205 (2001)

[Kri03] Krivine, J.-L.: Dependent choice, 'quote' and the clock. Theor. Comput. Sci. 308(1-3), 259–276 (2003)

[Kri09] Krivine, J.-L.: Realizability in classical logic. In: Interactive Models of Computation and Program Behaviour. Panoramas et synthèses, vol. 27, pp. 197–229. Société Mathématique de France (2009)

[Kri10] Krivine, J.-L.: Realizability algebras: a program to well order R. CoRR, abs/1005.2395 (2010)

[Miq07] Miquel, A.: Classical program extraction in the calculus of constructions. In: Duparc, J., Henzinger, T.A. (eds.) CSL 2007. LNCS, vol. 4646, pp. 313–327. Springer, Heidelberg (2007)

[Miq10] Miquel, A.: Existential witness extraction in classical realizability and via a negative translation. Logical Methods for Computer Science (2010)

[OS08] Oliva, P., Streicher, T.: On Krivine's realizability interpretation of classical second-order arithmetic. Fundam. Inform. 84(2), 207–220 (2008)

Tree Automata, (Dis-)Equality Constraints and Term Rewriting*
What's New?

Sophie Tison

University Lille 1, Mostrare project,
INRIA Lille Nord-Europe & LIFL (CNRS UMR8022)

Abstract. Connections between Tree Automata and Term Rewriting are now well known. Whereas tree automata can be viewed as a subclass of ground rewrite systems, tree automata are successfully used as decision tools in rewriting theory. Furthermore, applications, including rewriting theory, have influenced the definition of new classes of tree automata.

In this talk, we will first present a short and not exhaustive reminder of some fruitful applications of tree automata in rewriting theory. Then, we will focus on extensions of tree automata, specially tree automata with local or/and global (dis-)equality constraints: we will emphasize new results, compare different extensions, and sketch some applications.

Tree automata can be used in numerous ways as decision tools in rewriting theory. An ideal way is to encode the reducibility relation by a recognizable relation. Of course, recognizability of the reducibility relation is a very strong requirement which limits this approach to very restricted subclasses of term rewriting systems. A weaker requirement is that the rewrite relation preserves recognizability. This approach is a key point in reachability analysis and tree regular model checking. More generally, encoding set of descendants of terms and possibly sets of normal forms by tree automata provide canonical techniques to obtain decidability results and a lot of work has been done to characterize subclasses of term rewriting systems having "good recognizability properties".

Even powerful, these methods remain restricted. To enhance their power, numerous works have been developed, e.g. in reachability analysis, by using abstract interpretation or over-approximations. Other works have recently focused on using rewrite strategies. An other approach is the use of extended tree automata. E.g., equational tree automata have been proposed to handle equational theories end several extensions have been defined to take into account associativity.

This talk will focus on new results about extensions of tree automata with equality and disequality constraints.

---

* *This talk will be strongly inspired by recent works of several researchers, specially but not exclusively:* Emmanuel Filiot, Guillem Godoy, Florent Jacquemard, Jean-Marc Talbot, Camille Vacher.

L. Ong (Ed.): TLCA 2011, LNCS 6690, pp. 3–5, 2011.

From Local Constraints . . .

A typical example of language which is not recognized by a finite tree automaton is the set of ground instances of a non-linear term. This implies of course that the set of ground normal forms of a t.r.s. is not necessarily recognizable. An other point is that images by a non-linear morphism of a recognizable tree language are not necessarily recognizable. E.g. if a morphism associates $h(x)$ with $f(x, x)$ and a with b, the image by this morphism of $h^*(a)$ is the non-recognizable set of well-balanced terms over $\{f, b\}$. So, extending tree automata to handle non-linearity is very natural and in the early 80's, M. Dauchet and J. Mongy have proposed a new class for this purpose, by enriching tree automata rules by equality constraints. E.g. if a rule is associated with the constraint $1.1 = 1.2$, it can be applied at position p in t only if the subterms at positions $p.1.1$ and $p.1.2$ are equal. A more general class can be defined by allowing also disequality constraints. Unfortunately, emptiness is undecidable even when only equality constraints are allowed.

Several restrictions of this class have so been studied (see e.g. a survey in [2]). Let us remind two of them, which are of special interest for term rewriting. The first one, the class of automata with constraints between brothers, restricts equality and disequality tests to sibling positions. This class has good decidability and closure properties. E.g., this class allows to represent normal forms for left-shallow t.r.s. -but not for general ones - and it helped recently for providing new decidability results for normalization. The second one, the class of reduction automata, bounds, roughly speaking, the number of equality constraints. This class has provided the decidability of the reducibility theory and as a corollary a (new) way of deciding ground reducibility.

Recently, a strong work by Godoy & alt. [4] has given some new emphasis on these classes. Indeed, it defines some new subclasses having good properties. This enables them to decide whether the homomorphic image of a tree recognizable language is recognizable. As a corollary, they get a new simple proof of decidability of recognizability of the set of normal forms of a t.r.s..

. . . to Global Ones

A new approach has been recently proposed : adding constraints to perform (dis-)equality tests, but globally. The idea is to enrich the automaton by two relations, $=, \neq$, over the states. Roughly speaking, the run will be correct if the subterms associated with two "equal" (resp. "not equal") states are equal (resp. different). E.g. this approach enables to check that all the subterms rooted by a f are equal or to encode that every identifier is different. This approach has led to (almost) simultaneous definitions of classes by different researchers to different purposes: rigid tree automata [5], tree automata with global equality and disequality tests (TAGED) [3], tree automata with global constraints [1]. The second part of this talk will give an overview of these classes and sketch their links with other classes.

Here follows a very incomplete list of references the talk will rely upon:

References

1. Barguñó, L., Creus, C., Godoy, G., Jacquemard, F., Vacher, C.: The emptiness problem for tree automata with global constraints. In: Proceedings of the 25th Annual IEEE Symposium on Logic in Computer Science, LICS, pp. 263–272. IEEE Computer Society, Los Alamitos (2010)
2. Comon, H., Dauchet, M., Gilleron, R., Löding, C., Jacquemard, F., Lugiez, D., Tison, S., Tommasi, M.: Tree automata techniques and applications (2007), http://www.grappa.univ-lille3.fr/tata release (October 12, 2007)
3. Filiot, E., Talbot, J.-M., Tison, S.: Tree automata with global constraints. Int. J. Found. Comput. Sci. 21(4), 571–596 (2010)
4. Godoy, G., Giménez, O., Ramos, L., Àlvarez, C.: The hom problem is decidable. In: Schulman, L.J. (ed.) Proceedings of the 42nd ACM Symposium on Theory of Computing, STOC 2010, pp. 485–494 (2010)
5. Jacquemard, F., Klay, F., Vacher, C.: Rigid tree automata and applications. Inf. Comput. 209(3), 486–512 (2011)

Rewriting in Practice*

Ashish Tiwari

SRI International, Menlo Park, CA 94025
tiwari@csl.sri.com

The field of rewriting is broadly concerned with manipulating *representations* of *objects* so that we go from a larger representation to a *smaller* representation. The field of rewriting has contributed some fundamental results within the computer science discipline. This extended abstract explores a few impactful applications of rewriting in the areas of (a) design of algorithms, (b) formal modeling and analysis, and (c) term rewriting and theorem proving.

The theory of rewriting can be used as a basis to study the design and analysis of algorithms in at least two distinct ways. First, an algorithm can often be viewed as a set of rewrite rules. A *rewrite-based description* of algorithms cleanly separates the logical part of an algorithm from the implementation details. This applies to large class of algorithms including graph algorithms, sorting algorithms, and combinatorial algorithms. Second, rewriting provides a general paradigm for the design of algorithms. The (abstract) critical-pair completion algorithm is a generic procedure that can be instantiated in different domains to yield very important algorithms, such as, the algorithms implementing the union-find data structure [5], (associative-commutative) congruence closure [5], Gröbner basis algorithm [6,2,4], and the Simplex algorithm for linear constraints [8,9]. Apart from simplifying the correctness proofs, completion-based view of algorithms makes it possible to inherit certain optimizations.

The essence of the theory of rewriting, and in particular of the critical-pair completion procedure, can be described as: (a) add facts that make proofs of provable facts smaller and (b) delete facts that already have smaller proofs. This more general view of completion inspires the design of several other algorithms; most notably the algorithms used in saturation-based theorem proving [3]. In equational reasoning, proofs have a certain nice structure that admits interesting proof orderings, leading to algorithms such as standard and ordered completion [1]. For non-equational theories, proofs may not have a very nice structure and complexity of a proof is just the complexity of the facts used in the proof.

The view of rewriting as a paradigm for generating small facts is helpful not only for first-order theorem proving (where $\perp$ is the small fact of interest), but also for solving other problems, such as testing unsatisfiability of linear constraints [9] and nonlinear constraints [18], and generating (equational) invariants of a continuous dynamical system.

Rewriting provides both a language for formal modeling of systems, as well as a tool for simulating and analyzing the formal models [10]. A biochemical

* Research supported in part by the National Science Foundation under grant CSR-EHCS-0834810 and CSR-0917398.

reaction is naturally a rewrite rule [7,11,17]. Apart from the usual questions about reachability, there is also interest in characterizing certain kinds of *steady state* behaviors of rewrite systems that model biological processes. A steady state behavior is a cyclic derivation. Flux balance analysis is a commonly used technique for finding such steady state behaviors of Petrinets [15,19]. Systems Biology [13,14,16] also motivates a study of probabilistic rewrite systems, stochastic Petrinets, and Gillespie's stochastic simulation algorithm [12] for simulating certain timed stochastic rewrite systems. Biology also poses other challenges: *learning* rewrite rules or models from available data on rewrite derivations. The data could be of disease propagation in humans, and one could learn models of disease propagation and develop therapeutics based on the learned model.

References

1. Bachmair, L.: Canonical Equational Proofs. Birkhäuser, Basel (1991)
2. Bachmair, L., Ganzinger, H.: Buchberger's algorithm: A constraint-based completion procedure. In: Jouannaud, J.-P. (ed.) CCL 1994. LNCS, vol. 845. Springer, Heidelberg (1994)
3. Bachmair, L., Ganzinger, H.: Rewrite-based equational theorem proving with selection and simplification. J. of Logic and Computation 4, 217–247 (1994)
4. Bachmair, L., Tiwari, A.: D-bases for polynomial ideals over commutative noetherian rings. In: Ganzinger, H. (ed.) RTA 1996. LNCS, vol. 1103, pp. 113–127. Springer, Heidelberg (1996)
5. Bachmair, L., Tiwari, A., Vigneron, L.: Abstract congruence closure. J. of Automated Reasoning 31(2), 129–168 (2003)
6. Buchberger, B.: A critical-pair completion algorithm for finitely generated ideals in rings. In: Börger, E., Rödding, D., Hasenjaeger, G. (eds.) Rekursive Kombinatorik 1983. LNCS, vol. 171, pp. 137–161. Springer, Heidelberg (1984)
7. Danos, V., Feret, J., Fontana, W., Harmer, R., Krivine, J.: Rule-based modelling of cellular signalling. In: Caires, L., Vasconcelos, V.T. (eds.) CONCUR 2007. LNCS, vol. 4703, pp. 17–41. Springer, Heidelberg (2007)
8. Detlefs, D., Nelson, G., Saxe, J.B.: Simplify: A theorem prover for program checking. J. of the ACM 52(3), 365–473 (2005)
9. Dutertre, B., de Moura, L.: A fast linear-arithmetic solver for DPLL(T). In: Ball, T., Jones, R.B. (eds.) CAV 2006. LNCS, vol. 4144, pp. 81–94. Springer, Heidelberg (2006)
10. Clavel, M., et al.: Maude: Specification and Programming in Rewriting Logic. In: SRI International, Menlo Park, CA (1999), http://maude.csl.sri.com/manual/
11. Hlavacek, W.S., et al.: Rules for modeling signal-transduction systems. Sci. STKE 344 (2006), PMID: 16849649
12. Gillespie, D.T.: A general method for numerically simulating the stochastic time evolution of coupled chemical reactions. J. Comp. Physics 22, 403–434 (1976)
13. Kitano, H.: Systems biology: A brief overview. Science 295, 1662–1664 (2002)
14. Lincoln, P., Tiwari, A.: Symbolic systems biology: Hybrid modeling and analysis of biological networks. In: Alur, R., Pappas, G.J. (eds.) HSCC 2004. LNCS, vol. 2993, pp. 660–672. Springer, Heidelberg (2004)
15. Orth, J.D., Thiele, I., Palsson, B.O.: What is flux balance analysis? Nature Biotechnology 28, 245–248 (2010)

16. Priami, C.: Algorithmic systems biology. CACM 52(5), 80–88 (2009)
17. Talcott, C.L.: Pathway logic. In: Bernardo, M., Degano, P., Tennenholtz, M. (eds.) SFM 2008. LNCS, vol. 5016, pp. 21–53. Springer, Heidelberg (2008)
18. Tiwari, A.: An algebraic approach for the unsatisfiability of nonlinear constraints. In: Ong, L. (ed.) CSL 2005. LNCS, vol. 3634, pp. 248–262. Springer, Heidelberg (2005)
19. Tiwari, A., Talcott, C., Knapp, M., Lincoln, P., Laderoute, K.: Analyzing pathways using SAT-based approaches. In: Anai, H., Horimoto, K., Kutsia, T. (eds.) Ab 2007. LNCS, vol. 4545, pp. 155–169. Springer, Heidelberg (2007)

Combining Proofs and Programs

Stephanie Weirich

University of Pennsylvania, Philadelphia, USA
sweirich@cis.upenn.edu

Programming languages based on dependent type theory promise two great advances: *flexibility* and *security*. With the type-level computation afforded by dependent types, algorithms can be more generic, as the type system can express flexible interfaces via programming. Likewise, type-level computation can also express data structure invariants, so that programs can be proved correct through type checking. Furthermore, despite these extensions, programmers already know everything. Via the Curry-Howard isomorphism, the language of type-level computation and the verification logic is the programming language itself.

There are two current approaches to the design of dependently-typed languages: Coq, Epigram, Agda, which grew out of the logics of proof assistants, require that all expressions terminate. These languages provide decidable type checking and strong correctness guarantees. In contrast, functional programming languages, like Haskell and Ωmega, have adapted the features dependent type theories, but retain a strict division between types and programs. These languages trade termination obligations for more limited correctness assurances.

In this talk, I present a work-in-progress overview of the TRELLYS project. TRELLYS is new core language, designed to provide a smooth path from functional programming to dependently-typed programming. Unlike traditional dependent type theories and functional languages, TRELLYS allows programmers to work with total and partial functions uniformly. The language itself is composed of two fragments that share a common syntax and overlapping semantics: a simple logical language that guarantees total correctness and an expressive call-by-value programming language that guarantees types safety but not termination.

Importantly, these two fragments interact. The logical fragment may soundly reason about effectful, partial functions. Program values may be used as evidence by the logic. We call this principle *freedom of speech*: whereas proofs themselves must terminate, they must be allowed to reason about any function a programmer might write. To retain consistency, the TRELLYS type system keeps track of where potentially non-terminating computations may appear, so that it can prevent them from being used as proofs.

Acknowledgements. TRELLYS is a collaborative project between the University of Pennsylvania, the University of Iowa and Portland State University supported by the National Science Foundation (award 0910510). This talk is based on joint work with Aaron Stump, Tim Sheard, Chris Casinghino, Vilhelm Sjöberg, Brent Yorgey, Harley D. Eades III, Garrin Kimmel, and Nathan Collins.

L. Ong (Ed.): TLCA 2011, LNCS 6690, p. 9, 2011.
© Springer-Verlag Berlin Heidelberg 2011

Higher-Order Dynamic Pattern Unification for Dependent Types and Records

Andreas Abel[1] and Brigitte Pientka[2]

[1] Institut für Informatik, Ludwig-Maximilians-Universität, München, Deutschland
andreas.abel@ifi.lmu.de
[2] School of Computer Science, McGill University, Montreal, Canada
bpientka@cs.mcgill.ca

Abstract. While higher-order pattern unification for the λ^Π-calculus is decidable and unique unifiers exists, we face several challenges in practice: 1) the pattern fragment itself is too restrictive for many applications; this is typically addressed by solving sub-problems which satisfy the pattern restriction eagerly but delay solving sub-problems which are non-patterns until we have accumulated more information. This leads to a dynamic pattern unification algorithm. 2) Many systems implement $\lambda^{\Pi\Sigma}$ calculus and hence the known pattern unification algorithms for λ^Π are too restrictive.

In this paper, we present a constraint-based unification algorithm for $\lambda^{\Pi\Sigma}$-calculus which solves a richer class of patterns than currently possible; in particular it takes into account type isomorphisms to translate unification problems containing Σ-types into problems only involving Π-types. We prove correctness of our algorithm and discuss its application.

1 Introduction

Higher-order unification is a key operation in logical frameworks, dependently-typed programming systems, or proof assistants supporting higher-order logic. It plays a central role in type inference and reconstruction algorithms, in the execution of programs in higher-order logic programming languages, and in reasoning about the totality of functions defined by pattern-matching clauses.

While full higher-order unification is undecidable [7], Miller [9] identified a decidable fragment of higher-order unification problems, called the *pattern* fragment. A pattern is a unification problem where all meta-variables (or logic variables) occurring in a term are applied to some distinct bound variables. For example, the problem $\lambda x\,y\,z.\,X\,x\,y = \lambda x\,y\,z.\,x\,(\mathsf{suc}\,y)$ falls into the pattern fragment, because the meta-variable X is applied to distinct bound variables x and y; the pattern condition allows us to solve the problem by a simple abstraction $X = \lambda x\,y.\,x\,(\mathsf{suc}\,y)$. This is not possible for non-patterns; examples for non-pattern problems can be obtain by changing the left hand side of the previous problem to $\lambda x\,y\,z.\,X\,x\,x\,y$ (non-linearity), $\lambda x\,y\,z.\,X\,(Y\,x)\,y$ (X applied to another meta-variable) or $\lambda x\,y\,z.\,X\,x\,(\mathsf{suc}\,y)$ (X applied to non-variable term).

In practice we face several challenges: First, the pattern fragment is too restrictive for many applications. Systems such as Twelf [12], Beluga [14], and

L. Ong (Ed.): TLCA 2011, LNCS 6690, pp. 10–26, 2011.
© Springer-Verlag Berlin Heidelberg 2011

Delphin [15] solve eagerly sub-problems which fall into the pattern fragment and delay sub-problems outside the pattern fragment until more information has been gathered which in turn simplifies the delayed sub-problems. The meta-theory justifying the correctness of such a strategy is largely unexplored and complex (an exception is the work by Reed [16]).

Second, we often want to consider richer calculi beyond the λ^Π-calculus. In Beluga and Twelf for example we use Σ-types to group assumptions together. In Agda [11] we support Σ-types in form of records with associated η-equality in its general form. Yet, little work has been done on extending the pattern fragment to handle also Σ-types. The following terms may be seen as equivalent: (a) $\lambda y_1.\lambda y_2. X\,(y_1, y_2)$, (b) $\lambda y.\,X\,(\text{fst }y)\,(\text{snd }y)$ and (c) $\lambda y_1.\lambda y_2. X\,y_1\,y_2$. Only the last term falls within the pattern fragment as originally described by Miller. However, the other two terms can be transformed such that they also fall into the pattern fragment: for term (a), we replace X with $\lambda y.\,X'\,(\text{fst }y)\,(\text{snd }y)$; for term (b), we unfold y which stands for a pair and replace y with (y_1, y_2).

In this paper, we describe a higher-order unification algorithm for the $\lambda^{\Pi\Sigma}$ calculus; our algorithm handles lazily η-expansion and we translate terms into the pure pattern fragment where a meta-variable is applied to distinct bound variables. The key insight is to take into account type isomorphisms for Σ, the dependently typed pairs: $\Pi z{:}(\Sigma x{:}A.B).C$ is isomorphic to $\Pi x{:}A.\Pi y{:}B.[(x,y)/z]C$, and a function $f{:}\Pi x{:}A.\Sigma y{:}B.C$ can be translated into two functions $f_1 : \Pi x{:}A.B$ and $f_2 : \Pi x{:}A.[f_1\,x/y]C$. These transformations allow us to handle a richer class of dependently-typed patterns than previously considered.

Following Nanevski et al. [10] and Pientka [13], our description takes advantage of modelling meta-variables as closures; instead of directly considering a meta-variable X at function type $\Pi \boldsymbol{x}{:}\boldsymbol{A}.B$ which is applied to $\boldsymbol{x}$, we describe them as contextual objects, i.e., objects of type B in a context $\boldsymbol{x}{:}\boldsymbol{A}$, which are associated with a delayed substitution for the local context $\boldsymbol{x}{:}\boldsymbol{A}$.[1] This allows us to give a high-level description and analysis following Dowek et al. [2], but not resorting to explicit substitutions; more importantly, it provides a logical grounding for some of the techniques such as "pre-cooking" and handles a richer calculus including Σ-types. Our work also avoids some of the other shortcomings; as pointed out by Reed [16], the algorithm sketched in Dowek et al. [2] fails to terminate on some inputs. We give a clear specification of the pruning which eliminates bound variable dependencies for the dependently typed case and show correctness of the unification algorithms in three steps: 1) we show it terminates, 2) we show that the transformations in our unification algorithm preserve types, and 3) that each transition neither destroys nor creates (additional) solutions.

Our work is to our knowledge the first comprehensive description of constraint-based higher-order pattern unification for the $\lambda^{\Pi\Sigma}$ calculus. It builds on and extends prior work by Reed [16] to handle Σ-types. Previously, Elliot [4] described unification for Σ-types in a Huet-style unification algorithm. While it is typically straightforward to incorporate η-expansions and lowering for meta-variables of Σ-type [18,11] , there is little work on extending the notion of Miller

---

[1] We write $\boldsymbol{x}{:}\boldsymbol{A}$ for a vector $x_1{:}A_1, \ldots x_n{:}A_n$.

patterns to be able to handle meta-variables which are applied to projections of bound variables. Fettig and Löchner [5] describe a higher-order pattern unification algorithm with finite products in the simply typed lambda-calculus. Their approach does not directly exploit isomorphisms on types, but some of the ideas have a similar goal: for example abstractions $\lambda x.\,\mathsf{fst}\,x$ is translated into $\lambda(x_1, x_2).\,\mathsf{fst}\,(x_1\,,\,x_2)$ which in turn normalizes to $\lambda(x_1, x_2).x_1$ to eliminate projections. Duggan [3] also explores extended higher-order patterns for products in the simply-typed setting; he generalizes Miller's pattern restriction for the simply-typed lambda-calculus by allowing repeated occurrences of variables to appear as arguments to meta-variables, provided such variables are prefixed by distinct sequences of projections.

Our work has been already tested in practice. Some of the ideas described in this paper are incorporated into the implementation of the dependently-typed Beluga language; in Beluga, Σ-types occur in a restricted form, i.e., only variable declarations in contexts can be of Σ-type and there is no nesting of Σ-types.

Due to space restrictions, most proofs have been omitted; they can be found in an extended version of this article on the authors' homepages.

2 $\lambda^{\Pi\Sigma}$-Calculus

In this paper, we are considering an extension of the $\lambda^{\Pi\Sigma}$-calculus with meta-variables. Its grammar is mostly straightforward. We use x, y, z for bound variables to distinguish them from meta-variables u, v, and w.

Sorts	s	$::= \mathsf{type} \mid \mathsf{kind}$
Atomic types	P, Q	$::= \mathbf{a}\,M$
Types	A, B, C, D	$::= P \mid \Pi x{:}A.B \mid \Sigma x{:}A.B$
Kinds	κ	$::= \mathsf{type} \mid \Pi x{:}A.\kappa$
(Rigid) heads	H	$::= \mathbf{a} \mid \mathbf{c} \mid x$
Projections	π	$::= \mathsf{fst} \mid \mathsf{snd}$
Evaluation contexts	E	$::= \bullet \mid E\,N \mid \pi\,E$
Neutral terms	R	$::= E[H] \mid E[u[\sigma]]$
Normal terms	M, N	$::= R \mid \lambda x.M \mid (M\,,\,N)$
Substitutions	σ, τ	$::= \cdot \mid \sigma, M$
Variable substitutions	ρ, ξ	$::= \cdot \mid \rho, x$
Contexts	Ψ, Φ, Γ	$::= \cdot \mid \Psi, x{:}A$
Meta substitutions	θ, η	$::= \cdot \mid \theta, \hat{\Psi}.M/u$
Meta contexts	Δ	$::= \cdot \mid \Delta, u{:}A[\Psi]$

Meta-variables are characterized as a closure $u[\sigma]$ which is the use of the meta-variable u under the suspended explicit substitution σ. The term $\lambda x\,y\,z.\,X\,x\,y$ with the meta-variable X which has type $\Pi x{:}A.\Pi y{:}B.C$ is represented in our calculus as $\lambda x\,y\,z.\,u[x, y]$ where u has type $C[x{:}A, y{:}B]$ and $[x, y]$ is a substitution with domain $x{:}A, y{:}B$ and the range x, y, z. Instead of an abstraction, we can directly replace u with a closed object $x, y.\,x\,(\mathsf{suc}\,y)$. This eliminates the need to craft a λ-prefix for the instantiation of meta-variables, avoids spurious reductions, and provides simple justifications for techniques such as lowering.

In general, the meta-variable u stands for a contextual object $\hat{\Psi}.M$ where $\hat{\Psi}$ describes the ordinary bound variables which may occur in M. This allows us to rename the free variables occurring in M if necessary. We use the following convention: If the meta-variable u is associated with the identity substitution, we simply write u instead of $u[\text{id}]$. A meta-variable u has the contextual type $A[\Psi]$ thereby characterizing an object of type A in the context Ψ. Our grammar and our subsequent typing rules enforce that objects are β-normal.

A signature Σ is a collection of declarations, which take one of the forms: $\mathbf{a} : \kappa$ (type family declaration) or $\mathbf{c} : A$ (constructor declaration). Because variable substitutions ρ play a special role in the formulation of our unification algorithm, we recognize them as a subclass of general substitutions σ. Identity substitutions id_Φ are defined recursively by $\text{id.} = (\cdot)$ and $\text{id}_{\Phi,x:A} = (\text{id}_\Phi, x)$. The subscript Φ is dropped when unambiguous. If Φ is a sub-context of Ψ (in particular if $\Psi = \Phi$) then id_Φ is a well-formed substitution in Ψ, i.e., $\Psi \vdash \text{id}_\Phi : \Phi$ holds (see Fig. 1). We write $\hat{\Phi}$ for the list of variables $\text{dom}(\Phi)$ in order of declaration.

We write $E[M]$ for plugging term M into the hole $\bullet$ of evaluation context E. This will be useful when describing the unification algorithm, since we often need to have access to the head of a neutral term. In the λ^Π-calculus, this is often achieved using the spine notation [1] simply writing $H\,M_1 \ldots M_n$. Evaluation contexts are the proper generalization of spines to projections .

Occurrences and free variables. If α, β are syntactic entities such as evaluation contexts, terms, or substitutions, $\alpha, \beta ::= E \mid R \mid M \mid \sigma$, we write $\alpha\{\beta\}$ if β is a part of α. If we subsequently write $\alpha\{\beta'\}$ then we mean to replaceme the indicated occurrence of β by β'. We say an occurrence is *rigid* if it is not part of a delayed substitution σ of a meta-variable, otherwise it is termed *flexible*. For instance, in $\mathbf{c}\,(u[y_1])\,(x_1\,x_2)\,(\lambda z.\,z\,x_3\,v[y_2, w[y_3]])$ there are rigid occurrences of $x_{1..3}$ and flexible occurrences of $y_{1..3}$. The meta-variables u, v appear in a rigid and w in a flexible position. A rigid occurrence is *strong* if it is not in the evaluation context of a free variable. In our example, only x_2 does *not* occur strongly rigidly. Following Reed [16] we indicate rigid occurrences by $\alpha\{\beta\}^{\text{rig}}$ and strongly rigid occurrences by $\alpha\{\beta\}^{\text{srig}}$.

We denote the set of free variables of α by $\mathsf{FV}(\alpha)$ and the set of free meta variables by $\mathsf{FMV}(\alpha)$. A superscript ^{rig} indicates to count only the rigid variables.

Typing. We rely on a bi-directional type system to guarantee that well-typed terms are in β-normal form (Fig. 1). The typing rules are devided into rules which check that an object has a given type ($\Leftarrow$ judgments) and rules which synthesize a type for a given object ($\Rightarrow$ judgments). We have record types $\Sigma x{:}A.\,B$ but no record kinds $\Sigma x{:}A.\,\kappa$. Our typing rules ensure that terms are in β-normal form, but they need not be η-long. The judgment $A =_\eta C$ (rules omitted) compares A and C modulo η, i.e., modulo $R = \lambda x.\,R\,x\ (x \notin \mathsf{FV}(R))$ and $R = (\mathsf{fst}\,R\,,\,\mathsf{snd}\,R)$.

Hereditary substitution and meta-substitution. For α a well-typed entity in context Ψ and $\Delta; \Phi \vdash \sigma : \Psi$ a well-formed substitution, we facilitate a simultaneous substitution operation $[\sigma]_\Psi(\alpha)$ that substitutes the terms in σ for the

Neutral Terms/Types $\boxed{\Delta; \Psi \vdash R \Rightarrow A}$

$$\frac{\Sigma(\mathbf{a}) = \kappa}{\Delta; \Psi \vdash \mathbf{a} \Rightarrow \kappa} \quad \frac{\Sigma(\mathbf{c}) = A}{\Delta; \Psi \vdash \mathbf{c} \Rightarrow A} \quad \frac{\Psi(x) = A}{\Delta; \Psi \vdash x \Rightarrow A} \quad \frac{u{:}A[\Phi] \in \Delta \quad \Delta; \Psi \vdash \sigma \Leftarrow \Phi}{\Delta; \Psi \vdash u[\sigma] \Rightarrow [\sigma]_\Phi A}$$

$$\frac{\Delta; \Psi \vdash R \Rightarrow \Pi x{:}A.B \quad \Delta; \Psi \vdash M \Leftarrow A}{\Delta; \Psi \vdash R\,M \Rightarrow [M/x]_A B}$$

$$\frac{\Delta; \Psi \vdash R \Rightarrow \Sigma x{:}A.B}{\Delta; \Psi \vdash \mathsf{fst}\,R \Rightarrow A} \quad \frac{\Delta; \Psi \vdash R \Rightarrow \Sigma x{:}A.B}{\Delta; \Psi \vdash \mathsf{snd}\,R \Rightarrow [\mathsf{fst}\,R/x]_A B}$$

Normal Terms $\boxed{\Delta; \Psi \vdash M \Leftarrow A}$

$$\frac{\Delta; \Psi \vdash R \Rightarrow A \quad A =_\eta C}{\Delta; \Psi \vdash R \Leftarrow C}$$

$$\frac{\Delta; \Psi, x{:}A \vdash M \Leftarrow B}{\Delta; \Psi \vdash \lambda x.M \Leftarrow \Pi x{:}A.B} \quad \frac{\Delta; \Psi \vdash M \Leftarrow A \quad \Delta; \Psi \vdash N \Leftarrow [M/x]_A B}{\Delta; \Psi \vdash (M\,,\,N) \Leftarrow \Sigma x{:}A.B}$$

Substitutions $\boxed{\Delta; \Psi \vdash \sigma \Leftarrow \Psi'}$

$$\frac{}{\Delta; \Psi \vdash \cdot \Leftarrow \cdot} \quad \frac{\Delta; \Psi \vdash \sigma \Leftarrow \Psi' \quad \Delta; \Psi \vdash M \Leftarrow [\sigma]_{\Psi'} A}{\Delta; \Psi \vdash \sigma, M \Leftarrow \Psi', x{:}A}$$

LF Types and Kinds $\boxed{\Delta; \Psi \vdash A \Leftarrow s}$

$$\frac{\Delta; \Psi \vdash P \Rightarrow \mathsf{type}}{\Delta; \Psi \vdash P \Leftarrow \mathsf{type}} \quad \frac{\Delta; \Psi \vdash A \Leftarrow \mathsf{type} \quad \Delta; \Psi, x{:}A \vdash B \Leftarrow \mathsf{type}}{\Delta; \Psi \vdash \Sigma x{:}A.B \Leftarrow \mathsf{type}}$$

$$\frac{}{\Delta; \Psi \vdash \mathsf{type} \Leftarrow \mathsf{kind}} \quad \frac{\Delta; \Psi \vdash A \Leftarrow \mathsf{type} \quad \Delta; \Psi, x{:}A \vdash B \Leftarrow s}{\Delta; \Psi \vdash \Pi x{:}A.B \Leftarrow s}$$

Meta-Substitutions $\boxed{\Delta \vdash \theta \Leftarrow \Delta'}$

$$\frac{\text{for all } u{:}A[\Phi] \in \Delta' \text{ and } \hat{\Phi}.M/u \in \theta \quad \Delta; [\![\theta]\!]\Phi \vdash M \Leftarrow [\![\theta]\!]A}{\Delta \vdash \theta \Leftarrow \Delta'}$$

Meta-Context $\boxed{\vdash \Delta \ \mathsf{mctx}}$

$$\frac{\text{for all } u{:}A[\Psi] \in \Delta \quad \Delta \vdash \Psi \ \mathsf{ctx} \quad \Delta; \Psi \vdash A \Leftarrow \mathsf{type}}{\vdash \Delta \ \mathsf{mctx}}$$

Fig. 1. Typing rules for LF with meta-variables

variables as listed by Ψ in α and produces a β-normal result. Such an operation exists for well-typed terms, since $\lambda^{\Pi\Sigma}$ is normalizing. A naive implementation just substitutes and then normalizes, a refined implementation, called *hereditary substitution* [19], proceeds by resolving created redexes on the fly through new substitutions. Details can be found in Nanevski et al. [10], but do not concern us much here. Single substitution $[N/x]_A(\alpha)$ is conceived as a special case of simultaneous substitution. The type annotation A and the typing information in Ψ allow hereditary substitution to be defined by structural recursion; if no ambiguity arises, we may omit indices Ψ and A from substitutions.

The meta-substitution operation is written as $[\![\hat{\Psi}.M/u]\!](N)$ and the simultaneous meta-substitution is written as $[\![\theta]\!](N)$. When we apply $\hat{\Psi}.M/u$ to $u[\sigma]$ we first substitute $\hat{\Psi}.M$ for u in the substitution σ to obtain σ'. Subsequently, we continue to apply σ' to M hereditarily to obtain M'.

The typing rules ensure that the type of the instantiation $\hat{\Psi}.M$ and the type of u agree, i.e. we can replace u which has type $A[\Psi]$ with a normal term M if M has type A in the context Ψ. Because of α-conversion, the variables that are substituted at different occurrences of u may be different, and we write $\hat{\Psi}.M$ where $\hat{\Psi}$ binds all the free variables in M. We can always appropriately rename the bound variable in $\hat{\Psi}$ such that they match the domain of the postponed substitution σ'. This complication can be eliminated in an implementation of the calculus based on de Bruijn indexes.

3   Constraint-Based Unification

We define the unification algorithm using rewrite rules which solve constraints incrementally. Constraints K and sets of constraints $\mathcal{K}$ are defined as follows:

Constraint	$K ::= \top \mid \bot$	trivial constraint and inconsistency
	$\mid \ \Psi \vdash M = N : C$	unify term M with N
	$\mid \ \Psi \mid R{:}A \vdash E = E'$	unify evaluation context E with E'
	$\mid \ \Psi \vdash u \leftarrow M : C$	solution for u found
C. sets	$\mathcal{K} ::= K \mid \mathcal{K} \wedge K$	modulo laws of conjunction.

Our basic constraints are of the form $\Psi \vdash M = N : C$. The type annotation $\Psi \vdash C$ serves two purposes: First, it is necessary to ensure that all substitutions created and used in our transformations can be properly annotated and hence we can use the fact that their application will terminate and produce again normal forms. Second, the type annotations in the context Ψ are necessary to eliminate Σ-types. For both purposes, simple types, i.e., the dependency-erasure of $\Psi \vdash C$ would suffice. However, we keep dependency in this presentation to scale this work from $\lambda^{\Pi\Sigma}$ to non-erasable dependent types such as in Agda.

A unification problem is described by $\Delta \Vdash \mathcal{K}$ where Δ contains the typings of all the meta variables in $\mathcal{K}$. A meta-variable u is *solved*, if there is a constraint $\Psi \vdash u \leftarrow M : C$ in $\mathcal{K}$; otherwise we call u *active*. A solved metavariable does not appear in any other constraints nor in any type in Δ (nor in its solution M).

Intuitively, a set of constraints is well-formed if each constraint $\Psi \vdash M = N : C$ is well typed. Unfortunately, this is complicated by the fact that we may delay working on some sub-terms; to put it differently, we can work on subterms in an arbitrary order. Yet, the type of an equation may depend on the solvability of another postponed equation. Consider for example tuples. If (M_1, M_2) and (N_1, N_2) both have type $\Sigma x{:}A.B$, then M_1 and N_1 have type A. However, types may get out of sync when we consider M_2 and N_2. M_2 has type $[M_1/x]B$ while N_2 has type $[N_1/x]B$, and we only know that their types agree, if we know that M_1 is equal to N_1. Similar issues arise for function types and applications. Following Reed [16] we adopt here a weaker typing invariant, namely typing modulo constraints.

3.1 Typing Modulo

For all typing judgments $\Delta; \Psi \vdash J$ defined previously, we define $\Delta; \Psi \vdash_{\mathcal{K}} J$ by the same rules as for $\Delta; \Psi \vdash J$ except replacing syntactic equality $=$ with $=_{\mathcal{K}}$. We write $\hat{\Psi}.M =_{\mathcal{K}} N$ if for any ground meta-substitution θ that is a ground solution for $\mathcal{K}$, we have $[\![\theta]\!]M =_{\eta} [\![\theta]\!]N$. To put it differently, if we can solve $\mathcal{K}$, we can establish that M is equal to N. Following Reed [16], substitution and meta-substitution preserve typing modulo.

Intuitively, a unification problem $\Delta \Vdash \mathcal{K}$ is well-formed if all constraints $(\Psi \vdash M = N : C) \in \mathcal{K}$ are well-typed modulo $\mathcal{K}$, i.e., $\Delta; \Psi \vdash_{\mathcal{K}} M \Leftarrow C$ and $\Delta; \Psi \vdash_{\mathcal{K}} N \Leftarrow C$. We will come back to this later when we prove correctness of our algorithm, but it is helpful to keep the typing invariant in mind when explaining the transitions in our algorithm.

Decomposition of functions

$$\Psi \vdash \lambda x.M = \lambda x.N : \Pi x{:}A.\,B \qquad \mapsto_d \quad \Psi, x{:}A \vdash M = N : B$$

$$\Psi \vdash \lambda x.M = R : \Pi x{:}A.\,B \qquad \mapsto_d \quad \Psi, x{:}A \vdash M = R\,x : B$$

$$\Psi \vdash R = \lambda x.M : \Pi x{:}A.\,B \qquad \mapsto_d \quad \Psi, x{:}A \vdash R\,x = M : B$$

Decomposition of pairs

$$\Psi \vdash (M_1\,,M_2) = (N_1\,,N_2) : \Sigma x{:}A.\,B \qquad \mapsto_d \quad \Psi \vdash M_1 = N_1 : A \wedge \Psi \vdash M_2 = N_2 : [M_1/x]B$$

$$\Psi \vdash (M_1\,,M_2) = R : \Sigma x{:}A.\,B \qquad \mapsto_d \quad \Psi \vdash M_1 = \mathsf{fst}\,R : A \wedge \Psi \vdash M_2 = \mathsf{snd}\,R : [M_1/x]B$$

$$\Psi \vdash R = (M_1\,,M_2) : \Sigma x{:}A.\,B \qquad \mapsto_d \quad \Psi \vdash \mathsf{fst}\,R = M_1 : A \wedge \Psi \vdash \mathsf{snd}\,R = M_2 : [\mathsf{fst}\,R/x]B$$

Decomposition of neutrals

$$\Psi \vdash E[H] = E'[H] : C \qquad \mapsto_d \quad \Psi \mid H : A \vdash E = E' \text{ where } \Psi \vdash H \Rightarrow A$$

$$\Psi \vdash E[H] = E'[H'] : C \qquad \mapsto_d \quad \bot \text{ if } H \neq H'$$

Decomposition of evaluation contexts

$$\Psi \mid R : A \vdash \bullet = \bullet \qquad \mapsto_d \quad \top$$

$$\Psi \mid R : \Pi x{:}A.\,B \vdash E[\bullet\,M] = E'[\bullet\,M'] \qquad \mapsto_d \quad \Psi \vdash M = M' : A \wedge \Psi \mid R\,M : [M/x]B \vdash E = E'$$

$$\Psi \mid R : \Sigma x{:}A.\,B \vdash E[\mathsf{fst}\,\bullet] = E'[\mathsf{fst}\,\bullet] \qquad \mapsto_d \quad \Psi \mid \mathsf{fst}\,R : A \vdash E = E'$$

$$\Psi \mid R : \Sigma x{:}A.\,B \vdash E[\mathsf{snd}\,\bullet] = E'[\mathsf{snd}\,\bullet] \qquad \mapsto_d \quad \Psi \mid \mathsf{snd}\,R : [\mathsf{fst}\,R/x]B \vdash E = E'$$

$$\Psi \mid R : \Sigma x{:}A.\,B \vdash E[\pi\,\bullet] = E'[\pi'\,\bullet] \qquad \mapsto_d \quad \bot \text{ if } \pi \neq \pi'$$

Orientation

$$\Psi \vdash M = u[\sigma] : C \text{ with } M \neq v[\tau] \qquad \mapsto_d \quad \Psi \vdash u[\sigma] = M : C$$

η-Contraction

$$\Psi \vdash u[\sigma\{\lambda x.\,R\,x\}] = N : C \qquad \mapsto_e \quad \Psi \vdash u[\sigma\{R\}] = N : C$$

$$\Psi \vdash u[\sigma\{(\mathsf{fst}\,R, \mathsf{snd}\,R)\}] = N : C \qquad \mapsto_e \quad \Psi \vdash u[\sigma\{R\}] = N : C$$

Eliminating projections

$$\begin{array}{l} \Psi_1, x : \Pi\boldsymbol{y}{:}\boldsymbol{A}.\,\Sigma z{:}B.C, \Psi_2 \\ \quad \vdash u[\sigma\{\pi\,(x\,\boldsymbol{M})\}] = N : D \\ \text{where } \pi \in \{\mathsf{fst}, \mathsf{snd}\} \end{array} \quad \mapsto_p \quad \begin{array}{l} \Psi_1, x_1 : \Pi\boldsymbol{y}{:}\boldsymbol{A}.\,B,\; x_2 : \Pi\boldsymbol{y}{:}\boldsymbol{A}.\,[(x_1\,\boldsymbol{y})/z]C, \Psi_2 \\ \quad \vdash u[[\tau]\sigma] = [\tau]N : [\tau]D \\ \text{where } \tau = [\lambda\boldsymbol{y}.\,(x_1\,\boldsymbol{y}, x_2\,\boldsymbol{y})/x] \end{array}$$

Fig. 2. Local simplification

Local simplification

$$\Delta \Vdash \mathcal{K} \wedge K \mapsto \Delta \Vdash \mathcal{K} \wedge \mathcal{K}' \qquad \text{if } K \mapsto_m \mathcal{K}' \quad (m \in \{\mathsf{d}, \mathsf{e}, \mathsf{p}\})$$

Instantiation (notation)

$$\Delta \Vdash \mathcal{K} + (\Phi \vdash u \leftarrow M : A) \quad = \quad \llbracket \theta \rrbracket \Delta \Vdash \llbracket \theta \rrbracket \mathcal{K} \wedge \llbracket \theta \rrbracket \Phi \vdash u \leftarrow M : \llbracket \theta \rrbracket A$$
$$\text{where } \theta = \hat{\Phi}.M/u$$

Lowering

$$\Delta \Vdash \mathcal{K} \qquad\qquad\qquad \mapsto \Delta, v{:}B[\Phi, x{:}A] \Vdash \mathcal{K}$$
$$u{:}(\Pi x{:}A.B)[\Phi] \in \Delta \text{ active} \qquad + \Phi \vdash u \leftarrow \lambda x.v : \Pi x{:}A.\,B$$

$$\Delta \Vdash \mathcal{K} \qquad\qquad\qquad \mapsto \Delta, u_1{:}A[\Phi], u_2{:}([u_1/x]_A B)[\Phi] \Vdash \mathcal{K}$$
$$u{:}(\Sigma x{:}A.B)[\Phi] \in \Delta \text{ active} \qquad + \Phi \vdash u \leftarrow (u_1\,,\,u_2) : \Sigma x{:}A.\,B$$

Flattening Σ-types

$$\Delta \Vdash \mathcal{K} \quad (u{:}A[\Phi] \in \Delta \text{ active}) \quad \mapsto \Delta, v{:}([\sigma^{-1}]A)[\Phi'] \Vdash \mathcal{K} \;+\; \Phi \vdash u \leftarrow v[\sigma] : A$$
$$\Phi = \Phi_1, x : \Pi y{:}\boldsymbol{A}.\,\Sigma z{:}B.\,C,\;\Phi_2 \qquad \Phi' = \Phi_1, x_1 : \Pi y{:}\boldsymbol{A}.\,B,\; x_2 : \Pi y{:}\boldsymbol{A}.\,[x_1\,\boldsymbol{y}/z]C,\;\Phi_2$$
$$\sigma^{-1} = [\lambda \boldsymbol{y}.\,(x_1\,\boldsymbol{y}\,,\,x_2\,\boldsymbol{y})/x] \qquad \sigma = [\lambda \boldsymbol{y}.\,\mathsf{fst}\,(x\,\boldsymbol{y})/x_1, \lambda \boldsymbol{y}.\,\mathsf{snd}\,(x\,\boldsymbol{y})/x_2]$$

Pruning

$$\Delta \Vdash \mathcal{K} \qquad\qquad\qquad \mapsto \Delta' \Vdash \llbracket \eta \rrbracket \mathcal{K}$$
$$(\Psi \vdash u[\rho] = M : C) \in \mathcal{K} \qquad \text{if } \Delta \vdash \mathsf{prune}_\rho\,M \Rightarrow \Delta';\; \eta \text{ and } \eta \neq \mathsf{id}$$

Same meta-variable

$$\Delta \Vdash \mathcal{K} \wedge \Psi \vdash u[\rho] = u[\xi] : C \qquad \mapsto \Delta, v{:}A[\Phi_0] \Vdash \mathcal{K} + \Phi \vdash u \leftarrow v[\mathsf{id}_{\Phi_0}] : A$$
$$u{:}A[\Phi] \in \Delta \qquad\qquad\qquad \text{if } \rho \cap \xi : \Phi \Rightarrow \Phi_0$$

Failing occurs check

$$\Delta \Vdash \mathcal{K} \wedge \Psi \vdash u[\rho] = M : C \qquad \mapsto \bot \text{ if } \mathsf{FV}^{\mathsf{rig}}(M) \not\subseteq \rho$$

$$\Delta \Vdash \mathcal{K} \wedge \Psi \vdash u[\rho] = M : C \qquad \mapsto \bot \text{ if } M = M'\{u[\xi]\}^{\mathsf{srig}} \neq u[\xi]$$

Solving (with successful occurs check)

$$\Delta \Vdash \mathcal{K} \wedge \Psi \vdash u[\rho] = M : C \qquad \mapsto \Delta \Vdash \mathcal{K} + \Phi \vdash u \leftarrow M' : A$$
$$(u{:}A[\Phi]) \in \Delta;\; u \notin \mathsf{FMV}(M) \qquad \text{if } M' = [\rho/\hat{\Phi}]^{-1} M \text{ exists}$$

Fig. 3. Unification steps

3.2 A Higher-Order Dynamic Pattern Unification Algorithm for Dependent Types and Records

The higher-order dynamic pattern unification algorithm is presented as rewrite rules on the set of constraints $\mathcal{K}$ in meta variable context Δ. The *local simplification rules* (Figure 2) apply to a single constraint, decomposing it and molding it towards a pattern by η-contraction and projection elimination. Decomposition of a neutral term is defined using evaluation contexts to have direct access to the head.

The other *unification steps* (Figure 3) work on a meta-variable and try to find an instantiation for it. We write $\Delta \Vdash \mathcal{K} + \Phi \vdash u \mathrel{ot} M : A$ for instantiating the eta-variable u with the term M in the meta-context Δ and in the constraints $\mathcal{K}$. This abbreviation is defined under the heading **Instantiation** in Figure 3. Lowering rules transform a meta-variable of higher type to one of lower type. Flattening Σ-types concentrates on a meta-variable $u{:}A[\Phi]$ and eliminates

Σ-types from the context Φ. The combination of the flattening Σ-types transition and the eliminating projections transition allow us to transform a unification problem into one which resembles our traditional pattern unification problem. The pruning transition is explained in detail in Section 3.4 and unifying a meta-variable with itself is discussed in Section 3.5.

Our algorithm can deal with a larger class of patterns where we typically require that meta-variables are associated with a linear substitution. To motivate our rules, let us consider some problems $\Psi \vdash u[\sigma] = M : C$ that fall outside of the Miller pattern fragment, meaning that σ is not a list of disjoint variables. We may omit types and/or context if appropriate.

η-contraction. $u[\lambda x.\, y\, (\mathsf{fst}\, x\,,\, \mathsf{snd}\, x)] = M$
  Solved by contracting the l.h.s. to $u[y]$.

Eliminating projections. $y : \Pi x{:}A.\, \Sigma z{:}B.\, C \vdash u[\lambda x.\, \mathsf{fst}\, (y\, x)] = M$
  Applying substitution $\tau = [\lambda x.\, (y_1\, x\,,\, y_2\, x)/y]$ yields problem $y_1 : \Pi x{:}A.\, B$, $y_2 : \Pi x{:}A.\, [y_1\, x/z]C \vdash u[\lambda x.\, y_1\, x] = [\tau]M$ which is solved by η-contraction, provided $y_2 \notin \mathsf{FV}([\tau]M)$.

Lowering. $u : (\Sigma x{:}A.\, B)[\Phi] \Vdash \mathsf{fst}\, (u[y]) = \mathsf{fst}\, y$
  This equation determines only the first component of the tuple u. Thus, decomposition into $u[y] = y$, which also determines the second component, loses solutions. Instead we replace u by a pair (u_1, u_2) of meta-variables of *lower* type, yielding $u_1 : A[\Phi], u_2 : ([u_1/x]B)[\Phi] \Vdash u_1[y] = \mathsf{fst}\, y$.

Flattening Σ-types. $u : P[z : \Pi x{:}A.\, \Sigma y{:}B.\, C] \Vdash u[\lambda x.\, (z_1\, x\,,\, z_2\, x)] = M$
  By splitting z into two functions z_1, z_2 we arrive at $u : P[z_1 : \Pi x{:}A.\, B,\ z_2 : \Pi x{:}A.\, [z_1\, x/y]C] \Vdash u[\lambda x.\, z_1\, x,\ \lambda x.\, z_2\, x] = M$ and continue with η-contraction.

Solving in spite of non-linearity. $u[x, x, z] = \mathsf{suc}\, z$
  The non-linear occurrence of x on the l.h.s. can be ignored since x is not free on the r.h.s. We can solve this constraint by by $u[x, y, z] = \mathsf{suc}\, z$.

Pruning. $u[x] = \mathsf{suc}(v[x, y])$ and $v[x, \mathsf{zero}] = \mathsf{f}(x, \mathsf{zero})$
  Since u depends only on x, necessarily v cannot depend on y. We can prune away the second parameter of v by setting $v[x, y] = v'[x]$. This turns the second constraint into the pattern $v'[x] = \mathsf{f}(x, \mathsf{zero})$, yielding the solution $u[x] = \mathsf{suc}(\mathsf{f}(x, \mathsf{zero}))$.

  Note that pruning is more difficult in case of nested meta-variable. If instead $u[x] = \mathsf{suc}(v[x, w[y]])$ then there are two cases: either v does not depend on its second argument or w is constant. Pruning as we describe it in this article cannot be applied to this case; Reed [16] proceeds here by replacing y by a placeholder "$\_$". Once w gets solved the placeholder might occur as argument to v, where it can be pruned. If the placeholder appears in a rigid position, the constraints have no solution.

Pruning and non-linearity. $u[x, x] = v[x]$ and $u'[x, x] = v'[x, y]$
  Even though we cannot solve for u due to the non-linear x, pruning x from v could lose solutions. However, we can prune y from v' since only x can occur in $v'[x, y]$.

Failing occurs check. $u[x] = \mathsf{suc}\, y$

Pruning y fails because it occurs rigidly. The constraint set has no solution.

Same meta-variable. $u[x, y, x, z] = u[x, y, y, x]$

Since variables x, y, z are placeholders for arbitrary *open* well-typed terms, of which infinitely many exists for every type, the above equation can only hold if u does not depend on its 3rd and 4th argument. Thus, we can solve by $u[x, y, z, x'] = v[x, y]$ where $[x, y]$ is the *intersection* of the two variable environments $[x, y, x, z]$ and $[x, y, y, x]$.

Recursive occurrence. $u[x, y, x] = \mathsf{suc}\, u[x, y, y]$

Here, u has a *strong* rigid occurrence in its own definition. Even though not in the pattern fragment, this only has an infinite solution: consider the instance $u[z, z, z] = \mathsf{suc}\, u[z, z, z]$. Consequently, the occurs check signals unsolvability. [17, p. 105f] motivates why only *strong* rigid recursive occurrences force unsolvability. For instance, $f : \mathsf{nat} \to \mathsf{nat} \vdash u[f] = \mathsf{suc}\, (f\,(u[\lambda x.\,\mathsf{zero}]))$ has solution $u[f] = \mathsf{suc}\, (f\,(\mathsf{suc}\, \mathsf{zero}))$ in spite of a rigid occurrence of u in its definition.

If u occurs flexibly in its own definition, like in $u[x] = v[u[x]]$, we cannot proceed until we know more of v. Using the other constraints, we might manage to prune v's argument, arriving at $u[x] = v[]$, or find the solution of v directly; in these cases, we can revisit the constraint on u.

The examples suggest a *strategy* for implementation: Lowering can be integrated triggered by decomposition to resolve eliminations of a meta variable $E[u[\sigma]]$. After decomposition we have a set of $u[\sigma] = M$ problems. We try to turn the σs into variable substitutions by applying η-contraction, and where this gets stuck, elimination of projections and Σ-flattening. Solution of constraints $u[\rho] = M$ can then be attempted by pruning, where a failing occurs check signals unsolvability.

3.3   Inverting Substitutions

A most general solution for a constraint $u[\sigma] = M$ can only be hoped for if σ is a variable substitution. For instance $u[\mathsf{true}] = \mathsf{true}$ admits already two different solutions $u[x] = x$ and $u[x] = \mathsf{true}$ that are pure λ-terms. In a language with computation such as Agda infinitely more solutions are possible, because $u[x]$ could be defined by cases on x and the value of $u[\mathsf{false}]$ is completely undetermined.

But even constraints $u[\rho] = M$ can be ambiguous if the variable substitution ρ is not linear, i.e., no bijective variable renaming. For example, $u[x, x] = x$ has solutions $u[x, y] = x$ and $u[x, y] = y$. Other examples, like $u[x, x, z] = z$, which has unique solution $u[x, y, z] = z$, suggest that we can ignore non-linear variable occurrences as long as they do not occur on the r.h.s. Indeed, if we define a variable substitution ρ to be *invertible for term* M if there is exactly one M' such that $[\rho]M' = M$, then linearity is a sufficient, but not necessary condition. However, it is necessary that ρ must be linear if restricted to the free variables of (β-normal!) M. Yet instead of computing the free variables of M, checking that ρ is invertible, inverting ρ and applying the result to M, we can directly try to invert the effect of the substitution ρ on M.

For a variable substitution $\Psi \vdash \rho \Leftarrow \Phi$ and a term/substitution $\alpha ::= M \mid R \mid \tau$ in context Ψ, we define the partial operation $[\rho/\hat{\Phi}]^{-1}\alpha$ by

$$[\rho/\hat{\Phi}]^{-1}x \quad = y \qquad \text{if } x/y \in \rho/\hat{\Phi} \text{ and there is no } z \neq y \text{ with } x/z \in \rho/\hat{\Phi},$$
$$\text{undefined otherwise}$$
$$[\rho/\hat{\Phi}]^{-1}c \quad = c$$
$$[\rho/\hat{\Phi}]^{-1}(u[\tau]) = u[\tau'] \quad \text{where } \tau' = [\rho/\hat{\Phi}]^{-1}\tau$$

and homeomorphic in all other cases.

We can show by induction on α, that inverse substitution $[\rho/\hat{\Phi}]^{-1}\alpha$ is correct, preserves well-typedness and commutes with meta substitutions.

3.4   Pruning

If the constraint $u[\sigma] = M$ has a solution θ, then $[\![\theta]\!]\sigma]\theta(u) = [\![\theta]\!]M$, and since θ is closed ($\mathsf{FMV}(\theta) = \emptyset$), we have $\mathsf{FV}(\sigma) \supseteq \mathsf{FV}([\![\theta]\!]M)$. Thus, if $\mathsf{FV}(M) \not\subseteq \mathsf{FV}(\sigma)$ we can try to find a most general meta-substitution η which *prunes* the free variables of M that are not in the range of σ, such that $\mathsf{FV}([\![\eta]\!]M) \subseteq \mathsf{FV}(\sigma)$. For instance, in case $u[x] = \mathsf{suc}\, v[x,y]$, the meta-substitution $x,y.\, v'[x]/v$ does the job. However, pruning fails either if pruned variables occur rigidly, like in $u[x] = \mathsf{c}\, y\, v[x,y]$ (constraint unsolvable), or if the flexible occurrence is under another meta variable, like in $u[x] = v[x, w[x,y]]$. Here, two minimal pruning substitutions $\eta_1 = x,y.\, v'[x]/v$ and $\eta_2 = x,y.\, w'[x]/w$ exist which are not instances of each other—applying pruning might lose solutions.

We restrict pruning to situations $u[\rho] = M$ where ρ is a variable substitution. This is because we view pruning as a preparatory step to inverting ρ on M—which only makes sense for variable substitutions. Also, we do not consider partial pruning, as in pruning y from v in the situation $u[x] = v[x,y,w[x,y]]$, obtaining $u[x] = v'[x, w[x,y]]$. Such extensions to pruning are conceivable, but we have no data indicating that they strengthen unification significantly in practice. We employ the following judgments to define pruning (see Fig. reffig:prune):

$$\mathsf{prune\_ctx}_\rho(\tau\,/\,\Psi_1) \Rightarrow \Psi_2 \quad \text{prune } \tau \text{ such that } \mathsf{FV}^{\mathsf{rig}}([\tau]\mathsf{id}_{\Psi_2}) \subseteq \rho$$
$$\Delta \vdash \mathsf{prune}_\rho M \Rightarrow \Delta'; \eta \quad \text{prune } M \text{ such that } \mathsf{FV}([\![\eta]\!]M) \subseteq \rho.$$

When pruning substitution τ with domain Ψ_1 we look at each term M in τ which substitutes for an $x{:}A$ of Ψ_1. If M has a rigid occurrence of a variable $y \notin \rho$, we discard the entry $x{:}A$ from the domain Ψ_1, thus, effectively removing M from τ. If M has no occurrence of such an y we keep $x{:}A$. However, since we might have removed prior entries from Ψ_1 we need to ensure A is still well-formed, by validating that its free variables are bound in the pruned context. Finally, if M has a flexible occurrence of a $y \notin \rho$, pruning fails. Examples:

1. $\mathsf{prune\_ctx}_x(\mathsf{c}\, x, y \quad /\ x'{:}A, y'{:}B) \Rightarrow x'{:}A$
2. $\mathsf{prune\_ctx}_y(\mathsf{c}\, x, y \quad /\ x'{:}A, y'{:}B) \Rightarrow y'{:}B$
3. $\mathsf{prune\_ctx}_y(\mathsf{c}\, x, u[y]\ /\ x'{:}A, y'{:}B) \Rightarrow y'{:}B$
4. $\mathsf{prune\_ctx}_y(u[x], y \quad /\ x'{:}A, y'{:}B) \qquad \text{fails}$

$\boxed{\mathsf{prune\_ctx}_\rho(\tau \ / \ \Psi_1) \Rightarrow \Psi_2}$ Prune $\tau : \Psi_1$, returning a sub-context Ψ_2 of Ψ_1.

$$\frac{}{\mathsf{prune\_ctx}_\rho(\cdot \ / \ \cdot) \Rightarrow \cdot} \qquad \frac{\mathsf{prune\_ctx}_\rho(\tau \ / \ \Psi_1) \Rightarrow \Psi_2 \quad \mathsf{FV}^{\mathrm{rig}}(M) \not\subseteq \rho}{\mathsf{prune\_ctx}_\rho(\tau, M \ / \ \Psi_1, x{:}A) \Rightarrow \Psi_2}$$

$$\frac{\mathsf{prune\_ctx}_\rho(\tau \ / \ \Psi_1) \Rightarrow \Psi_2 \quad \mathsf{FV}(M) \subseteq \rho \quad \mathsf{FV}(A) \subseteq \hat{\Psi}_2}{\mathsf{prune\_ctx}_\rho(\tau, M \ / \ \Psi_1, x{:}A) \Rightarrow \Psi_2, x{:}A}$$

$\boxed{\Delta \vdash \mathsf{prune}_\rho M \Rightarrow \Delta'; \ \eta}$ Prune M, returning $\Delta' \vdash \eta \Leftarrow \Delta$.

$$\frac{v{:}B[\Psi_1] \in \Delta \quad \mathsf{prune\_ctx}_\rho(\tau \ / \ \Psi_1) \Rightarrow \Psi_2 \quad \Psi_2 \neq \Psi_1 \quad \mathsf{FV}(B) \subseteq \hat{\Psi}_2 \quad \eta = \hat{\Psi}_1.v'[\mathsf{id}_{\Psi_2}]/v}{\Delta \vdash \mathsf{prune}_\rho(v[\tau]) \Rightarrow \llbracket \eta \rrbracket(\Delta, v'{:}B[\Psi_2]); \ \eta}$$

$$\frac{v{:}B[\Psi_1] \in \Delta \quad \mathsf{prune\_ctx}_\rho(\tau \ / \ \Psi_1) \Rightarrow \Psi_1}{\Delta \vdash \mathsf{prune}_\rho(v[\tau]) \Rightarrow \Delta; \ \mathsf{id}_\Delta} \qquad \frac{x \in \rho}{\Delta \vdash \mathsf{prune}_\rho x \Rightarrow \Delta; \ \mathsf{id}_\Delta}$$

$$\frac{}{\Delta \vdash \mathsf{prune}_\rho c \Rightarrow \Delta; \ \mathsf{id}_\Delta} \qquad \frac{\Delta \vdash \mathsf{prune}_\rho R \Rightarrow \Delta_1; \ \eta_1 \quad \Delta_1 \vdash \mathsf{prune}_\rho(\llbracket \eta_1 \rrbracket M) \Rightarrow \Delta_2; \ \eta_2}{\Delta \vdash \mathsf{prune}_\rho(R\,M) \Rightarrow \Delta_2; \ \llbracket \eta_2 \rrbracket \eta_1}$$

$$\frac{\Delta \vdash \mathsf{prune}_\rho M \Rightarrow \Delta'; \ \eta}{\Delta \vdash \mathsf{prune}_\rho(\pi\,M) \Rightarrow \Delta'; \ \eta} \qquad \frac{\Delta \vdash \mathsf{prune}_{\rho,x} M \Rightarrow \Delta'; \ \eta}{\Delta \vdash \mathsf{prune}_\rho(\lambda x.\,M) \Rightarrow \Delta'; \ \eta}$$

$$\frac{\Delta \vdash \mathsf{prune}_\rho M \Rightarrow \Delta_1; \ \eta_1 \quad \Delta_1 \vdash \mathsf{prune}_\rho(\llbracket \eta_1 \rrbracket N) \Rightarrow \Delta_2; \ \eta_2}{\Delta \vdash \mathsf{prune}_\rho(M\,,\,N) \Rightarrow \Delta_2; \ \llbracket \eta_2 \rrbracket \eta_1}$$

Fig. 4. Pruning

Pruning a term M with respect to ρ ensures that all rigid variables of M are in the range of ρ (see variable rule). Also, for each rigid occurrence of a meta-variable $v[\tau]$ in M we try to prune the substitution τ. If τ is already pruned, we leave v alone; otherwise, if the domain Ψ_1 of τ shrinks to Ψ_2 then we replace $v : B[\Psi_1]$ by a new meta-variable $v' : B[\Psi_2]$ with domain Ψ_2. However, we need to ensure that the type B still makes sense in Ψ_2; otherwise, pruning fails.

Lemma 1 (Soundness and completeness of pruning)

1. *If $\Delta \vdash_\mathcal{K} \Psi_1$ ctx and $\mathsf{prune\_ctx}_\rho(\tau \ / \ \Psi_1) \Rightarrow \Psi_2$ then $\Delta \vdash_\mathcal{K} \Psi_2$ ctx and $\mathsf{FV}(\llbracket \tau \rrbracket \mathsf{id}_{\Psi_2}) \subseteq \rho$. Additionally, if $x \in \Psi_1 \setminus \Psi_2$ then $\mathsf{FV}^{\mathrm{rig}}(\llbracket \tau \rrbracket x) \not\subseteq \rho$.*
2. *If $\Delta \vdash \mathsf{prune}_\rho M \Rightarrow \Delta'; \ \eta$ then $\Delta' \vdash_\mathcal{K} \eta \Leftarrow \Delta$ and $\mathsf{FV}(\llbracket \eta \rrbracket M) \subseteq \rho$. Also, if θ solves $\Psi \vdash u[\rho] = M_0\{M\}^{\mathrm{rig}} : C$ then there is some θ' such that $\theta = \llbracket \theta' \rrbracket \eta$.*

In an implementation, we may combine pruning with inverse substitution and the occurs check. Since we already traverse the term M for pruning, we may also check whether $[\rho/\hat{\Phi}]^{-1}M$ exists and whether u occurs in M.

3.5   Unifying Two Identical Existential Variables

Any solution $\hat{\Phi}.N/u$ for a meta variable $u : A[\Phi]$ with constraint $u[\rho] = u[\xi]$ must fulfill $[\rho]N = [\xi]N$, which means that $[\rho]x = [\xi]x$ for all $x \in \mathsf{FV}(N)$. This means that u can only depend on those of its variables in Φ that are mapped to the same term by ρ and ξ. Thus, we can substitute u by $\hat{\Phi}.v[\rho']$ where ρ' is the *intersection* of substitutions ρ and ξ. Similarly to context pruning, we obtain ρ' as $[\rho]\mathrm{id}_{\Phi'}$, which identical to $[\xi]\mathrm{id}_{\Phi'}$, where Φ' is a subcontext of Φ mentioning only the variables that have a common image under ρ and ξ. This process is given as judgement $\boxed{\rho \cap \xi : \Phi \Rightarrow \Phi'}$ with the following rules:

$$\overline{\cdot \cap \cdot : \cdot \Rightarrow \cdot}$$

$$\frac{\rho \cap \xi : \Phi \Rightarrow \Phi'}{(\rho, y) \cap (\xi, y) : (\Phi, x{:}A) \Rightarrow (\Phi', x{:}A)} \qquad \frac{\rho \cap \xi : \Phi \Rightarrow \Phi' \quad z \neq y}{(\rho, z) \cap (\xi, y) : (\Phi, x{:}A) \Rightarrow \Phi'}$$

Lemma 2 (Soundness of intersection). *If $\Delta; \Psi \vdash_{\mathcal{K}} \rho, \xi \Leftarrow \Phi$ and $\rho \cap \xi : \Phi \Rightarrow \Phi'$, then $\Delta \vdash_{\mathcal{K}} \Phi'$ ctx and $\Delta; \Phi \vdash_{\mathcal{K}} \mathrm{id}_{\Phi'} \Leftarrow \Phi'$ and $[\rho]\mathrm{id}_{\Phi'} = [\xi]\mathrm{id}_{\Phi'}$.*

Although we have restricted intersection to variable substitutions, it could be extended to meta-ground substitutions, i.e., substitutions that do not contain any meta-variables.

4   Correctness

Theorem 1 (Termination). *The algorithm terminates and results in one of the following states:*

- *A solved state where only assignments $\Psi \vdash u \leftarrow M : A$ remain.*
- *A stuck state, i.e., no transition rule applies.*
- *Failure $\bot$.*

Proof. Let the size $|M|$ of a term be as usual the number of nodes and leaves in its tree representation, with the exception that we count λ-nodes twice. This modification has the effect that $|\lambda x.M| + |R| > |M| + |R\,x|$, hence, an η-expanding decomposition step also decreases the sum of the sizes of the involved terms [6]. We define the size $|A[\Phi]|$ of a type A in context Φ by $|P[\Phi]| = 1 + \sum_{A \in \Phi} |A[]|$, $|(\Pi x{:}A.\,B)[\Phi]| = 1 + |B[\Phi, x{:}A]|$ and $|(\Sigma x{:}A.\,B)[\Phi]| = 1 + |A[\Phi]| + |B[\Phi]|$. The size of a type can then be obtained as $|A| = |A[]|$ and the size of a context as $|\Phi| = \sum_{A \in \Phi} |A|$. The purpose of this measure is to give Σ-types a large weight that can "pay" for *flattening*.

Let the weight of a solved constraint be 0, whereas the weight $|K|$ for a constraint $\Psi \vdash M = M' : C$ be the ordinal $(|M| + |M'|)\omega + |\Psi|$ if a decomposition step can be applied, and simply $|\Psi|$ else. Similarly, let the weight of constraint $\Phi \mid R{:}A \vdash E = E'$ be $(|E| + |E'|)\omega + |\Psi|$. Finally, let the weight $|\Delta \Vdash \mathcal{K}|$ of a unification problem be the ordinal

$$\sum_{u:A[\Phi] \in \Delta \text{ active}} |A[\Phi]|\omega^2 + \sum_{K \in \mathcal{K}} |K|.$$

By inspection of the transition rules we can show that each unification step reduces the weight of the unification problem. $\qquad\square$

4.1 Solutions to Unification

A solution to a set of equations $\mathcal{K}$ is a meta-substitution θ for all the meta-variables in Δ s.t. $\Delta' \vdash \theta \Leftarrow \Delta$ and

1. for every $\Psi \vdash u \leftarrow M : A$ in $\mathcal{K}$ we have $\hat{\Psi}.M/u \in \theta$,
2. for all equations $\Psi \vdash M = N : A$ in $\mathcal{K}$, we have $[\![\theta]\!]M = [\![\theta]\!]N$.

A *ground* solution to a set of equations $\mathcal{K}$ can be obtained from a solution to $\mathcal{K}$ by applying a grounding meta-substitution θ' where $\cdot \vdash \theta' \Leftarrow \Delta'$ to the solution θ. We write $\theta \in \mathsf{Sol}(\Delta \Vdash \mathcal{K})$ for a ground solution to the constraints $\mathcal{K}$.

Before, we prove that transitions preserve solutions, we first prove that there always exists a meta-substitution relating the original meta-variable context Δ_0 to the meta-variable context Δ_1 we transition to. It is useful to state this property in isolation, although it is also folded into theorem 2.

Lemma 3. *If* $\Delta_0 \Vdash \mathcal{K}_0 \mapsto \Delta_1 \Vdash \mathcal{K}_1$ *then there exists a meta-substitution* θ *s.t.* $\Delta_1 \vdash_{\mathcal{K}_1} \theta \Leftarrow \Delta_0$.

We also observe that if we start in a state $\Delta_0 \Vdash \mathcal{K}_0$ and transition to a state $\Delta_1 \Vdash \mathcal{K}_1$ the meta-variable context strictly grows, i.e., $\mathsf{dom}(\Delta_0) \subseteq \mathsf{dom}(\Delta_1)$. We subsequently show that if we have a solution for $\Delta_0 \Vdash \mathcal{K}_0$, then transitioning to a new state $\Delta_1 \Vdash \mathcal{K}_1$ will not add any additional solutions nor will it destroy some solution we may already have. In other words, any additional constraints which may be added in $\Delta_1 \Vdash \mathcal{K}_1$ are consistent with the already existing solution.

Theorem 2 (Transitions preserve solutions). *Let* $\Delta_0 \Vdash \mathcal{K}_0 \mapsto \Delta_1 \Vdash \mathcal{K}_1$.

1. *If* $\theta_0 \in \mathsf{Sol}(\Delta_0 \Vdash \mathcal{K}_0)$ *then there exists a meta-substitution* θ' *s.t.* $\Delta_1 \vdash_{\mathcal{K}_1} \theta' \Leftarrow \Delta_0$ *and a solution* $\theta_1 \in \mathsf{Sol}(\Delta_1 \Vdash \mathcal{K}_1)$ *such that* $[\![\theta_1]\!]\theta' = \theta_0$.
2. *If* $\theta_1 \in \mathsf{Sol}(\Delta_1 \Vdash \mathcal{K}_1)$ *then* $[\![\theta_1]\!]\mathsf{id}_{\Delta_0} \in \mathsf{Sol}(\Delta_0 \Vdash \mathcal{K}_0)$.

4.2 Transitions Preserve Types

Our goal is to prove that if we start with a well-typed unification problem our transitions preserve the type, i.e., we can never reach an ill-typed state and hence, we cannot generate a solution which may contain an ill-typed term. In the statement below it is again important to note that the meta-context strictly grows, i.e., $\Delta_0 \subseteq \Delta_1$ and that there always exists a meta-substitution θ which maps Δ_0 to Δ_1. Moreover, since transitions preserve solutions, if we have a solution for $\mathcal{K}_0$ there exists a solution for $\mathcal{K}_1$.

Lemma 4 (Transitions preserve typing). *Let* $\Delta_0 \Vdash \mathcal{K}_0 \mapsto \Delta_1 \Vdash \mathcal{K}_1$ *and* $\Delta_1 \vdash_{\mathcal{K}_1} \theta \Leftarrow \Delta_0$.

1. If $A =_{\mathcal{K}_0} B$, then $A =_{\mathcal{K}_1} B$.
2. If $\Delta_0; \Psi \vdash_{\mathcal{K}_0} M \Leftarrow A$ then $\Delta_1; [\![\theta]\!]\Psi \vdash_{\mathcal{K}_1} [\![\theta]\!]M \Leftarrow [\![\theta]\!]A$.
3. If $\Delta_0; \Psi \vdash_{\mathcal{K}_0} R \Rightarrow A$ then $\Delta_1; [\![\theta]\!]\Psi \vdash_{\mathcal{K}_1} [\![\theta]\!]R \Rightarrow A'$ and $[\![\theta]\!]A =_{\mathcal{K}_1} A'$.

Next, we define when a set of equations which constitute a unification problem are well-formed using the judgment $\Delta_0 \Vdash_{\mathcal{K}_0} \mathcal{K}$ wf, which states that each equation $\Psi \vdash M = N : A$ must be well-typed modulo the equations in $\mathcal{K}_0$, i.e., $\Delta_0; \Psi \vdash_{\mathcal{K}_0} M \Leftarrow A$ and $\Delta_0; \Psi \vdash_{\mathcal{K}_0} N \Leftarrow A$. We simply write $\Delta_0 \Vdash \mathcal{K}$ wf to mean $\Delta_0 \Vdash_{\mathcal{K}} \mathcal{K}$ wf.

Lemma 5 (Equations remain well-formed under meta-substitutions)
If $\Delta_0 \Vdash \mathcal{K}$ wf and $\Delta_1 \vdash_{[\![\theta]\!]\mathcal{K}} \theta \Leftarrow \Delta_0$ then $\Delta_1 \Vdash [\![\theta]\!]\mathcal{K}$ wf.

Lemma 6 (Well-formedness of equations is preserved by transitions)
If $\Delta_0 \Vdash \mathcal{K}_0 \mapsto \Delta_1 \Vdash \mathcal{K}_1$ and $\Delta_0 \Vdash_{\mathcal{K}_0} \mathcal{K}$ wf then $\Delta_1 \vdash_{\mathcal{K}_1} \mathcal{K}$ wf.

Theorem 3 (Unification preserves types)
If $\Delta_0 \Vdash \mathcal{K}_0$ wf and $\Delta_0 \Vdash \mathcal{K}_0 \mapsto \Delta_1 \Vdash \mathcal{K}_1$ then $\Delta_1 \Vdash \mathcal{K}_1$ wf.

5   Conclusion

We have presented a constraint-based unification algorithm which solves higher-order patterns dynamically and showed its correctness. We have extended higher-order pattern unification to handle Σ-types; this has been an open problem so far, and it is of practical relevance:

1. In LF-based systems such as Beluga, Twelf or Delphin, a limited form of Σ-types arises due to context blocks: Σ-types are used to introduce several assumptions simultaneously. For Beluga, the second author has implemented the flattening of context blocks and it works well in type reconstruction.
2. In dependently typed languages such as Agda, Σ-types, or, more generally, record types, are commonly used but unification has not been adapted to records. McBride [8, p.6] gives a practial example where the unification problem $T\,(\text{fst}\,\gamma)\,(\text{snd}\,\gamma) = T'\,\gamma$ appears, which is not solved by Agda at this point. With the techniques described in this article we will be able to solve such problems and make systems such as Agda more robust.

Correctness of our unification constraint solver is proved using *typing modulo* [16]. This is possible since we have no constraints on the type level and we are dealing with terms whose normalization via hereditary substitutions can be defined by recursion on their type. Even in the presence of unsolvable constraints, which lead to ill-typed terms, normalization is terminating. This does not scale to Agda which has large eliminations and unification on the type level; there, ill-typed terms may lead to divergence of type reconstruction. A solution has been described by Norell [11]: unsolved constraints block normalization, thus guaranteeing termination of the unification algorithm. The idea has been implemented in Agda 2 and can be extended to Σ-types and the unification rules described in this article.

Acknowledgments. We thank Jason Reed for his insightful work and his explanations given via email. Thanks to Conor McBride for in-depth comments on this text and his suggestions for improvement. We also acknowledge the anonymous referees who have given constructive critique on this article.

References

1. Cervesato, I., Pfenning, F.: A linear spine calculus. Journal of Logic and Computation 13(5), 639–688 (2003)
2. Dowek, G., Hardin, T., Kirchner, C., Pfenning, F.: Unification via explicit substitutions: The case of higher-order patterns. In: Joint International Conference Logic Programming, pp. 259–273. MIT Press, Cambridge (1996)
3. Duggan, D.: Unification with extended patterns. Theoretical Computer Science 206(1-2), 1–50 (1998)
4. Elliott, C.: Extensions and Applications of Higher-Order Unification. PhD thesis, School of Computer Science, Carnegie Mellon University (1990)
5. Fettig, R., Löchner, B.: Unification of higher-order patterns in a simply typed lambda-calculus with finite products and terminal type. In: Ganzinger, H. (ed.) RTA 1996. LNCS, vol. 1103, pp. 347–361. Springer, Heidelberg (1996)
6. Goguen, H.: Justifying algorithms for $\beta\eta$ conversion. In: Sassone, V. (ed.) FOSSACS 2005. LNCS, vol. 3441, pp. 410–424. Springer, Heidelberg (2005)
7. Goldfarb, W.D.: The undecidability of the second-order unification problem. Theoretical Computer Science 13, 225–230 (1981)
8. McBride, C.: Outrageous but meaningful coincidences: Dependent type-safe syntax and evaluation. In: ACM SIGPLAN Workshop on Genetic Programming (WGP 2010), pp. 1–12. ACM, New York (2010)
9. Miller, D.: Unification of simply typed lambda-terms as logic programming. In: Eighth International Logic Programming Conference, pp. 255–269. MIT Press, Cambridge (1991)
10. Nanevski, A., Pfenning, F., Pientka, B.: Contextual modal type theory. ACM Transactions on Computational Logic 9(3), 1–49 (2008)
11. Norell, U.: Towards a practical programming language based on dependent type theory. PhD thesis, Department of Computer Science and Engineering, Chalmers University of Technology, Technical Report 33D (2007)
12. Pfenning, F., Schürmann, C.: System description: Twelf - A meta-logical framework for deductive systems. In: Ganzinger, H. (ed.) CADE 1999. LNCS (LNAI), vol. 1632, pp. 202–206. Springer, Heidelberg (1999)
13. Pientka, B.: Tabled higher-order logic programming. PhD thesis, Department of Computer Science, Carnegie Mellon University, CMU-CS-03-185 (2003)
14. Pientka, B., Dunfield, J.: Beluga: a framework for programming and reasoning with deductive systems (System Description). In: Giesl, J., Hähnle, R. (eds.) IJCAR 2010. LNCS, vol. 6173, pp. 15–21. Springer, Heidelberg (2010)
15. Poswolsky, A., Schürmann, C.: System description: Delphin—a functional programming language for deductive systems. In: International Workshop on Logical Frameworks and Meta-Languages: Theory and Practice (LFMTP 2008). ENTCS, vol. 228, pp. 135–141. Elsevier, Amsterdam (2009)
16. Reed, J.: Higher-order constraint simplification in dependent type theory. In: International Workshop on Logical Frameworks and Meta-Languages: Theory and Practice, LFMTP 2009 (2009)

17. Reed, J.: A Hybrid Logical Framework. PhD thesis, School of Computer Science, Carnegie Mellon University (2009)
18. Schack-Nielson, A., Schürmann, C.: Pattern unification for the lambda calculus with linear and affine types. In: International Workshop on Logical Frameworks and Meta-Languages: Theory and Practice (LFMTP 2010). EPTCS, vol. 34, pp. 101–116 (2010)
19. Watkins, K., Cervesato, I., Pfenning, F., Walker, D.: A concurrent logical framework I: Judgements and properties. Technical report, School of Computer Science, Carnegie Mellon University, Pittsburgh (2003)

Classical Call-by-Need and Duality

Zena M. Ariola[1], Hugo Herbelin[2], and Alexis Saurin[2]

[1] University of Oregon
ariola@cs.uoregon.edu
[2] Laboratoire PPS, équipe πr^2, CNRS, INRIA & Université Paris Diderot
{herbelin,saurin}@pps.jussieu.fr

Abstract. We study call-by-need from the point of view of the duality between call-by-name and call-by-value. We develop sequent-calculus style versions of call-by-need both in the minimal and classical case. As a result, we obtain a natural extension of call-by-need with control operators. This leads us to introduce a call-by-need $\lambda\mu$-calculus. Finally, by using the dualities principles of $\overline{\lambda}\mu\tilde{\mu}$-calculus, we show the existence of a new call-by-need calculus, which is distinct from call-by-name, call-by-value and usual call-by-need theories.

Keywords: call-by-need, lazy evaluation, duality of computation, sequent calculus, $\lambda\mu$-calculus, classical logic, control, subtraction connective.

Introduction

The theory of call-by-name λ-calculus [9,8] is easy to define. Given the syntax of λ-calculus $M ::= x \mid \lambda x.M \mid M\,M$, the reduction semantics is entirely determined by the β-reduction rule $(\lambda x.M)\,N \rightarrow_\beta M[x \leftarrow N]$ in the sense that:

- for any closed term M, either M is a value $\lambda x.N$ or M is a β-redex and for all $M \twoheadrightarrow V$, there is standard path $M \longmapsto V'$ made only of β-redexes at the head, together with the property that $V' \twoheadrightarrow V$ using internal β-reductions;
- the observational closure of β induces a unique rule η that fully captures observational equality over finite normal terms (Böhm theorem);
- the extension with control, typically done à la Parigot [30], is relatively easy to get by adding just two operational rules and an observational rule (though the raw version of Böhm theorem fails [15,35]).

The theory of call-by-value λ-calculus, as initiated by Plotkin [31], has a similar property with respect to the β_v rule (the argument of β being restricted to a variable or a $\lambda x.M$ only), but the observational closure is noticeably more complex: it at least includes the rules unveiled by Moggi [27] as was shown by Sabry and Felleisen [34]. Extensions of standardization and Böhm theorem for call-by-value are more delicate than in call-by-name [24,33].

Comparatively, call-by-need λ-calculus, though at the core of programming languages such as Haskell [16], is at a much less advanced stage of development.

L. Ong (Ed.): TLCA 2011, LNCS 6690, pp. 27–44, 2011.

The first approach to call-by-need as a calculus goes back to the 90's with the works of Ariola *et al.* [3] and Maraist *et al.* [26] for whom the concern was the characterization of standard weak-head reduction. Our paper aims at studying call-by-need in terms that are familiar to the study of call-by-name and call-by-value. In particular, we will address the question of adding control to call-by-need and the question of what is the dual of call-by-need along the lines of the duality between call-by-name and call-by-value [20,36,11]. Call-by-need is close to call-by-value in the sense that only values are substituted, but call-by-need is also close to call-by-name in the sense that only those terms that are bound to needed variables are evaluated. In particular, with respect to evaluation of pure closed terms, the call-by-name and call-by-need calculi are not distinguishable. In order to tackle the problem of developing a classical version of call-by-need, we first study how to formulate (minimal) call-by-need in the sequent calculus setting [22] (while current call-by-need calculi are based on natural deduction [32]). An advantage of a sequent calculus presentation of a calculus is that its extension to the classical case does not require the introduction of new rules but simply the extension of existing ones [7].

Curien and Herbelin [11] designed a calculus that provides an appealing computational interpretation of proofs in sequent calculus, while providing at the same time a syntactic duality between terms, *i.e.*, *producers*, and evaluation contexts, *i.e.*, *consumers*, and between the call-by-name and call-by-value reduction strategies. By giving priority to the producer one obtains call-by-value, whereas by giving priority to the consumer one obtains call-by-name. In this paper, we present how call-by-need fits in the duality of computation. Intuitively, call-by-need corresponds to focusing on the consumer to the point where the producer is *needed*. The focus goes then to the producer till a value is reached. At that point, the focus returns to the consumer. We call this calculus *lazy call-by-value*, it is developed in Section 2 and 3. In addition to the properties of confluence and standardization, we show its correctness with respect to the call-by-name sequent calculus [11]. In Section 4, we develop the natural deduction presentation of call-by-need. The reduction theory is contained in the one of Maraist *et al.* [26] and extends the one of Ariola *et al.* [3]. Interestingly, the sequent calculus has suggested an alternative standard reduction which consists of applying some axioms (*i.e.*, *lift* and *assoc*) eagerly instead of lazily. In Section 5, we show that the natural deduction and sequent calculus call-by-need are in reduction correspondence. In Section 6, we extend the minimal sequent calculus call-by-need with control, in both sequent calculus and natural deduction form. The calculi still enjoy confluence and standardization. The sequent calculus presentation of call-by-need naturally leads to a dual call-by-need, which corresponds to focusing on the producer and going to the consumer on a need basis. We call this calculus lazy call-by-name. In Section 7, we show how the dual call-by-need is obtained by dualizing the lazy call-by-value extended with the subtraction connective. We conclude and discuss our future work in Section 8. We start next with an overview of the duality of computation.

1   The Duality of Computation

Curien and Herbelin [11] provided classical sequent calculus with a term assignment, which is called the $\overline{\lambda}\mu\tilde{\mu}$ calculus. In $\overline{\lambda}\mu\tilde{\mu}$ there are two *dual* syntactic categories: terms which *produce* values and contexts which *consume* values. The interaction between a producer v and a consumer e is rendered by a command written as $\langle v \| e \rangle$, which is the computational counterpart of a sequent calculus cut. Contexts can be seen as evaluation contexts, that is, commands with a hole, written as $\square$, standing for the term whose computation is to be done next: $\langle \square \| e \rangle$. Thus, a command $\langle v \| e \rangle$ can be seen as filling the hole of the evaluation context e with v. Dually, terms can also be seen as commands with a *context* hole, standing for the context in which the term shall be computed. The duality of terms and contexts is also reflected at the variable level. One has two distinct sets of variables. The usual term variables $(x, y, \cdots)$ and the context variables $(\alpha, \beta, \cdots)$, which correspond to continuation variables. The set of terms, in addition to variables and lambda abstractions, contains a term of the form $\mu\alpha.c$, where c is a command, after Parigot's $\lambda\mu$-calculus [30]. The μ construct is similar to Felleisen's $\mathcal{C}$ control operator [18,19,17] and one can approximatively read $\mu\alpha.c$ as $\mathcal{C}(\lambda\alpha.c)$ (see [4] for a detailed analysis of the differences). Whereas the μ construct allows one to give a name to a context, so as to invoke it later, the dual construct, named $\tilde{\mu}$, allows one to name terms. One can read $\tilde{\mu}x.c$ as let $x = \square$ in c. Given a context e, $v \cdot e$ is also a context, which corresponds to an applicative context of the form $e[\square v]$. The grammar of $\overline{\lambda}\mu\tilde{\mu}$ and its reduction theory are given below:

$$c ::= \langle v \| e \rangle \qquad v ::= x \mid \lambda x.v \mid \mu\alpha.c \qquad e ::= \alpha \mid \tilde{\mu}x.c \mid v \cdot e$$

$$
\begin{array}{lll}
(\beta) & \langle \lambda x.v \| v' \cdot e \rangle & \rightarrow \langle v' \| \tilde{\mu}x.\langle v \| e \rangle \rangle \\
(\mu) & \langle \mu\alpha.c \| e \rangle & \rightarrow c[\alpha \leftarrow e] \\
(\tilde{\mu}) & \langle v \| \tilde{\mu}x.c' \rangle & \rightarrow c'[x \leftarrow v]
\end{array}
$$

The reduction theory can be seen as consisting of structural reduction rules, μ and $\tilde{\mu}$, as well as logical reduction rules (here, only β, the rule corresponding to implication).

The calculus is not confluent due to a critical pair between μ and $\tilde{\mu}$:

$$\langle z \| \beta \rangle \leftarrow_\mu \langle \mu\alpha.\langle z \| \beta \rangle \| \tilde{\mu}x.\langle y \| \beta \rangle \rangle \rightarrow_{\tilde{\mu}} \langle y \| \beta \rangle$$

To regain confluence one can impose a strategy on how to resolve the critical pair $\mu/\tilde{\mu}$. By giving priority to the $\tilde{\mu}$ rule one captures *call-by-name*, whereas by giving priority to the μ rule one captures *call-by-value*. More generally, one can describe various ways to specialize the pair $\mu/\tilde{\mu}$ as reduction rules parametrized by sets $\mathcal{E}$ and $\mathcal{V}$, which denote sets of contexts and terms, respectively:

$$
\begin{array}{lll}
(\mu_\mathcal{E}) & \langle \mu\alpha.c \| e \rangle \rightarrow c[\alpha \leftarrow e] & \text{if } e \in \mathcal{E} \\
(\tilde{\mu}_\mathcal{V}) & \langle v \| \tilde{\mu}x.c' \rangle \rightarrow c'[x \leftarrow v] & \text{if } v \in \mathcal{V}
\end{array}
$$

This presentation with parametric rules is inspired by the work of Ronchi and Paolini on parametric λ-calculus [33]. A strategy corresponds to specifying which contexts and terms can be duplicated or erased. For call-by-name, $\mathcal{V}$ is instantiated by V_n below but there are two possible choices E_n^{basic} and E_n for instantiating $\mathcal{E}$.

$$V_n ::= x \mid \lambda x.v \mid \mu\alpha.c \qquad E_n^{basic} ::= \alpha \mid v \cdot \tilde{\mu}x.c \qquad E_n ::= \alpha \mid v \cdot E_n$$

In both cases, this captures the fact that in call-by-name every term can be substituted for a term variable, while only specific contexts can be substituted for a context variable. In particular, the context $\tilde{\mu}x.c$ is not substitutable. The difference between the basic strategy and the second strategy is reminiscent of the difference between Plotkin's call-by-name continuation-passing-style semantics [31] and Lafont-Reus-Streicher's one [25]: the first one is not compatible with η but the second is. In the rest of the paper, we will consider the second strategy and the reduction rules corresponding to V_n and E_n are denoted as μ_n and $\tilde{\mu}_n$, respectively. We refer to E_n as an applicative context since it consists of a list of arguments. For call-by-value, the instantiations are V_v and E_v:

$$V_v ::= x \mid \lambda x.v \qquad E_v ::= \alpha \mid v \cdot E_v \mid \tilde{\mu}x.c$$

capturing the fact that only restricted terms (values) are substituted for a term variable, while every context can be substituted for a context variable. The resulting reduction rules are denoted as μ_v and $\tilde{\mu}_v$, respectively. Notice also that full non-deterministic $\overline{\lambda}\mu\tilde{\mu}$ corresponds to choosing μ_v together with $\tilde{\mu}_n$. As discussed next, call-by-need $\overline{\lambda}\mu\tilde{\mu}$-calculus will be defined with respect to another choice of parameters.

In addition to the instantiations of the structural rules $\mu_{\mathcal{E}}$ and $\tilde{\mu}_{\mathcal{V}}$, the calculi developed in the rest of the paper will contain rules for evaluating connectives. We will only consider implication, except in Section 7 where subtraction will also be added. We will also consider the following instances of the usual extensionality rules for μ and $\tilde{\mu}$ in $\overline{\lambda}\mu\tilde{\mu}$:

$$\begin{aligned}
(\eta_\mu^{\mathcal{V}}) \; &\mu\alpha.\langle v\|\alpha\rangle \rightarrow v & v \in \mathcal{V} \text{ and } \alpha \text{ not free in } v \\
(\eta_{\tilde{\mu}}^{\mathcal{E}}) \; &\tilde{\mu}x.\langle x\|e\rangle \rightarrow e & e \in \mathcal{E} \text{ and } x \text{ not free in } e
\end{aligned}$$

We denote the corresponding call-by-value and call-by-name instantiations as $\eta_\mu^v/\eta_{\tilde{\mu}}^v$ and $\eta_\mu^n/\eta_{\tilde{\mu}}^n$, respectively.

2   Call-by-Need and Duality

As we did for call-by-name and call-by-value, we have to specify the parametric sets used for call-by-need, that is, which terms and contexts can be substituted for term and context variables. Since call-by-need avoids duplication of work, it is natural to restrict the set $\mathcal{V}$ to V_v, thus allowing substitution of variables and lambda abstractions only. In order to specify which contexts are substitutable, it is important to notice that the goal of the structural rules is to bring to the top of a term the redex to be perfomed [2]. Thus, one should allow the reduction of $\langle \mu\alpha.\langle I\|I \cdot \alpha\rangle\|v \cdot \alpha\rangle$ (I stands for $\lambda x.x$) to $\langle I\|I \cdot v \cdot \alpha\rangle$ since the applicative redex (*i.e.*, the one underlined) is *needed* in order to continue the computation. This implies that E_n should be part of $\mathcal{E}$. That however is not enough. One would also want to reduce $\langle \mu\alpha.\langle I \| I \cdot \alpha \rangle \| \tilde{\mu}x.\langle x \| \alpha \rangle \rangle$ to $\langle I \| I \cdot \tilde{\mu}x.\langle x \| \alpha \rangle \rangle$. This however does not imply that $\tilde{\mu}x.c$ should be part of $\mathcal{E}$ since that would unveil an unwanted redex, such as in $\langle \mu\alpha.\langle I \| I \cdot \alpha \rangle \| \tilde{\mu}x.\langle z \| \alpha \rangle \rangle$. The only time we want to allow a change of focus from the consumer to the producer is when the producer is *needed*, which means that the variable bound to the producer occurs in the hole of a context; x is needed in $\langle x \| E_n \rangle$ but it is not needed in $\langle x \| \tilde{\mu}y.\langle z \| y \cdot \alpha \rangle \rangle$. This notion will still not capture a situation such as $\langle \mu\alpha.\langle I \| I \cdot \alpha \rangle \| \tilde{\mu}x.\langle v_1 \| \tilde{\mu}y.\langle x \| E_n \rangle \rangle \rangle$, since the needed variable is buried under the binding for y. This motivates the introduction of the notion of a call-by-need meta-context, which is simply a command hole surrounded by $\tilde{\mu}$-bindings:

$$C_l^{\tilde{\mu}} ::= \Box \mid \langle \mu\alpha.c \| \tilde{\mu}z.C_l^{\tilde{\mu}} \rangle$$

A variable x is needed in a command c, if c is of the form $C_l^{\tilde{\mu}}[\langle x \| E_n \rangle]$.

We have so far determined that $\mathcal{E}$ contains the call-by-name applicative contexts and contexts of the form $\tilde{\mu}x.C_l^{\tilde{\mu}}[\langle x \| E_n \rangle]$. This would allow the reduction of $\langle \mu\alpha.\langle I \| I \cdot \alpha \rangle \| \tilde{\mu}f.\langle f \| f \cdot \alpha \rangle \rangle$ to $\langle I \| I \cdot \tilde{\mu}f.\langle f \| f \cdot \alpha \rangle \rangle$. The problem is that the call-by-name applicative context considered so far does not contain a $\tilde{\mu}$. This is necessary to capture sharing. For example, in the above term $\langle I \| I \cdot \tilde{\mu}f.\langle f \| f \cdot \alpha \rangle \rangle$, the $\tilde{\mu}f$ captures the sharing of II. We need however to be careful about which $\tilde{\mu}$ we allow in the notion of applicative context. For example, we should disallow contexts such as $I \cdot \tilde{\mu}f.\langle z \| f \cdot \alpha \rangle$ since they might cause unwanted computation. Indeed, in the following reduction the application of I to I is computed while it is not needed to derive the result:

$$\langle I \| I \cdot \tilde{\mu}f.\langle z \| f \cdot \alpha \rangle \rangle \rightarrow_\beta \langle I \| \tilde{\mu}x.\langle x \| \tilde{\mu}f.\langle z \| f \cdot \alpha \rangle \rangle \rangle \rightarrow_{\tilde{\mu}_v} \langle I \| \tilde{\mu}f.\langle z \| f \cdot \alpha \rangle \rangle \rightarrow_{\tilde{\mu}_v} \langle z \| I \cdot \alpha \rangle.$$

This implies that a context $\tilde{\mu}x.c$ is allowed in an applicative context only if c demands x.

We are ready to instantiate the structural and extensional rules; $\mathcal{V}$ and $\mathcal{E}$ are instantiated as follows:

$$V_v ::= x \mid \lambda x.v \qquad E_l ::= F \mid \tilde{\mu}x.C_l^{\tilde{\mu}}[\langle x \| F \rangle] \quad \text{with} \quad F ::= \alpha \mid v \cdot E_l$$

resulting in reduction rules that we will denote as μ_l, $\tilde{\mu}_v$ and η_μ^v.

3   Minimal Call-by-Need in Sequent Calculus Form ($\overline{\lambda}_{mlv}$)

A classical sequent calculus naturally provides a notion of control. However, one can restrict the calculus to be control-free by limiting the set of continuation variables to a single variable, conventionally written $\star$, which is linearly used. This corresponds to the restriction to minimal logic [1]. We introduce next the lazy call-by-value calculus, $\overline{\lambda}_{mlv}$.

Definition 1. *The syntax of $\overline{\lambda}_{mlv}$ is defined as follows:*

$$
\begin{array}{lll}
\text{command} & c & ::= \langle v \| e \rangle \\
\text{term} & v & ::= x \mid \lambda x.v \mid \mu\star.c \\
\text{context} & e & ::= E \mid \tilde{\mu}x.c \\
\text{yielding context} & E & ::= F \mid \tilde{\mu}x.C[\langle x\|F\rangle] \\
\text{applicative context} & F & ::= \star \mid v \cdot E \\
\text{meta-context} & C & ::= \square \mid \langle \mu\star.c\|\tilde{\mu}z.C\rangle
\end{array}
$$

The reduction of $\overline{\lambda}_{mlv}$, written as $\rightarrow_{mlv}$, denotes the compatible closure of β, μ_l, $\tilde{\mu}_v$ and η_μ^v; the relation $\twoheadrightarrow_{mlv}$ denotes the reflexive and transitive closure of $\rightarrow_{mlv}$ while $=_{mlv}$ denotes its reflexive, symmetric and transitive closure. The notion of weak head standard reduction is defined as:

$$
\frac{c \rightarrow_\beta c'}{C[c] \mapsto_{mlv} C[c']} \qquad \frac{c \rightarrow_{\mu_l} c'}{C[c] \mapsto_{mlv} C[c']} \qquad \frac{c \rightarrow_{\tilde{\mu}_v} c'}{C[c] \mapsto_{mlv} C[c']}
$$

The notation $\mapsto\!\!\!\rightarrow_{mlv}$ stands for the reflexive and transitive closure of $\mapsto_{mlv}$. A weak head normal form (whnf) is a command c such that for no c', $c \mapsto_{mlv} c'$.

Notice how in the lazy call-by-value calculus, the standard redex does not necessarily occur at the top level. In $\langle v_1\|\tilde{\mu}x_1.\langle v_2\|\tilde{\mu}x_2.\langle \lambda x.v\|s \cdot \star\rangle\rangle\rangle$, the standard redex is buried under the bindings for x_1 and x_2, which is why the standard reduction refers to the meta-context. This however can be solved simply by going to a calculus with explicit substitutions, which would correspond to an abstract machine we are currently investigating. Some more discussions on this topic are available in section 8. Note that in a term of the form $\langle \lambda z.v\|\tilde{\mu}x.\langle x\|\tilde{\mu}y.\langle y\|\star\rangle\rangle\rangle$, the substitution for y is not the standard redex, and in

$$
\langle \mu\star.\langle I\|I \cdot \star\rangle\|\tilde{\mu}x.\langle x\|\tilde{\mu}y.\langle y\|\star\rangle\rangle\rangle \qquad \underline{\langle \mu\star.\langle V\|\tilde{\mu}y.\langle y\|\star\rangle\rangle\|\tilde{\mu}x.\langle x\|\star\rangle\rangle}
$$

the standard redex is the underlined one. The η_μ^v rule is not needed for standard reduction. The η_μ^v rule turns a computation into a value, allowing for example the reduction: $\langle \mu\star.\langle V\|\star\rangle\|\tilde{\mu}x.\langle y\|x \cdot \star\rangle\rangle \rightarrow \langle V\|\tilde{\mu}x.\langle y\|x \cdot \star\rangle\rangle \rightarrow \langle y\|V \cdot \star\rangle$, which is not standard; in fact, the starting term is already in whnf.

Proposition 1 (Confluence). *$\rightarrow_{mlv}$ is confluent.*

Indeed, the only critical pair in $\overline{\lambda}_{mlv}$ is between η_μ^v and μ_l and it trivially converges since both rules produce the same resulting command.

Remark 1. In $\overline{\lambda}_{mlv}$ the duplicated redexes are all disjoint. This was not the situation in λ_{need} [26], where the *assoc* rule could have duplicated a *lift* redex. This does not happen in $\overline{\lambda}_{mlv}$ because the contexts are moved all at once, as described in the example below, which mimics the situation in λ_{need}.

$$
\begin{array}{ccc}
\langle \mu\star\langle \mu\star.\langle z\|\tilde{\mu}y.\langle y\|\star\rangle\rangle\|\tilde{\mu}x.\langle x\|\star\rangle\rangle\|N \cdot \star\rangle & \rightarrow_{\mu_n} & \langle \mu\star.\langle z\|\tilde{\mu}y.\langle y\|\tilde{\mu}x.\langle x\|\star\rangle\rangle\rangle\|N \cdot \star\rangle \\
\downarrow_{\mu_n} & & \downarrow_{\mu_n} \\
\langle \mu\star.\langle z\|\tilde{\mu}y.\langle y\|\star\rangle\rangle\|\tilde{\mu}x.\langle x\|N \cdot \star\rangle\rangle & \rightarrow_{\mu_n} & \langle z\|\tilde{\mu}y.\langle y\|\tilde{\mu}x.\langle x\|N \cdot \star\rangle\rangle\rangle
\end{array}
$$

The needed constraint breaks the property that commands in weak head normal form are of the form $\langle x \| E \rangle$ or $\langle \lambda x.v \| \star \rangle$ (a property that holds for $\overline{\lambda}\mu\tilde{\mu}$ in call-by-name or call-by-value).

Definition 2. *Let $\boldsymbol{x}$ be a sequence of variables. $c_{\boldsymbol{x}}$ is defined by the grammar:*

$$c_{\boldsymbol{x}} ::= \langle \mu\star.c \| \tilde{\mu}y.c_{y\boldsymbol{x}} \rangle \mid \langle \lambda x.v \| \star \rangle \mid \langle z \| F \rangle \qquad z \notin \boldsymbol{x}.$$

Note that in $c_{\boldsymbol{x}}$, $\boldsymbol{x}$ records the variables which are $\tilde{\mu}$-bound to a computation on the path from the top of the term to the current position.

Proposition 2. *A command c is in weak head normal form iff it is in c_{ϵ}, where ϵ denotes the empty sequence of variables.*

Indeed, any command in c_{ϵ} is in whnf. Conversily, if c is in whnf, there is no c' such that $c \mapsto_{mlv} c'$ by definition, which means that c must be either of the form $C[\langle \lambda x.v \| \star \rangle]$ or $C[\langle z \| F \rangle]$ with z not bound by a $\tilde{\mu}$, otherwise said it must be in c_{ϵ}.

$\langle x \| \star \rangle$ is in whnf, however it is not of the form c_x since it demands variable x. Neither $\langle y \| \tilde{\mu}x.c \rangle$ nor $\langle \mu\star.c \| \tilde{\mu}x.\langle x \| \star \rangle \rangle$ are in whnf. A whnf is either of the form $C[\langle x \| F \rangle]$ or $C[\langle \lambda x.v \| \star \rangle]$. It is easy to see that standard redexes are mutually exclusive. For instance, a command c which had a standard β redex cannot have a μ_l or $\tilde{\mu}_v$ redex. Hence:

Proposition 3 (Unique Decomposition). *A command c is either a whnf or there exists a unique meta-context C and redex c' such that c is of the form $C[c']$.*

Using standard techniques (commutation of inner reductions with standard reduction), the following easily comes by induction:

Proposition 4 (Standardization). *Given a command c and a whnf c', if $c \twoheadrightarrow_{mlv} c'$ then there exists a whnf c'' such that $c \mapsto\!\!\!\twoheadrightarrow_{mlv} c''$ and $c'' \twoheadrightarrow_{mlv} c'$.*

3.1   Soundness and Completeness of $\overline{\lambda}_{mlv}$

The call-by-need calculus in natural deduction form is observationally equivalent to call-by-name. We show next that the same holds for call-by-need in sequent calculus form. To that end, we first review Curien and Herbelin call-by-name sequent calculus, called $\overline{\lambda}\mu\tilde{\mu}_T$ (after Danos *et al*'s LKT [12,13]). $\overline{\lambda}\mu\tilde{\mu}_T$ restricts the syntax of legal contexts capturing the intuition that according to the call-by-name continuation passing style, the continuation follows a specific pattern. The syntax of $\overline{\lambda}\mu\tilde{\mu}_T$ becomes:

$$c ::= \langle v \| e \rangle \qquad v ::= V_n \qquad e ::= \tilde{\mu}x.c \mid E_n$$

Notice that whereas $v \cdot \tilde{\mu}x.c$ is a legal context in $\overline{\lambda}\mu\tilde{\mu}$, it is not legal in $\overline{\lambda}\mu\tilde{\mu}_T$. The reduction theory of $\overline{\lambda}\mu\tilde{\mu}_T$ consists of β, μ_n and $\tilde{\mu}_n$.

The $\overline{\lambda}_{mlv}$ calculus is sound and complete with respect to the minimal restriction of $\overline{\lambda}\mu\tilde{\mu}_T$. We first need to translate $\overline{\lambda}_{mlv}$ terms to $\overline{\lambda}\mu\tilde{\mu}_T$ terms by giving a name to the $\tilde{\mu}$-term contained in a linear context. The translation, written as $(.)^\circ$, is defined as follows (the interesting cases of the translation are the last two cases), with $n \geq 0$:

$$x^\circ = x$$
$$(\lambda x.v)^\circ = \lambda x.v^\circ$$
$$(\mu\star.c)^\circ = \mu\star.c^\circ$$
$$(\langle v\|w_1\cdot\ldots w_n\cdot\star\rangle)^\circ = \langle v^\circ\|w_1^\circ\cdot\ldots w_n^\circ\cdot\star\rangle$$
$$(\langle v\|\tilde{\mu}x.c\rangle)^\circ = \langle v^\circ\|\tilde{\mu}x.c^\circ\rangle$$
$$(\langle v\|w_0\cdot\ldots w_n\cdot\tilde{\mu}x.c\rangle)^\circ = \langle \mu\star.\langle v^\circ\|w_0^\circ\cdot\ldots w_n^\circ\cdot\star\rangle\|\tilde{\mu}x.c^\circ\rangle$$

we then have the following properties:

Lemma 1. *If c is a command in $\overline{\lambda}_{mlv}$, then c° is a command in $\overline{\lambda}\mu\tilde{\mu}_T$.*

The previous lemma holds since the syntactical constraint on $\overline{\lambda}\mu\tilde{\mu}_T$ commands is that a context is either of the form $\tilde{\mu}x.c$ or it is a stack a terms pushed on top of $\star$: the translation precisely achieves this goal.

Proposition 5. *(i) Given a $\overline{\lambda}_{mlv}$ term v, $v =_{mlv} v^\circ$.*
  (ii) Given terms v and w in $\overline{\lambda}\mu\tilde{\mu}_T$, v and w are also in $\overline{\lambda}_{mlv}$ and we have:
  (a) $v =_{mlv} w$ then $v =_{\overline{\lambda}\mu\tilde{\mu}_T} w$;
  (b) $v =_{\overline{\lambda}\mu\tilde{\mu}_T} \langle\lambda x.w\|\star\rangle$ then $v =_{mlv} C[\langle\lambda x.w'\|\star\rangle]$ for some C and w'.

Indeed, $\overline{\lambda}_{mlv}$ theory restricted to the call-by-name syntax of $\overline{\lambda}\mu\tilde{\mu}_T$ is included in $\overline{\lambda}\mu\tilde{\mu}_T$ theory.

Intermezzo 2. Soundness can also be shown with respect to the $\overline{\lambda}\mu\tilde{\mu}$ calculus without the need of doing a translation, since the $\overline{\lambda}\mu\tilde{\mu}$ calculus does not impose any restrictions on the context. This however requires extending the $\tilde{\mu}$ rule to $\langle v\|v_1\cdots v_n.\tilde{\mu}x.c\rangle \to c[x = \mu\star.\langle v\|v_1\cdots v_n.\star\rangle]$. The rule is sound for call-by-name extended with the eta rule, called $\eta^R_{\to}$ in [23], given as $y = \lambda x.\alpha.\langle y\|x.\alpha\rangle$. We have:

$$\langle v\|w\cdot\tilde{\mu}x.c\rangle =_{\tilde{\mu}} \langle v\|\tilde{\mu}y.\langle y\|w\cdot\tilde{\mu}x.c\rangle\rangle$$
$$=_{\eta^R_{\to}} \langle v\|\tilde{\mu}y.\langle\lambda z.\mu\star.\langle y\|z\cdot\star\rangle\|w\cdot\tilde{\mu}x.c\rangle\rangle$$
$$=_{\to} \langle v\|\tilde{\mu}y.\langle w\|\tilde{\mu}z.\langle\mu\star.\langle y\|z\cdot\star\rangle\|\tilde{\mu}x.c\rangle\rangle\rangle$$
$$=_{\tilde{\mu}} \langle v\|\tilde{\mu}y.\langle w\|\tilde{\mu}z.c[x = \mu\star.\langle y\|z\cdot\star\rangle]\rangle\rangle$$
$$=_{\tilde{\mu}} \langle v\|\tilde{\mu}y.c[x = \mu\star.\langle y\|w\cdot\star\rangle]\rangle$$
$$=_{\tilde{\mu}} c[x = \mu\star.\langle v\|w\cdot\star\rangle]$$

4   Minimal Call-by-Need in Natural Deduction (λ_{need})

We now present a natural deduction counterpart to $\overline{\lambda}_{mlv}$.

Definition 3. *The syntax of λ_{need} is defined as follows:*

term	M	$::= V \mid M_{nv}$
value	V	$::= x \mid \lambda x.M$
computation	M_{nv}	$::= MM \mid \text{let } x = M \text{ in } N$
applicative context	C_{ap}	$::= \Box \mid C_{ap}M$
needed context	C	$::= C_{ap} \mid \text{let } x = M_{nv} \text{ in } C \mid \text{let } x = C_{ap}M \text{ in } C[x]$

Reduction in λ_{need}, written as $\rightarrow_{need}$, is the compatible closure of the following rules:

(β)	$(\lambda x.N)M$	$\rightarrow$ let $x = M$ in N
($lift$)	(let $x = M$ in P)N	$\rightarrow$ let $x = M$ in PN
($deref_v$)	let $x = V$ in M	$\rightarrow M[x \leftarrow V]$
($assoc$)	let $z = ($let $x = M$ in $N)$ in $C[z]$	$\rightarrow$ let $x = M$ in let $z = N$ in $C[z]$

The relation $\twoheadrightarrow_{need}$ denotes the reflexive and transitive closure of $\rightarrow_{need}$. The notion of weak head standard reduction is defined as:

$$\frac{M \rightarrow_{\beta,lift} N}{C_{\beta l}[M] \mapsto_{need} C_{\beta l}[N]} \qquad \frac{M \rightarrow_{deref_v,assoc} N}{C_{da}[M] \mapsto_{need} C_{da}[N]}$$

where

$$C_{\beta l} ::= C_{ap} \mid \text{let } x = M_{nv} \text{ in } C_{\beta l} \mid \text{let } x = C_{ap} \text{ in } C[x]$$
$$C_{da} ::= \square \mid \text{let } x = M_{nv} \text{ in } C_{da}$$

The notation $\mapsto\!\!\!\rightarrow_{need}$ stands for the reflexive and transitive closure of $\mapsto_{need}$. A weak head normal form (whnf) is a term M such that for no N, $M \mapsto_{need} N$.

Unlike the calculi defined by Maraist *et al.* [26] and Ariola *et al.* [3], the *deref$_v$* rule follows the call-by-value discipline since it substitutes a value for each occurrence of the bound variable, even if the variable is not needed. The rule is derivable in the calculus of Maraist *et al.* using garbage collection. The *assoc* rule is more constrained than in the calculus of Maraist *et al.* since it performs the flattening of the bindings on a demand basis. The *assoc* requires the variable to appear in the hole of a context C, whose definition does not allow a hole to be bound to a let variable. For example, let $x = \square$ in x and let $x = \square$ in let $y = x$ in y are not C contexts. This restriction is necessary to make sure that in a term of the form

$$\text{let } x = (\text{let } z = N \text{ in } P) \text{ in let } y = x \text{ in } y$$

the standard redex is the substitution for y and not the *assoc* redex. The *assoc* rule is more general than in [3], since it does not require the binding for z to be an answer (*i.e.*, an abstraction surrounded by bindings). The *lift* rule is the same as in [26], it is more general than the corresponding rule in [3] since the left-hand side of the application is not restricted to be an answer. The calculi in [26] and [3] share the same standard reduction. For example, in the terms:

$$(\text{let } y = M \text{ in } (\lambda x.x)y)P \qquad \text{let } y = (\text{let } z = N \text{ in } (\lambda x.x)y) \text{ in } y$$

$(\lambda x.x)y$ is the standard redex. Our standard reduction differs. The above terms correspond to a *lift* and *assoc* redex, respectively. Moreover, our standard reduction is also defined for open terms. Thus, the following terms:

$$(\text{let } y = xz \text{ in } y)P \qquad \text{let } y = (\text{let } z = xP \text{ in } z) \text{ in } y$$

instead of being of the form $C[x]$, reduce further. The standard reduction requires different closure operations to avoid the interference between reductions. In

$$\text{let } z = (\text{let } x = V \text{ in } N) \text{ in } z \qquad \text{let } y = (\text{let } z = (\text{let } x = M \text{ in } N) \text{ in } P) \text{ in } y$$

the standard redex is the (outermost) *assoc*, and in let $x = II$ in let $y = x$ in y, the $deref_v$ is the standard redex.

Proposition 6 (Confluence). $\rightarrow_{need}$ *is confluent.*

This is because all critical pairs converge.

Proposition 7 (Unique Decomposition). *A term M is either a whnf or there exists a unique $C_{\beta l}$ such that M is of the form $C_{\beta l}[P]$, where P is a β or lift redex, or there exists a unique C_{da} such that M is of the form $C_{da}[P]$, where P is a $deref_v$ or assoc redex.*

The previous proposition essentially relies on the facts that $C[x]$ is a whnf and that $C_{da} \subset C \subset C_{\beta l}$.

Proposition 8 (Standardization). *Given a term M and whnf N, if $M \twoheadrightarrow_{need} N$ then there exists a whnf N' such that $M \longmapsto_{need} N'$ and $N' \twoheadrightarrow_{need} N$.*

Definition 4. *Let $\boldsymbol{x}$ be a sequence of variables. $M_{\boldsymbol{x}}$ is defined as:*

$$M_{\boldsymbol{x}} ::= \lambda x.N \quad | \text{ let } y = N_{nv} \text{ in } M_{y\boldsymbol{x}}$$
$$| \, zN_1 \cdots N_n \, | \text{ let } y = zNN_1 \cdots N_n \text{ in } C[y] \qquad z \notin \boldsymbol{x}$$

Proposition 9. *A term M is in whnf iff it is in M_ϵ (with ϵ the empty sequence).*

4.1   Soundness and Completeness of λ_{need}

Our calculus is sound and complete for evaluation to an answer (*i.e.*, an abstraction or a let expression whose body is an answer) with respect to the standard reduction of the call-by-need calculi defined in [26] and [3], denoted by $\longmapsto_{mow}^{af}$.

Proposition 10. *Let M be a term and A be an answer.*

- *If $M \longmapsto_{need} A$ then there exists an answer A' such that $M \longmapsto_{mow}^{af} A'$;*
- *If $M \longmapsto_{mow}^{af} A$ then there exists an answer A' such that $M \longmapsto_{need} A'$.*

Indeed, the discussion at the beginning of the section evidences that $\rightarrow_{need}$ is contained in $\rightarrow_{mow}$ and contains $\rightarrow_{af}$ and the result follows from the fact that standard reductions of $\rightarrow_{mow}$ and $\rightarrow_{af}$ coincide.

5   Correspondence between $\overline{\lambda}_{mlv}$ and λ_{need}

The calculi $\overline{\lambda}_{mlv}$ and λ_{need} are in reduction correspondence for the following translations from λ_{need} to $\overline{\lambda}_{mlv}$ and vice-versa:

Definition 5. *Given a term M in λ_{need}, a term v, a context e and a command c in $\overline{\lambda}_{mlv}$, translations $M^{\triangleright}$, $M_e^{\triangleright}$, $v^{\triangleleft}$, $e^{\triangleleft}$ and $c^{\triangleleft}$ are defined as follows:*

$$
\begin{aligned}
x^{\triangleright} &= x \\
(\lambda x.M)^{\triangleright} &= \lambda x.M^{\triangleright} \\
(MN)^{\triangleright} &= \mu\star.(MN)_{\star}^{\triangleright} \\
(\text{let } x = M \text{ in } N)^{\triangleright} &= \mu\star.(\text{let } x = M \text{ in } N)_{\star}^{\triangleright} \\
(MN)_e^{\triangleright} &= M_{N^{\triangleright}\cdot e}^{\triangleright} \\
(\text{let } x = M \text{ in } N)_e^{\triangleright} &= \begin{cases} M_{\tilde{\mu}x.N_e^{\triangleright}}^{\triangleright} & N \equiv C[x] \\ \langle M^{\triangleright} \| \tilde{\mu}x.N_e^{\triangleright} \rangle & \text{otherwise} \end{cases} \\
V_e^{\triangleright} &= \langle V^{\triangleright} \| e \rangle
\end{aligned}
$$

$$
\begin{aligned}
\langle v \| e \rangle^{\triangleleft} &= e^{\triangleleft}[v^{\triangleleft}] \\
x^{\triangleleft} &= x \\
(\lambda x.v)^{\triangleleft} &= \lambda x.v^{\triangleleft} \\
(\mu\star.c)^{\triangleleft} &= c^{\triangleleft} \\
\star^{\triangleleft} &= \square \\
(v \cdot E)^{\triangleleft} &= E^{\triangleleft}[\square v^{\triangleleft}] \\
(\tilde{\mu}x.c)^{\triangleleft} &= \text{let } x = \square \text{ in } c^{\triangleleft}
\end{aligned}
$$

We first illustrate the correspondence on an example.

Example 3. Consider the following λ_{need} reduction, where I stands for $\lambda y.y$ and M for $(\lambda f.fI(fI))((\lambda z.\lambda w.zw)(II))$:

$$
\begin{array}{llll}
M \to_\beta \text{ let} & \to_\beta \text{ let} & \to_{assoc} \text{ let} \\
\quad f = (\lambda z.\lambda w.zw)(II) & \quad f = \text{ let} & \quad z = II \\
\quad \text{in } fI(fI) & \qquad z = II & \quad \text{in let} \\
& \qquad \text{in } \lambda w.zw & \qquad f = \lambda w.zw \\
& \quad \text{in } fI(fI) & \qquad \text{in } fI(fI)
\end{array}
$$

We have $M_\star^{\triangleright} = \langle \lambda f.\mu\star.\langle f \| I \cdot (fI)^{\triangleright} \cdot \star \rangle \| \mu\star.\langle \lambda z.\lambda w.(zw)^{\triangleright} \| (II)^{\triangleright} \cdot \star \rangle \cdot \star \rangle$. The first β step is simulated by the following $\overline{\lambda}_{mlv}$ reduction, where we underline the redex to be contracted unless it occurs at the top:

$$
\begin{aligned}
&\langle \lambda f.\mu\star.\langle f \| I \cdot (fI)^{\triangleright} \cdot \star \rangle \| \mu\star.\langle \lambda z.\lambda w.(zw)^{\triangleright} \| (II)^{\triangleright} \cdot \star \rangle \cdot \star \rangle \to_\beta \\
&\langle \mu\star.\langle \lambda z.\lambda w.(zw)^{\triangleright} \| (II)^{\triangleright} \cdot \star \rangle \| \tilde{\mu}f.\underline{\langle \mu\star.\langle f \| I \cdot (fI)^{\triangleright} \cdot \star \rangle \| \star \rangle} \rangle \to_{\mu_l} \\
&\langle \mu\star.\langle \lambda z.\lambda w.(zw)^{\triangleright} \| (II)^{\triangleright} \cdot \star \rangle \| \tilde{\mu}f.\langle f \| I \cdot (fI)^{\triangleright} \cdot \star \rangle \rangle \to_{\mu_l} \\
&\langle \lambda z.\lambda w.(zw)^{\triangleright} \| (II)^{\triangleright} \cdot \tilde{\mu}f.\langle f \| I \cdot (fI)^{\triangleright} \cdot \star \rangle \rangle
\end{aligned}
$$

The second μ_l step corresponds to moving the redex in the context let $f = \square$ in $C[f]$ at the top. The simulation of the second β step leads to:

$$
\langle (II)^{\triangleright} \| \tilde{\mu}z.\langle \lambda w.(zw)^{\triangleright} \| \tilde{\mu}f.\langle f \| I \cdot (fI)^{\triangleright} \cdot \star \rangle \rangle \rangle
$$

The *assoc* corresponds to an identity in $\overline{\lambda}_{mlv}$.

Notice that the restriction on the *assoc* rule is embedded in the sequent calculus. The simulation of a non restricted *assoc* would require a generalization of the μ_l rule. For example, the simulation of the reduction:

$$
\text{let } x = (\text{let } y = II \text{ in } y) \text{ in } 0 \to \text{ let } y = II \text{ in let } x = y \text{ in } 0
$$

would require equating the following terms:

$$
\langle \mu\star.\langle I \| I \cdot \tilde{\mu}y.\langle y \| \star \rangle \rangle \| \tilde{\mu}x.\langle 0 \| \star \rangle \rangle = \langle \mu\star.\langle I \| I \cdot \star \rangle \| \tilde{\mu}y.\langle y \| \tilde{\mu}x.\langle 0 \| \star \rangle \rangle \rangle
$$

However, those should not be equated to $\langle I \| I \cdot \tilde{\mu}y.\langle y \| \tilde{\mu}x.\langle 0 \| \star \rangle \rangle \rangle$. That would correspond to relaxing the restriction of E_l in the μ_l rule, and has the problem of bringing the redex II to the top and thus becoming the standard redex.

Proposition 11 (Simulation). *Call-by-need reduction in natural deduction and sequent calculus form are in reduction correspondence:*

(i) $M \twoheadrightarrow_{need} M^{\triangleright\triangleleft}$

(ii) $c \twoheadrightarrow_{mlv} c^{\triangleleft\triangleright}$

(iii) If $M \rightarrow_{need} N$ *then* $M^{\triangleright} \twoheadrightarrow_{mlv} N^{\triangleright}$

(iv) If $c \rightarrow_{mlv} c'$ *then* $c^{\triangleleft} \twoheadrightarrow_{need} c'^{\triangleleft}$

Remark 4. Note that the translation $(\_)^{\triangleright}_e$ of a let expression depends on the bound variable being needed or not. The choice of this *optimized* translation was required to preserve reduction. Indeed, otherwise, to simulate the *assoc* reduction one would need an expansion in addition to a reduction.

6   Classical Call-by-Need in Sequent Calculus ($\overline{\lambda}_{lv}$) and Natural Deduction Form ($\lambda\mu_{need}$)

Defining sequent calculus classical call-by-need, called $\overline{\lambda}_{lv}$, requires extending the applicative context and the μ construct to include a generic continuation variable[1]. The syntax of $\overline{\lambda}_{lv}$ becomes:

$$
\begin{array}{ll}
c ::= \langle v \| e \rangle & E ::= F \mid \tilde{\mu}x.C[\langle x \| F \rangle] \\
v ::= x \mid \lambda x.v \mid \mu\alpha.c & F ::= \alpha \mid v \cdot E \\
e ::= E \mid \tilde{\mu}x.c & C ::= \Box \mid \langle \mu\alpha.c \| \tilde{\mu}z.C \rangle
\end{array}
$$

Reduction, weak head standard reduction (written as $\rightarrow_{lv}$ and $\mapsto_{lv}$, respectively) and weak head normal form (whnf) are defined as in the minimal case by replacing $\star$ with any context variable α. For example, a term of the form $\langle \mu\alpha.\langle x \| \beta \rangle \| \tilde{\mu}x.\langle y \| y \cdot \delta \rangle \rangle$ is in weak head normal form.

Unique decomposition, confluence and standardization extend to the classical case. Once control is added to the calculus, call-by-need and call-by-name are observationally distinguishable, as witnessed by the example given in the next section. It is important to notice that the bindings are not part of the captured context. For example, in the following command, the redex II written as $\mu\alpha.\langle \lambda x.x \| (\lambda x.x) \cdot \alpha \rangle$ will be executed only once. Whereas, if the bindings were part of the captured context then that computation would occur twice.

$$\langle II \| \tilde{\mu}z.\langle \mu\alpha.\langle \lambda x.\mu\beta.\langle z \| (\mu\delta.\langle \lambda x.x \| \alpha \rangle) \cdot \beta \rangle \| \alpha \rangle \| \tilde{\mu}f.\langle f \| z \cdot \gamma \rangle \rangle \rangle\rangle$$

Unlike the sequent calculus setting, to extend minimal natural deduction to the classical case, we need to introduce two new constructs: the capture of a continuation and the invocation of it, written as $\mu\alpha.J$ and $[\alpha]M$, where J stands for a jump (*i.e.*, an invocation of a continuation). The reduction semantics makes use of the notion of *structural substitution*, which was first introduced in [30] and is written as $J[\alpha \leftarrow [\alpha]F]$ indicating that each occurrence of $[\alpha]M$ in J is replaced by $[\alpha]F[M]$, where F is the context captured by a continuation which is either

---

[1] To reduce closed commands one can introduce a constant named **tp** as in [5], or one can encode the top-level using subtraction (see Section 7).

$\Box M$ or let $x = \Box$ in $C[x]$. The benefits of structural substitution are discussed in [4]. In addition to *lift*, *assoc*, $deref_v$ and β, the reduction theory includes the following reduction rules:

$$
\begin{array}{llll}
(\mu_{ap}) & (\mu\alpha.J)M & \to \mu\alpha.J[\alpha \leftarrow [\alpha](\Box M)] \\
(\mu_{let}) & \text{let } x = \mu\alpha.J \text{ in } C[x] & \to \mu\alpha.J[\alpha \leftarrow [\alpha](\text{let } x = \Box \text{ in } C[x])] \\
(\mu_{lift}) & \text{let } x = M_{nv} \text{ in } \mu\alpha.[\beta]N & \to \mu\alpha.[\beta](\text{let } x = M_{nv} \text{ in } N) \\
(\mu_{base}) & [\beta]\mu\alpha.J & \to J[\alpha \leftarrow \beta]
\end{array}
$$

The relation $\to_{\mu need}$ denotes the compatible closure of $\to$, and $\twoheadrightarrow_{\mu need}$ denotes the reflexive and transitive closure of $\to_{\mu need}$. The weak head standard reduction is defined as follows:

$$
\frac{M \to_{\beta, lift, \mu_{ap}} N}{[\alpha]C_{\beta l}[M] \mapsto_{\mu need} [\alpha]C_{\beta l}[N]} \qquad \frac{M \to_{deref_v, assoc, \mu_{let}, \mu_{lift}} N}{[\alpha]C_{da}[M] \mapsto_{\mu need} [\alpha]C_{da}[N]} \qquad \frac{J \to_{\mu_{base}} J'}{J \mapsto_{\mu need} J'}
$$

The notation $\longmapsto_{\mu need}$ stands for the reflexive and transitive closure of $\mapsto_{\mu need}$. Note that we only reduce jumps. A jump J is in weak head normal form if for no J', $J \mapsto_{\mu need} J'$. A weak head normal form (whnf) is a term M such that, either M is $\mu\alpha.J$ with J in weak head normal form or, for no J, $[\alpha]M \mapsto_{\mu need} J$. For example, let $x = \mu\alpha.[\beta]P$ in yx is in whnf.

Proposition 12 (Confluence). $\to_{\mu need}$ *is confluent.*

Proposition 13 (Standardization). *Given a term M and whnf N, if $M \twoheadrightarrow_{\mu need} N$ then there exists a whnf N' such that $M \longmapsto_{\mu need} N'$ and $N' \twoheadrightarrow_{\mu need} N$.*

The translation between classical call-by-need in natural deduction and sequent calculus form is modified in the following way to cover the classical constructs:

$$
\begin{array}{llll}
(\mu\alpha.J)^{\triangleright} & = \mu\alpha.J^{\triangleright} & ([\alpha]M)^{\triangleright} & = M_{\alpha}^{\triangleright} \\
\alpha_{\alpha} & = \alpha & F_{\alpha} & = \Box \\
\alpha_{v \cdot E} & = \alpha_E & F_{v \cdot E} & = F_E[\Box v^{\triangleleft}] \\
\alpha_{\tilde{\mu}x.\langle v \| e \rangle} & = \alpha_e & F_{\tilde{\mu}x.\langle v \| e \rangle} & = \text{let } x = \Box \text{ in } F_e[v^{\triangleleft}] \\
(\mu\alpha.c)^{\triangleleft} & = \mu\alpha.c^{\triangleleft} & e^{\triangleleft} & = [\alpha_e]F_e
\end{array}
$$

Proposition 14 (Equational correspondence). *Classical call-by-need in natural deduction and sequent calculus form are in equational correspondence:*
(i) $M =_{\mu need} M^{\triangleright\triangleleft}$
(ii) $c =_{lv} c^{\triangleleft\triangleright}$
(iii) If $M =_{\mu need} N$ then $M^{\triangleright} =_{lv} N^{\triangleright}$
(iv) If $c =_{lv} c'$ then $c^{\triangleleft} =_{\mu need} c'^{\triangleleft}$

Notice that the main reason for having only equational correspondence instead of a more precise reduction correspondence is the fact that, in $\lambda\mu_{need}$, μ_{ap} can be applied atomically $(\mu\alpha.J)N_1 \ldots N_n \to_{\mu_{ap}} (\mu\alpha.J[\alpha \leftarrow [\alpha]\Box N_1])N_2 \ldots N_n$ while in $\overline{\lambda}_{lv}$ the whole applicative context $\Box N_1 \ldots N_n$ is moved at once. In particular, the following holds: if $c \to_{lv} c'$ then $c^{\triangleleft} \twoheadrightarrow_{\mu need} c'^{\triangleleft}$.

7   Dual Classical Call-by-Need in Sequent Calculus Form ($\overline{\lambda}_{ln}^{-}$)

In call-by-need, the focus is on the consumer and goes to the producer on a need basis. This suggests a dual call-by-need which corresponds to focusing on the producer and going to the consumer on a need basis. To that end, we first extend the classical call-by-need calculus of the previous section, $\overline{\lambda}_{lv}$, with the dual of the implication, the subtraction connective, and then build the dual classical call-by-need calculus by using duality constructions typical from $\overline{\lambda}\mu\tilde{\mu}$-calculi.

While μ and $\tilde{\mu}$ constructs are dual of each other, implicative constructions $\lambda x.t$ and $v \cdot E$ currently have no dual in $\overline{\lambda}_{lv}$. We extend $\overline{\lambda}_{lv}$ by adding constructions for the subtraction connective [10]. Subtraction was already considered in the setting of $\overline{\lambda}\mu\tilde{\mu}$ in Curien *et al.* [11]. We follow the notation introduced by Herbelin in his habilitation thesis [23]. Terms are extended with the construction $v - e$ and contexts with $\tilde{\lambda}\alpha.e$. The corresponding reduction is:

$$(-)\qquad \langle v - e \| \tilde{\lambda}\alpha.e' \rangle \rightarrow \langle \mu\alpha.\langle v \| e' \rangle \| e \rangle$$

We can now present the classical call-by-need calculus extended with subtraction, $\overline{\lambda}_{lv}^{-}$. The structural rules are obtained by instantiating $\mathcal{V}$ and $\mathcal{E}$ as:

$$V_v^{-} = x \mid \lambda x.t \mid (V_v^{-} - e)$$

$$E_l^{-} = F^{-} \mid \tilde{\mu}x.C_l^{\tilde{\mu}}[\langle x \| F^{-} \rangle] \text{ with } F^{-} = \alpha \mid v \cdot E_l^{-} \mid \tilde{\lambda}\alpha.e$$

The syntax for the language with subtraction is finally as follows (with $c = \langle v \| e \rangle$):

command	c	$::= \langle t \| e \rangle$
term	v	$::= V \mid \mu\alpha.c$
linear term	V	$::= x \mid \lambda x.v \mid V - e$
context	e	$::= E \mid \tilde{\mu}x.c$
yielding context	E	$::= F \mid \tilde{\mu}x.C[\langle x \| F \rangle]$
linear context	F	$::= \alpha \mid v \cdot E \mid \tilde{\lambda}\alpha.e$
meta-context	C	$::= \Box \mid \langle \mu\alpha.c \| \tilde{\mu}x.C \rangle$

Using the duality principles developed in [11], we obtain $\overline{\lambda}_{ln}^{-}$ by dualizing $\overline{\lambda}_{lv}^{-}$: The syntax of the calculus is obtained by dualizing $\overline{\lambda}_{lv}^{-}$ syntax and its reductions are also obtained by duality: (β) and $(-)$ are dual of each other while μ_l and $\tilde{\mu}_v$ are respectively turned into:

- the μ-reduction associated with set $E_n^{-} ::= \alpha \mid v \cdot E_n^{-} \mid \tilde{\lambda}\alpha.e$, written μ_n
- the $\tilde{\mu}$-reduction associated with set $V_l^{-} ::= W \mid \mu\alpha.C_l^{\mu}[\langle W \| \alpha \rangle]$, with $W ::= x \mid \lambda x.t \mid V - e$ (and C_l^{μ} being the dual of $C_l^{\tilde{\mu}}$), written $\tilde{\mu}_l$.

Since only linear contexts are substituted for context variables, as in call-by-name, but only on a needed basis, we call the resulting calculus lazy call-by-name. Its syntax is given as follows:

$$
\begin{array}{lll}
command & c & ::= \langle t \| e \rangle \\
term & v & ::= V \mid \mu\alpha.c \\
yielding\ term & V & ::= W \mid \mu\alpha.C[\langle W \| \alpha \rangle] \\
linear\ term & W & ::= x \mid \lambda x.v \mid V - e \\
context & e & ::= E \mid \tilde{\mu}x.c \\
linear\ context & E & ::= \alpha \mid v \cdot E \mid \tilde{\lambda}\alpha.e \\
meta\text{-}context & C & ::= \Box \mid \langle \mu\alpha.C \| \tilde{\mu}x.c \rangle
\end{array}
$$

The four theories can be discriminated by the following command:

$$
\begin{aligned}
c \equiv \langle \mu\alpha.\langle \lambda x.\mu\_.\langle \lambda y.x \| \alpha \rangle \| \alpha \rangle \\
\| \tilde{\mu}f.\langle \mu\beta.\langle f \| t \cdot \beta \rangle \\
\| \tilde{\mu}x_1.\langle \mu\gamma.\langle f \| s \cdot \gamma \rangle \\
\| \tilde{\mu}x_2.\langle x_1 \| x_2 \cdot x_2 \cdot \delta \rangle \rangle \rangle \rangle
\end{aligned}
$$

We call c_1 the command obtained by instantiating t and s to $\lambda x.\lambda y.x$ and $\lambda x.\lambda y.y$, respectively. Then c_1 evaluates to $\langle \lambda x.\lambda y.x \| \delta \rangle$ in lazy call-by-value and to $\langle \lambda x.\lambda y.y \| \delta \rangle$ in call-by-name. We call c_2 the command obtained by instantiating t and s to $\lambda f.\lambda x.\mu\alpha.\langle f \| x \cdot \alpha \rangle$ and $\lambda x.x$. We now consider c_3 to be $\langle \mu\gamma.c_2 \| \tilde{\mu}w.c_1 \rangle$, where w does not occur free in c_1 and γ does no occur free in c_2. In call-by-name and lazy call-by-value, c_3 evaluates as c_1, up to garbage collection. However, c_3 evaluates to $\langle \lambda f.\lambda x.\mu\alpha.\langle f \| x \cdot \alpha \rangle \| \delta \rangle$ in call-by-value, and to $\langle I \| \delta \rangle$ in lazy call-by-name, up to garbage collection. This can be generalized by the following example, where we assume that α_1 does not occur free in c and V, and that x_1 does not occur free in c' and E. If we define

$$
c_0 \triangleq \langle \mu\alpha_1.\langle \mu\alpha_2.\langle V \| \alpha_2 \rangle \| \tilde{\mu}y.c \rangle \| \tilde{\mu}x_1.\langle \mu\beta.c' \| \tilde{\mu}x_2.\langle x_2 \| E \rangle \rangle \rangle
$$

then
$$
\begin{aligned}
c_0 &\twoheadrightarrow_n c'[\beta \leftarrow E[x_2 \leftarrow \mu\beta.c']] \\
c_0 &\twoheadrightarrow_v c[y \leftarrow V[\alpha_2 \leftarrow \tilde{\mu}y.c]] \\
c_0 &\twoheadrightarrow_{ln} \langle \mu\alpha_1.c[y \leftarrow \mu\alpha_2.\langle V \| \alpha_2 \rangle] \| \tilde{\mu}x_1.\langle \mu\beta.c' \| \tilde{\mu}x_2.\langle x_2 \| E \rangle \rangle \rangle \\
c_0 &\twoheadrightarrow_{lv} \langle \mu\alpha_1.\langle \mu\alpha_2.\langle V \| \alpha_2 \rangle \| \tilde{\mu}y.c \rangle \| \tilde{\mu}x_1.c'[\beta \leftarrow \tilde{\mu}x_2.\langle x_2 \| E \rangle]\rangle
\end{aligned}
$$

8   Conclusions and Future Work

The advantage of studying evaluation order in the context of sequent calculus has shown its benefits: extending the calculus (both syntax and reduction theory) to the classical case simply corresponds to going from one context variable to many. The study has also suggested how to provide a call-by-need version of Parigot's $\lambda\mu$-calculus, and in the minimal case, has led to a new notion of standard reduction, which applies the *lift* and *assoc* rule eagerly. In the minimal case, the single context variable, called $\star$, could be seen as the constant **tp** discussed in [6,5]. In the cited work, it is also presented how delimited control can be captured by extending **tp** to a dynamic variable named $\widehat{tp}$. This suggests that one could use $\widehat{tp}$ instead of **tp** to represent computations also in the minimal setting. Since evaluation goes under a $\widehat{tp}$, it means that one would obtain a different notion of standard reduction, which would correspond to the one of Ariola *et al.* [3] and Maraist *et al.* [26].

A benefit of sequent calculus over natural deduction in both call-by-name and call-by-value is that the standard redex in the sequent calculus always occurs at the top of the command. In other words, there is no need to perform an unbounded search to reach the standard redex [2]: this search is embedded in the structural reduction rules. However, this does not apply to our call-by-need sequent calculus: the standard redex can be buried under an arbitrary number of bindings. This can be easily solved by considering a calculus with explicit substitutions. A command now becomes $\langle v \| e \rangle \tau$, where τ is a list of declarations. For example, the critical pair will be solved as: $\langle \mu\alpha.c \| \tilde{\mu}x.c' \rangle \tau \rightarrow c'[x = \mu\alpha.c]\tau$ and the switching of context is realized by the rule: $\langle x \| E \rangle \tau_0[x := \mu\alpha.c]\tau_1 \rightarrow c[\alpha := \tilde{\mu}x.\langle x \| E \rangle \tau_0]\tau_1$. This will naturally lead us to developing abstract machines, which will be compared to the abstract machines of Garcia *et al.* [21] and Danvy *et al.* [14], inspired by natural deduction.

We have related the lazy call-by-value with subtraction to its dual. We plan to provide a simulation of lazy call-by-value in lazy call-by-name and vice-versa, without the use of subtraction. We are also interested in devising a complete set of axioms with respect to a classical extension of the call-by-need continuation-passing style of Okasaki *et al.* [29]. A natural development will then be to extend our lazy call-by-value and lazy call-by-name with delimited control. Following a suggestion by Danvy, we will investigate connections between our lazy call-by-name calculus and a calculus with futures [28]. At last, we want to better understand the underlying logic or type system.

Aknowledgements: The authors wish to thank Olivier Danvy and the anonymous referees for valuable comments and suggestions. Zena M. Ariola has been supported by NSF grant CCF-0917329. This research has been developed under INRIA Équipe Associée SEMACODE.

References

1. Ariola, Z., Herbelin, H.: Minimal classical logic and control operators. In: Baeten, J.C.M., Lenstra, J.K., Parrow, J., Woeginger, G.J. (eds.) ICALP 2003. LNCS, vol. 2719, Springer, Heidelberg (2003)
2. Ariola, Z.M., Bohannon, A., Sabry, A.: Sequent calculi and abstract machines. Transactions on Programming Languages and Systems (TOPLAS) 31(4), 275–317 (2009)
3. Ariola, Z.M., Felleisen, M.: The call-by-need lambda calculus. J. Funct. Program. 7(3), 265–301 (1997)
4. Ariola, Z.M., Herbelin, H.: Control reduction theories: the benefit of structural substitution. J. Funct. Program. 18(3), 373–419 (2008)
5. Ariola, Z.M., Herbelin, H., Sabry, A.: A proof-theoretic foundation of abortive continuations. Higher Order Symbol. Comput. 20, 403–429 (2007)
6. Ariola, Z.M., Herbelin, H., Sabry, A.: A type-theoretic foundation of delimited continuations. Higher Order Symbol. Comput. 22, 233–273 (2009)

7. Barendregt, H., Ghilezan, S.: Lambda terms for natural deduction, sequent calculus and cut elimination. J. Funct. Program. 10, 121–134 (2000)
8. Barendregt, H.P.: The Lambda Calculus: Its Syntax and Semantics. North Holland, Amsterdam (1984)
9. Church, A.: A set of postulates for the foundation of logic. Annals of Mathematics 2(33), 346–366 (1932)
10. Crolard, T.: Subtractive Logic. Theoretical Computer Science 254(1-2), 151–185 (2001)
11. Curien, P.-L., Herbelin, H.: The duality of computation. In: International Conference on Functional Programming, pp. 233–243 (2000)
12. Danos, V., Joinet, J.-B., Schellinx, H.: LKQ and LKT: sequent calculi for second order logic based upon dual linear decompositions of the classical implication. In: Advances in Linear Logic, vol. 222, pp. 211–224. Cambridge University Press, Cambridge (1995)
13. Danos, V., Joinet, J.-B., Schellinx, H.: A new deconstructive logic: Linear logic. J. Symb. Log. 62(3), 755–807 (1997)
14. Danvy, O., Millikin, K., Munk, J., Zerny, I.: Defunctionalized interpreters for call-by-need evaluation. In: Blume, M., Kobayashi, N., Vidal, G. (eds.) FLOPS 2010. LNCS, vol. 6009, pp. 240–256. Springer, Heidelberg (2010)
15. David, R., Py, W.: Lambda-mu-calculus and Böhm's theorem. J. Symb. Log. 66(1), 407–413 (2001)
16. Fasel, J.H., Hudak, P., Peyton Jones, S., Wadler, P. (eds.): Haskell special issue. SIGPLAN Notices 27(5) (May 1992)
17. Felleisen, M., Friedman, D.: Control operators, the secd machine, and the lambda-calculus. In: Formal Description of Programming Concepts-III, pp. 193–217. North-Holland, Amsterdam (1986)
18. Felleisen, M., Friedman, D., Kohlbecker, E.: A syntactic theory of sequential control. Theoretical Computer Science 52(3), 205–237 (1987)
19. Felleisen, M., Friedman, D., Kohlbecker, E., Duba, B.: Reasoning with continuations. In: First Symposium on Logic and Computer Science, pp. 131–141 (1986)
20. Filinski, A.: Declarative Continuations and Categorical Duality. Master thesis, DIKU, Danmark (August 1989)
21. Garcia, R., Lumsdaine, A., Sabry, A.: Lazy evaluation and delimited control. In: POPL 2009: Proceedings of the 36th Annual ACM SIGPLAN-SIGACT Symposium on Principles of Programming Languages, pp. 153–164. ACM, New York (2009)
22. Gentzen, G.: Investigations into logical deduction. In: Szabo, M. (ed.) Collected papers of Gerhard Gentzen, pp. 68–131. North-Holland, Amsterdam (1969)
23. Herbelin, H.: C'est maintenant qu'on calcule. In: Habilitation à diriger les reserches (2005)
24. Herbelin, H., Zimmermann, S.: An operational account of call-by-value minimal and classical λ-calculus in "natural deduction" form. In: Curien, P.-L. (ed.) TLCA 2009. LNCS, vol. 5608, pp. 142–156. Springer, Heidelberg (2009)
25. Lafont, Y., Reus, B., Streicher, T.: Continuations semantics or expressing implication by negation. Technical Report 9321, Ludwig-Maximilians-UniversitÃd't, MÃi-jnchen (1993)
26. Maraist, J., Odersky, M., Wadler, P.: The call-by-need λ-calculus. J. Funct. Program. 8(3), 275–317 (1998)
27. Moggi, E.: Computational λ-calculus and monads. In: Logic in Computer Science (1989)
28. Niehren, J., Schwinghammer, J., Smolka, G.: A concurrent lambda calculus with futures. Theor. Comput. Sci. 364, 338–356 (2006)

29. Okasaki, C., Lee, P., Tarditi, D.: Call-by-need and continuation-passing style. In: Lisp and Symbolic Computation, pp. 57–81. Kluwer Academic Publishers, Dordrecht (1993)
30. Parigot, M.: Lambda-mu-calculus: An algorithmic interpretation of classical natural deduction. In: Voronkov, A. (ed.) LPAR 1992. LNCS, vol. 624, pp. 190–201. Springer, Heidelberg (1992)
31. Plotkin, G.D.: Call-by-name, call-by-value and the lambda-calculus. Theoretical Comput. Sci. 1, 125–159 (1975)
32. Prawitz, D.: Natural Deduction, a Proof-Theoretical Study. Almquist and Wiksell, Stockholm (1965)
33. Ronchi Della Rocca, S., Paolini, L.: The Parametric λ-Calculus: a Metamodel for Computation. Texts in Theoretical Computer Science: An EATCS Series. Springer, Heidelberg (2004)
34. Sabry, A., Felleisen, M.: Reasoning about programs in continuation-passing style. Lisp and Symbolic Computation 6(3-4), 289–360 (1993)
35. Saurin, A.: Separation with streams in the λμ-calculus. In: Proceedings of 20th IEEE Symposium on Logic in Computer Science (LICS 2005), Chicago, IL, USA, June 26-29, pp. 356–365. IEEE Computer Society, Los Alamitos (2005)
36. Selinger, P.: Control categories and duality: on the categorical semantics of the lambda-mu calculus. Mathematical Structures in Computer Science 11(2), 207–260 (2001)

Homotopy-Theoretic Models of Type Theory

Peter Arndt[1] and Krzysztof Kapulkin[2]

[1] University of Oslo, Oslo, Norway
peter.arndt@mathematik.uni-regensburg.de
[2] University of Pittsburgh, Pittsburgh, PA, USA
krk56@pitt.edu

Abstract. We introduce the notion of a logical model category, which is a Quillen model category satisfying some additional conditions. Those conditions provide enough expressive power that one can soundly interpret dependent products and sums in it while also having a purely intensional interpretation of the identity types. On the other hand, those conditions are easy to check and provide a wide class of models that are examined in the paper.

1 Introduction

Starting with the Hofmann-Streicher groupoid model [HS98] it has become clear that there exist deep connections between Intensional Martin-Löf Type Theory ([ML72], [NPS90]) and homotopy theory. These connections have recently been studied very intensively. We start by summarizing this work — a more complete survey can be found in [Awo10].

It is well-known that *Identity Types* (or *Equality Types*) play an important role in type theory since they provide a relaxed and coarser notion of equality between terms of a given type. For example, assuming standard rules for type Nat one cannot prove that

$$n : \mathsf{Nat} \vdash n + 0 = n : \mathsf{Nat}$$

but there is still an inhabitant

$$n : \mathsf{Nat} \vdash p : \mathsf{Id}_{\mathsf{Nat}}(n + 0, \ n).$$

Identity types can be axiomatized in a usual way (as inductive types) by FORM, INTRO, ELIM, and COMP rules (see eg. [NPS90]). A type theory where we do not impose any further rules on the identity types is called *intensional*. One may be interested in adding the following *reflection rule*:

$$\frac{\Gamma \vdash p : \mathsf{Id}_A(a, b)}{\Gamma \vdash a = b : A} \ \ \mathsf{Id}\text{-}\mathrm{REFL}$$

Now, the Id would not be any coarser than the usual definitional equality. However, assuming the reflection rule destroys one of the most important properties of type theory, namely decidability of type-checking.

L. Ong (Ed.): TLCA 2011, LNCS 6690, pp. 45–60, 2011.

In order to interpret Martin-Löf Type Theory in the categorical setting, one has to have some notion of 'dependent products' and 'dependent sums'. As it was shown by Robert Seely [See84], locally cartesian closed categories (recall that $\mathbb{C}$ is a locally cartesian closed category if every slice of $\mathbb{C}$ is cartesian closed) provide a natural setting to interpret such operations. However, this interpretation forces the reflection rule to hold, that is if $p : \mathrm{Id}_A \longrightarrow A \times A$ is an interpretation of Id-type over A, it has to be isomorphic to the diagonal map $\Delta \colon A \longrightarrow A \times A$ in $\mathbb{C}/(A \times A)$.

For a semantics that does not force the reflection rule, one can pass to so-called *Quillen model categories*. Model categories, introduced by Daniel Quillen (cf. [Qui67]) give an abstract framework for homotopy theory which has found many applications e.g. in Algebraic Topoogy and Algebraic Geometry. The idea of interpreting type theory in model categories has been recently very intensively explored. In [AW09], [War08] Steve Awodey and Michael Warren showed that the Id-types can be purely intensionally interpreted as fibrant objects in a model category satisfying some additional conditions. Following this idea Nicola Gambino and Richard Garner in [GG08] defined a weak factorization system in the classifying category of a given type theory.

Another direction is to build higher categorical structures out of type theory. An ∞-category has, apart from objects and morphisms, also 2-morphisms between morphisms, 3-morphisms between 2-morphisms, and so on. All this data is organized with various kinds of composition. The theory of higher-dimensional categories has been successfully studied by many authors (see for example [Bat98], [Lei04], [Lur09]) and subsequently been brought into type theory by Richard Garner and Benno van den Berg ([GvdB08]), Peter LeFanu Lumsdaine ([Lum08], [Lum10]) and Richard Garner ([Gar08]).

In this paper we make an attempt to obtain sound models of type theory with the type constructors Π and Σ within the model-categorical framework. In good cases i.e. when certain coherence conditions (see [GvdB10]) are satisfied, our notion of a model extends the well-known models for the Id-types. Following [Kap10] we propose a set of conditions on a model category that provide enough structure in order to interpret those type constructors. Such a model category will be called a *logical model category*. Our intention was to give conditions that on one hand will be easy to check but on the other hand, will provide a wide class of examples. It is important to stress that this paper presents only a part of the ongoing project [AK11] devoted to study of Π and Σ types in homotopy theoretic models of type theory. The details and direction of this project may be found in the last section.

This paper is organized as follows: Sections 2 and 3 provide a background on type theory and abstract homotopy theory, respectively. In Sect. 4 we define the notion of logical model category and show that such a category is a sound model of a theory with Π- and Σ-types. Next, within this section we give a range of examples of logical model categories. Finally, in Sect. 5 we sketch the directions of our further research in this project.

2   Background on Type Theory

2.1   Logical Framework of Martin-Löf Type Theory

In this section we will review some basic notions of type theory (cf. [NPS90]).
Martin-Löf Type Theory is a dependent type theory i.e. apart from simple types
and their terms it allows type dependency as in the judgement

$$x : A \vdash B(x) \text{ type}.$$

In this example B can be regarded as a family of types indexed over A.

There are some basic types as for example: $\mathbf{0}$, $\mathbf{1}$, Nat and some type-forming
operations. The latter can be divided into two parts:

- simple type-forming operations such as $A \longrightarrow B$, $A \times B$, and $A + B$.
- operations on dependent types such as $\Pi_{x:A}B(x)$, $\Sigma_{x:A}B(x)$, and $\mathrm{Id}_A(x, y)$.

The language of type theory consists of *hypothetical judgements* (or just judgements) of the following six forms:

1. $\Gamma \vdash A$ type
2. $\Gamma \vdash A = B$ type
3. $\Gamma \vdash a : A$
4. $\Gamma \vdash a = b : A$
5. $\vdash \Gamma$ cxt
6. $\vdash \Gamma = \Delta$ cxt

Judgements of the form 5. establish that Γ is a well-formed context. Γ is said
to be a well-formed context if Γ is a (possibly empty) sequence of the form
$(x_0 : A_0, x_1 : A_1(x), \dots, x_n : A_n(x_0, \dots, x_{n-1}))$ and

$$\Gamma \vdash A_0 \text{ type}$$

and for $i = 1, 2, \dots n$

$$\Gamma, x_0 : A_0, x_1 : A_1, \dots, x_{i-1} : A_{i-1} \vdash A_i(x_0, x_1, \dots, x_{i-1}) \text{ type}.$$

The deduction rules of Martin-Löf Type Theory can be divided into two parts:

- *structural* rules.
- rules governing the forms of types.

The structural rules are standard and may be found in the Appendix. The rules
governing the forms of types consist of the following:

- a *formation* rule, providing the conditions under which we can form a certain
  type.
- *introduction* rules, giving the canonical elements of a type. The set of introduction rules can be empty.
- an *elimination* rule, explaining how the terms of a type can be used in
  derivations.
- *computation* rules, reassuring that the introduction and elimination rules are
  compatible in some suitable sense. Each of the computation rules corresponds
  to some introduction rule.

2.2   Type Constructors Π and Σ

In this paper we will be interested only in two dependent type constructors: Π and Σ. Below we present the rules governing those constructors.

Π-types. One can think of Π as a type-theoretic counterpart under Curry-Howard isomorphism (see [SU06]) of the universal quantifier $\forall$ or as product of a family of sets indexed by a set.

$$\frac{\Gamma,\ x : A \vdash B(x)\ \text{type}}{\Gamma \vdash \Pi_{x:A}B(x)\ \text{type}}\ \Pi\text{-FORM}$$

$$\frac{\Gamma,\ x : A \vdash B(x)\ \text{type} \qquad \Gamma,\ x : A \vdash b(x) : B(x)}{\Gamma \vdash \lambda x : A.b(x) : \Pi_{x:A}B(x)}\ \Pi\text{-INTRO}$$

$$\frac{\Gamma \vdash f : \Pi_{x:A}B(x) \qquad \Gamma \vdash a : A}{\Gamma \vdash \text{app}(f,\ a) : B(a)}\ \Pi\text{-ELIM}$$

$$\frac{\Gamma,\ x : A \vdash B(x)\ \text{type} \qquad \Gamma,\ x : A \vdash b(x) : B(x) \qquad \Gamma \vdash a : A}{\Gamma \vdash \text{app}(\lambda x : A.b(x),\ a) = b(a) : B(a)}\ \Pi\text{-COMP}$$

Σ-types. The Σ-types correspond to the existential quantifier $\exists$.

$$\frac{\Gamma \vdash A\ \text{type} \qquad \Gamma,\ x : A \vdash B(x)\ \text{type}}{\Gamma \vdash \Sigma_{x:A}B(x)\ \text{type}}\ \Sigma\text{-FORM}$$

$$\frac{\Gamma \vdash A\ \text{type} \qquad \Gamma,\ x : A \vdash B(x)\ \text{type}}{\Gamma,\ x : A,\ y : B(x) \vdash \text{pair}(x,\ y) : \Sigma_{x:A}B(x)}\ \Sigma\text{-INTRO}$$

$$\frac{\Gamma,\ z : \Sigma_{x:A}B(x) \vdash C(z)\ \text{type} \qquad \Gamma,\ x : A,\ y : B(x) \vdash d(x,y) : C(\text{pair}(x,\ y))}{\Gamma,\ z : \Sigma_{x:A}B(x) \vdash \text{split}_d(z) : C(z)}\ \Sigma\text{-ELIM}$$

$$\frac{\Gamma,\ z : \Sigma_{x:A}B(x) \vdash C(z)\ \text{type} \qquad \Gamma,\ x : A,\ y : B(x) \vdash d(x,y) : C(\text{pair}(x,\ y))}{\Gamma,\ x : A,\ y : B(x) \vdash \text{split}_d(\text{pair}(x,y)) = d(x,y) : C(\text{pair}(x,\ y))}\ \Sigma\text{-COMP}$$

3   Background on Model Categories

In this section we will gather some notions and results from model category theory.

Definition 1. *Let $\mathbb{C}$ be a category. We say that $f\colon A \longrightarrow B$ has the left lifting property with respect to $g\colon C \longrightarrow D$ or equivalently that g has the right lifting property with respect to f (we write $f \pitchfork g$) if every commutative square $g \circ u = v \circ f$ as below has a diagonal filler i.e. a map $j\colon B \longrightarrow C$ making the diagram*

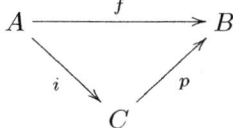

commute (i.e. $jf = u$ and $gj = v$).

For a collection $\mathcal{M}$ of morphisms in $\mathbb{C}$ we denote by $^{\pitchfork}\mathcal{M}$ (resp. $\mathcal{M}^{\pitchfork}$) the collection of maps having the left (resp. right) lifting property with respect to all maps in $\mathcal{M}$.

Definition 2. *A weak factorization system $(\mathcal{L}, \mathcal{R})$ on a category $\mathbb{C}$ consists of two collections of morphisms $\mathcal{L}$ (left maps) and $\mathcal{R}$ (right maps) in the category $\mathbb{C}$ such that:*

1. *Every map $f\colon A \longrightarrow B$ admits a factorization*

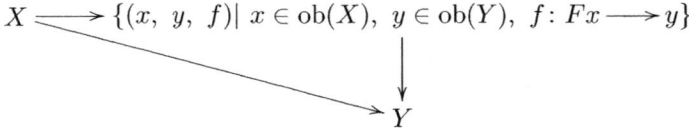

   where $i \in \mathcal{L}$ and $p \in \mathcal{R}$.
2. *$\mathcal{L}^{\pitchfork} = \mathcal{R}$ and $\mathcal{L} = {}^{\pitchfork}\mathcal{R}$.*

Examples 3. *The following are examples of weak factorization systems:*

1. *There is a weak factorization system in the category **Sets** with: $\mathcal{L} :=$ monomor-phisms and $\mathcal{R} :=$ epimorphisms. Note that the factorization and filling are not unique!*
2. *There is also a weak factorization system in the category **Gpd** of groupoids with: $\mathcal{L} :=$ injective equivalences and $\mathcal{R} :=$ fibrations . Recall that a functor is called* injective equivalence *if it is an equivalence of categories which is injective on objects. We factor a functor $F\colon X \longrightarrow Y$ in **Gpd** as*

$$X \Longrightarrow \{(x,\ y,\ f)\mid x \in \mathrm{ob}(X),\ y \in \mathrm{ob}(Y),\ f\colon Fx \longrightarrow y\}$$
$$\downarrow$$
$$Y$$

We now turn towards model categories. All results and notions given without reference can be found in [Hov99].

Definition 4. *A model category is a finitely complete and cocomplete category $\mathbb{C}$ equipped with three subcategories closed under retracts: $\mathcal{F}$ (fibrations), $\mathcal{C}$ (cofibrations), and $\mathcal{W}$ (weak equivalences) satisfying the following two conditions:*

1. *(Two-of-three) Given a commutative triangle*

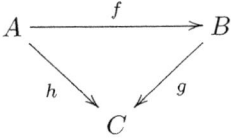

   if any two of f, g, h belong to $\mathcal{W}$, then so does the third.
2. *Both $(\mathcal{C}, \mathcal{F} \cap \mathcal{W})$ and $(\mathcal{C} \cap \mathcal{W}, \mathcal{F})$ are weak factorization systems.*

We will refer to model categories sometimes by the tuple $(\mathbb{C}, \mathcal{W}, \mathcal{C}, \mathcal{F})$ or, if no ambiguity arises, just by the underlying category $\mathbb{C}$.

From a model category $\mathbb{C}$ one has a functor into its associated *homotopy category* $\mathrm{Ho}(\mathbb{C})$, which is the initial functor mapping the weak equivalences to isomorphisms (which defines $\mathrm{Ho}(\mathbb{C})$ up to equivalence). A morphism which is both a fibration and a weak equivalence is called an *trivial fibration*. Similarly, a morphism which is both a cofibration and a weak equivalence is called a *trivial cofibration*.

Examples 5. *The following are examples of model categories:*

1. *On any complete and cocomplete category $\mathbb{C}$ one has the* discrete *model structure with $\mathcal{C} := \mathcal{F} := \mathrm{mor}\,\mathbb{C}$ and $\mathcal{W} := \mathrm{iso}\,\mathbb{C}$. This is the only model structure with $\mathcal{W} = \mathrm{iso}\,\mathbb{C}$.*
2. *The category* **Gpd** *of groupoids has a structure of a model category with: $\mathcal{F} :=$ fibrations, $\mathcal{C} :=$ functors injective on objects and $\mathcal{W} :=$ categorical equivalences*
3. *The category* **SSets** $:=$ **Sets**$^{\Delta^{\mathrm{op}}}$ *of simplicial sets (where Δ is the category of finite non-empty linearly ordered sets) has a standard model structure with $\mathcal{W} := \{$those morphisms inducing isomorphisms on all homotopy groups$\}$, $\mathcal{C} := \{$monomorphisms$\}$ and $\mathcal{F} := (\mathcal{W} \cap \mathcal{C})^{\pitchfork}$*

Definition 6. *An object A is called* fibrant *if the canonical map $A \longrightarrow 1$ is a fibration. Similarly, an object is called* cofibrant *if $0 \longrightarrow A$ is a cofibration.*

Definition/Proposition 7. *The following are equivalent for a pair of adjoint functors $L \colon \mathbb{C} \leftrightarrows \mathbb{D} \colon R$ between model categories:*

1. *L preserves cofibrations and trivial cofibrations.*
2. *R preserves fibrations and trivial fibrations.*

An adjoint pair satisfying these conditions is called a Quillen adjunction *and it induces an adjunction $\mathrm{Ho}(\mathbb{C}) \leftrightarrows \mathrm{Ho}(\mathbb{D})$ between the homotopy categories. It is called a* Quillen equivalence *if this induced adjunction is an equivalence of categories.*

Notation. Assume $\mathbb{C}$ is a complete category and $f\colon B \longrightarrow A$ is a morphism in $\mathbb{C}$. Following [GK09], by $\Delta_f\colon \mathbb{C}/A \longrightarrow \mathbb{C}/B$ we will denote a functor taking an object in the slice over A to its pullback along f.[1] This functor has a left adjoint denoted by Σ_f which takes an object in the slice over B and composes it with f. If Δ_f also has a right adjoint, it will be denoted by Π_f.

A model category interacts with its slice categories in the following way:

Proposition 8. *Let $\mathbb{C}$ be a model category and C an object in $\mathbb{C}$. Define that a morphism f in $\mathbb{C}/C$ is a fibration/cofibration/weak equivalence if it is a fibration/cofibration/weak equivalence in $\mathbb{C}$. Then $\mathbb{C}/C$ is a model category with the model structure described above. Furthermore for every morphism f the adjunction $\Sigma_f \dashv \Delta_f$ is a Quillen adjunction.*

Definition/Proposition 9. *The following are equivalent [Rez02, Prop. 2.7]:*

1. *Pullbacks of weak equivalences along fibrations are weak equivalences*
2. *For every weak equivalence $f\colon X \longrightarrow Y$ the induced Quillen adjunction $\mathbb{C}/X \leftrightarrows \mathbb{C}/Y$ is a Quillen equivalence*

If a model category satisfies these conditions it is called right *proper. From the second formulation one can deduce that right properness only depends on the class of weak equivalences, not the whole model structure (see also [Cis06, Cor. 1.5.21]). There also is the dual notion of left properness (pushouts of weak equivalences along cofibrations are weak equivalences).*

Proposition 10 ([Rez02], Rem 2.8)

1. *If all objects in a model category $\mathbb{C}$ are fibrant (resp. cofibrant) then $\mathbb{C}$ is right (resp. left) proper*
2. *If a model category $\mathbb{C}$ is right proper then so are its slice categories $\mathbb{C}/X$*

A further important property of model categories is the following:

Definition 11. *Let λ be a regular cardinal and $(\mathbb{C}, \mathcal{W}, \mathcal{C}, \mathcal{F})$ a model category. It is called λ-combinatorial if the underlying category is locally λ-presentable and there are sets I (resp. J) of morphisms between λ-presentable (resp. λ-presentable and cofibrant) objects such that $I^{\pitchfork} = \mathcal{F} \cap \mathcal{W}$ and $J^{\pitchfork} = \mathcal{F}$. A model category is* combinatorial *if it is λ-combinatorial for some λ.*

Examples 12.   1. *The category of sets with the discrete model structure is seen to be combinatorial taking $I := \{\{0,1\} \to \{0\}, \emptyset \to \{0\}\}$ and $J := \emptyset$*
2. *The category $\mathbf{SSets}$ with the standard model structure is combinatorial. Being a topos it is locally presentable, with the representable functors playing the role of generators of the category. Since the initial object is the constant functor with value the empty set and cofibrations are the monomorphisms, every object is cofibrant. One can take $I := \{\partial\Delta^n \to \Delta^n\}$ (the inclusions of the borders into full n-simplices) and $J := \{\Lambda_k^n \to \Delta^n\}$ (the inclusions of the k-th n-horns into n-simplices).*

---

[1] To avoid confusion, we denote the diagonal map by δ.

Combinatorial model categories serve as input for two constructions whose output are further combinatorial model categories:

Theorem 13 (see e.g. [Lur09],sect. A.2.8). *Let* $(\mathbb{C}, \mathcal{W}, \mathcal{C}, \mathcal{F})$ *be a combinatorial model category and* $\mathbb{D}$ *a small category. If on the functor category* $\mathbb{C}^{\mathbb{D}}$ *one defines the following classes of morphisms,*

$$\mathcal{C}_{inj} := \{morphisms \ which \ are \ objectwise \ in \ \mathcal{C}\}$$
$$\mathcal{W}_{\mathbb{C}^{\mathbb{D}}} := \{morphisms \ which \ are \ objectwise \ in \ \mathcal{W}\}$$
$$\mathcal{F}_{proj} := \{morphisms \ which \ are \ objectwise \ in \ \mathcal{F}\}$$

then one has:

- $(\mathbb{C}^{\mathbb{D}}, \mathcal{W}_{\mathbb{C}^{\mathbb{D}}}, \mathcal{C}_{inj}, (\mathcal{W}_{\mathbb{C}^{\mathbb{D}}} \cap \mathcal{C}_{inj})^{\pitchfork})$ *and* $(\mathbb{C}^{\mathbb{D}}, \mathcal{W}_{\mathbb{C}^{\mathbb{D}}}, {}^{\pitchfork}(\mathcal{W}_{\mathbb{C}^{\mathbb{D}}} \cap \mathcal{F}_{proj}), \mathcal{F}_{proj})$ *are combinatorial model category structures on* $\mathbb{C}^{\mathbb{D}}$, *called the injective and the projective structure, respectively.*
- *If* $(\mathbb{C}, \mathcal{W}, \mathcal{C}, \mathcal{F})$ *is left or right proper, then so are the above model structures.*

Examples 14. *1. The construction applied to the discrete model structure on the category of sets yields the discrete model structure on presheaves.*
2. For a small category $\mathbb{D}$ *the injective model structure on simplicial presheaves is an example for the above construction applied to the combinatorial left and right proper model category of simplicial sets, yielding the so-called global injective model structure on* **SSets**$^{\mathbb{D}}$.

Next we summarize *left Bousfield localization*. This is a technique to replace a given model structure on a category by another one, enlarging the class of weak equivalences, keeping the class of cofibrations and adjusting the class of fibrations accordingly. The applicability of this technique is only ensured when the model category is either *cellular* (for this see [Hir03]) or combinatorial.

In the following theorem we will use the *mapping space* $\mathbb{R}\operatorname{Hom}(X, Y) \in$ Ho(**SSets**) which one can associate to any two objects X, Y of a model category as e.g. in [Hov99, sec. 5.4] — for simplicial model categories this can be taken to be the simplicial Hom-set of morphisms between a cofibrant replacement of X and a fibrant replacement of Y.

Theorem 15 (J. Smith, proven in [Bar10]). *Let* $(\mathbb{C}, \mathcal{W}, \mathcal{C}, \mathcal{F})$ *be a left proper combinatorial model category and* H *a set of morphisms of* Ho$(\mathbb{C})$. *Define an object* $X \in \mathcal{M}$ *to be* H-local *if any morphism* $f \colon A \longrightarrow B$ *in* H *induces an isomorphism* $f^* \colon \mathbb{R}\operatorname{Hom}(B, X) \longrightarrow \mathbb{R}\operatorname{Hom}(A, X)$ *in* Ho(**SSets**). *Define* $\mathcal{W}_H$, *the class of* H-equivalences, *to be the class of morphisms* $f \colon A \longrightarrow B$ *which induce isomorphisms* $f^* \colon \mathbb{R}\operatorname{Hom}(B, X) \longrightarrow \mathbb{R}\operatorname{Hom}(A, X)$ *in* Ho(**SSets**) *for all* H-local objects X. *Then* $(\mathbb{C}, \mathcal{W}_H, \mathcal{C}, \mathcal{F}_H := (\mathcal{W}_H \cap \mathcal{C})^{\pitchfork})$ *is a left proper combinatorial model structure.*

Remark 16. 1. The fibrant objects in the localized model structure are exactly the H-local objects which are fibrant in the original model structure.
2. A shorter proof of the above theorem for the special case of simplicial sheaves can be found in [MV99], sect. 2.2

Example 17.   1. One can localize the discrete model structure on presheaves on a site taking H to be the set of morphisms of the following form: For each cover $\{A_i \to X | i \in I\}$ in the given Grothendieck topology take the canonical morphism $\mathrm{Coequ}(\coprod_{I \times I} \mathrm{Hom}(-, A_i \times_X A_j) \rightrightarrows \coprod_I \mathrm{Hom}(-, A_i)) \to X$. This yields a non-discrete model structure whose homotopy category is equivalent to the category of sheaves. Being H-local means in this case satisfying the descent condition for the covers given from the Grothendieck topology in question.

We still record a property of Bousfield localizations of simplicial sheaf categories which will be of interest:

Theorem 18 ([MV99], Thm 2.2.7). *Any Bousfield localization of a category of simplicial sheaves with injective model structure is right proper.*

Last, we gather some basic facts about *Cisinski model structures*. In the following we will use the terminology of small and large sets, small meaning to be contained in a Grothendieck universe.

Definition 19. *Let $\mathbb{C}$ be a topos. A set $\mathcal{W}$ of morphisms of $\mathbb{C}$ is called a* localizer *if*

1. *$\mathcal{W}$ has the two-of-three property (see 4.1)*
2. *$\mathcal{W}$ contains $(Mono\,\mathbb{C})^{\mathrm{fh}}$*
3. *$\mathcal{W}$ is closed under pushouts and transfinite compositions (i.e. for a chain of morphisms in $\mathcal{W}$ the canonical morphism from the domain of the first one to the colimit of the chain is again in $\mathcal{W}$)*

For any set of morphisms S there is a smallest localizer $\mathcal{W}(S)$ containing it, namely the intersection of all localizers containing S. A localizer is called accessible if it is generated by a small set.

Theorem 20 ([Cis02], Thm. 3.9). *For any accessible localizer $\mathcal{W}$ in a topos $\mathbb{C}$, the tuple $(\mathbb{C}, \mathcal{W}, Mono, (\mathcal{W} \cap Mono)^{\mathrm{fh}})$ is a model structure.*

A model structure arising in the above way is called a *Cisinski model structure*. Since the cofibrations are the monomorphisms and every morphism with domain an initial object is a monomorphism, every object in a Cisinski model category is cofibrant. Hence any such model structure is left proper (10.1). Right properness is adressed in the following statement:

Proposition 21 ([Cis02], Prop. 3.12 and Cor. 4.11). *Let $\mathbb{C}$ be a topos and $(X_i | i \in I)$ a small family of objects of $\mathbb{C}$. Then the localizer generated by the projections $\{Z \times X_i \to Z | Z \in \mathrm{ob}(\mathbb{C})\}$ is accessible and the corresponding model structure is right proper.*

Example 22. An example from mathematical practice of this last kind of model structure, obtained by "contracting" a family of objects, is the category $\mathbf{Sets}^{\Delta^{op} \times \mathbf{Sm}/S}$ of $\mathbf{SSets}$-valued functors on smooth schemes over a base S where one starts with the local injective model structure on $\mathbf{SSets}^{\mathbf{Sm}/S}$ and localizes by the set

$$\{! \times 1 : \mathbb{A}^1 \times \mathrm{Hom}(-, X) \longrightarrow \mathrm{Hom}(-, X) | X \in \mathbf{Sm}/S\}$$

4   Main Theorem and Examples

In this section we will define the notion of logical model category and show how one can interpret Π- and Σ-types in such a category.

Definition 23. *We say that $\mathbb{C}$ is a* logical model category *if $\mathbb{C}$ is a model category and in addition the following two conditions hold:*

1. *if $f\colon B \longrightarrow A$ is a fibration in $\mathbb{C}$, then there exists the right adjoint Π_f to the pullback functor Δ_f.*
2. *the class of trivial cofibrations is closed under pullback along a fibration.*

Clearly, one has the following corollary which provides a convenient way of checking that a model category is in fact a logical model category.

Corollary 24. *If $\mathbb{C}$ is a model model satisfying the following three conditions:*

1. *if $f\colon B \longrightarrow A$ is a fibration in $\mathbb{C}$, then there exists the right adjoint Π_f to the pullback functor Δ_f.*
2. *the class of cofibrations is closed under pullback along a fibration.*
3. *$\mathbb{C}$ is right proper.*

then $\mathbb{C}$ is a logical model category.

Given a logical model category $\mathbb{C}$ one may interpret the syntax of type theory as follows:

- contexts are interpreted as fibrant objects. In particular the empty context is interpreted as a terminal object in $\mathbb{C}$.
- a judgement $\Gamma \vdash A$ type is interpreted as a fibration $[\![\Gamma, \, x : A]\!] \longrightarrow [\![\Gamma]\!]$.
- a judgement $\Gamma \vdash a : A$ is interpreted as a section of $[\![\Gamma, \, x : A]\!] \longrightarrow [\![\Gamma]\!]$ i.e.

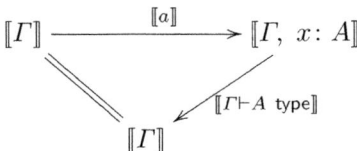

- substitution along $f\colon \Gamma \longrightarrow \Phi$ is interpreted by means of the pullback functor Δ_f.

Remark 25. The key point of this interpretation (i.e. homotopy theoretic) is that we allow only fibrant objects and fibrations as interpretations of contexts and

One can observe that the notion of model for type theory, which is a logical model category, is too strong. In fact, we do not need the whole model structure but only one weak factorization system. However, all the examples we have in mind are already model categories so introducing the notion of model in this way is not really a restriction from this point of view. We refer the reader to [AK11] where we study models in categories equipped with a single (algebraic) weak factorization system.

Theorem 26. *If* $\mathbb{C}$ *is a logical model category, then the above interpretation is sound.*

Proof. Because of the limited space we will not go into details and sketch only the main idea. We address the interested reader to [AK11].

The interpretation of Π- and Σ-types goes along the lines of [See84], that is Π-types are interpreted by means of the right adjoint and Σ-types are interpreted by means of the left adjoint to the pullback functor. The only new thing one has to show is that if $f: B \longrightarrow A$ is a fibration, then Π_f and Σ_f preserve fibrations (to validate the formation rules). In case of Σ_f it is clear since the class of fibrations is closed under composition. For Π_f we use theorem 7 combined with the condition 2. from the Definition 23.

Further details, including for example the coherence problems (as studied in details in [See84]) such as the Beck-Chevalley condition will be presented in [AK11]. ∎

There is a big class of examples:

Proposition 27.   *1. Any right proper Cisinski model structure admits a good interpretation of Π- and Σ-types, i.e. satisfies conditions 1 – 3.*
   2. Any left Bousfield localization of the category of sheaves on a site with the injective model structure is an example of this.

will

Proof. We proceed by checking the conditions of Corollary 24:

Condition 1: Toposes are locally cartesian closed.

Condition 2: The cofibrations of the injective model structure are the monomorphisms and by definition they stay the same after a left Bousfield localisation. The class of monomorphisms is closed under pullback.

Condition 3: Right properness is ensured by hypothesis in the first case and by 18 in the second. ∎

Proposition 28. *If* $\mathbb{C}$ *is a logical model category, then so is any slice category* $\mathbb{C}/X$.

Proof. By definition of the model structure in 8, the cofibrations, fibrations and weak equivalences in $\mathbb{C}/X$ are defined to be those of $\mathbb{C}$. Since pullbacks in a slice category are also pullbacks in the original category, trivial cofibrations are preserved under pullback along fibrations by hypothesis. The right adjoint of pullback along a fibration in $\mathbb{C}/X$ is the one from $\mathbb{C}$, the adjointness property follows from the one for $\mathbb{C}$ using that $(\mathbb{C}/X)/A \cong \mathbb{C}/A$ for every object $A \to X$ of $\mathbb{C}/X$. ∎

We now give some concrete examples of model categories satisfying conditions (1)–(3).

4.1   Groupoids

Our first example is the category **Gpd** of groupoids. It is well known (see [Gir64], [Con72]) that the right adjoint Π_f to the pullback functor exists in **Gpd** (as a subcategory of **Cat**) if and only if f is a so-called *Conduché fibration*. In particular, any fibration or opfibration is a Conduché fibration. Since we are interested only in taking Π_f for f being a fibration, the condition (1) is satisfied. Alternatively, as was pointed out by a referee, one can deduce satisfaction of (1) from the fact that we are only interested in the restriction of the pullback functor to fibrations and this restricted pullback functor always has a right adjoint (no matter what we are pulling back along). Conditions (2) and (3) also hold. It is standard to verify that a pullback of a functor injective on objects (resp. an equivalence of categories) is again injective on objects (an equivalence).

The model obtained above is known as the Hofmann-Streicher *groupoid model* which was the first intensional model of Martin-Löf Type Theory (see [HS98]).

The model structure on **Gpd** is a restriction of Joyal's model structure on the category **Cat**. Another interesting example of a model category that arises as a restriction from **Cat** is the model structure on the category **PreOrd** of preordered sets (i.e. equipped with a relation that is reflexive and transitive). It is again standard to verify that it a logical model category. Moreover, from the point of view of Id-types: even though the Id-type over A consists more information that the definitional equality, the proof-relevance principle is satisfied.

4.2   Extreme Examples

Discrete model structures: Any bicomplete, locally cartesian closed category endowed with the *discrete model structure* ($\mathcal{C} := \mathcal{F} := \mathrm{mor}\ \mathbb{C}$ and $\mathcal{W} := \mathrm{iso}\ \mathbb{C}$) satisfies conditions (1), (2) and (3): The right adjoint to the pullback functor exists by hypothesis, the cofibrations are trivially closed under pullbacks and since every object is fibrant right properness is ensured by the first criterion given in 10. The interpretation of type theories given in the last section then coincides with the usual extensional one.

Two indiscrete model structures on a Grothendieck topos: On any Grothendieck topos (automatically satisfying condition (1)) one has two model structures in which the weak equivalences are all morphisms: A Cisinski model structure, and one in which the cofibrations are the *complemented monomorphisms* and the fibrations are the split epimorphisms. Right properness of this latter model structure follows from the fact that the first model structure is right proper together with the fact that properness depends only on the weak equivalences (9). Stability of cofibrations under pullback is given by the fact that pulling back commutes with taking complements in a topos.

The minimal Cisinski model structure on Sets: There is a model structure on **Sets**, see [Cis02] Ex. 3.7, such that cofibrations are monomorphisms and weak equivalences are all morphisms except those whose domain is the empty set and whose codomain is not. The fibrations in this model structure are the epimorphisms and the maps whose domain is the empty set. It is the minimal Cisinski model structure, $\mathcal{W} := \mathcal{W}(\emptyset)$ and thus by 21, applied to the empty set of morphisms, is right proper.

4.3   Localized Structures on (Pre)sheaf Categories

Taking a category of presheaves on a site (it is locally cartesian closed, ensuring (1)) we can endow it with the discrete model structure. By 12 the discrete model structure on the category of sets is combinatorial, hence by 13 and 14(1) so is the one on presheaves. It is left and right proper since all objects are fibrant and cofibrant (by 10) and hence one can apply left Bousfield localization.

Localization by Grothendieck topologies: Examples from mathematical practice are ubiquitous and include the **Sets**-valued presheaves localized by a Grothendieck topology as in 17(3) — e.g. by the Grothendieck topologies for sheaf toposes of topological spaces, for the étale, crystalline, or Zariski toposes of schemes, Weil-étale toposes or classifying toposes of geometric theories.

Presheaves on Test categories: There is a theory of functor categories on test categories, which more or less by definition are categories supporting model structures which are Quillen equivalent to the standard model structure on topological spaces, see [Cis06]. The best known examples are cubical sets and simplicial sets, whose behaviour as a model for type theory has been described in detail in [Kap10] and cubical sets. To give an idea about how the further studies of these examples will be pursued we add a description of the Π-functor for **SSets**:

Let $p\colon X \longrightarrow B$ be an object in **SSets**$/B$ and $f\colon B \longrightarrow A$ a fibration of simplicial sets. Moreover let $a\colon \Delta^n \longrightarrow A$ be a cell inclusion and $\Delta_f(a) =: \overline{a}$. Then

$$\Pi_f(X,p)_n = \{h\colon \Delta^n \times_A B \longrightarrow X \,|\, p \circ h = \overline{a}\}.$$

Models for the $\mathbb{A}^1$-homotopy category: Let S be some base scheme and **Sm**$/S$ the category of smooth schemes over S. There are two Quillen equivalent Cisinski models for the $\mathbb{A}^1$-homotopy category, one model structure on the category of **Sets**-valued Nisnevich sheaves on **Sm**$/S$, and one on the category of **SSets**-valued Nisnevich sheaves on **Sm**$/S$. The latter is obtained by taking the injective model structure on **SSets**$^{\mathbf{Sm}/S}$ and localizing it by the class $\{! \times 1 \colon \mathbb{A}^1 \times \mathrm{Hom}(-, X) \longrightarrow \mathrm{Hom}(-, X) | X \in \mathbf{Sm}/S\}$, the former by a similar process described in [Voe98].

5   Future Research

In this section we show further directions that are of our interest in [AK11].

Completeness result and coherence conditions for Id-*types.* Theorem 26 provides only a soundness result for Π- and Σ-types. A natural question in this case would be the one of completeness with respect to some similar semantics. As shown in

[GG08] the semantics of Id-types is essentially complete in categories equipped with a weak factorization system. On the other hand, providing a sound and not too strict model of Id-types satisfying all the coherence conditions is still an open issue. One attempt to that can be found in [GvdB10]. Another approach which is worth considering would be to work with *fibration categories* which provide more structure than a single weak factorization system but are still weaker than model categories.

Further properties of Π-types. One may recognize in Π-COMP the standard β-rule from λ-calculus. This rule is by Theorem 26 satisfied in any logical model category. This is not the case for the other two rules, namely η- and ξ- (known as function extensionality). In [AK11] we will give a detailed treatment of the models where those rules are satisfied up to definitional equality or just up to a term of Id-type.

Extending the class of categories admitting an interpretation of Π-types One may try broaden the scope of our interpretation of Π-types in two directions: First, our definition of logical model category required the existence of a right adjoint to the pullback functor along a fibration. It seems enough, however, to require a right adjoint to the restriction of this pullback functor to the full subcategory whose objects are fibrations. Second, we required the right adjoint to take fibrations to fibrations. Alternatively one may try to force this property using the fibrant replacement functor in a model category. However, in both cases a careful consideration of coherence issues is required, while in the setting of this work the standard methods apply to give coherence.

Acknowledgements. We are very grateful to Thorsten Altenkirch, Steve Awodey, Richard Garner, Martin Hyland, Peter Lumsdaine, Markus Spitzweck, Thomas Streicher and Marek Zawadowski for many fruitful and interesting conversations. We are grateful to an anonymous referee for pointing out many useful suggestions. The first-named author would like to thank the Topology group at the University of Oslo for their warm hospitality which he enjoyed during the preparation of this work and in particular John Rognes for arranging financial support via the Norwegian Topology Project RCN 185335/V30. The second-named author would like to acknowledge the support of the Department of Mathematics at the University of Pittsburgh (especially Prof. Paul Gartside) and the A&S fellowship he was enjoying in the Spring Semester 2011 as well as the Department of Philosophy at Carnegie Mellon University and in particular his supervisor, Steve Awodey, who has been a constant help and a good friend. He dedicates this work to his Mother whose help and support for him when writing the paper went far beyond the obvious.

References

AK11.   Arndt, P., Kapulkin, C.: $\prod$- and $\sum$-types in homotopy theoretic models of type theory (work in progress)
AW09.   Awodey, S., Warren, M.A.: Homotopy theoretic models of identity types. Math. Proc. of the Cam. Phil. Soc. (2009)

Awo10.   Awodey, S.: Type theory and homotopy (2010) (preprint)

Bar10.   Barwick, C.: On left and right model categories and left and right bousfield localizations. Homology, Homotopy and Applications 12(2), 245–320 (2010)

Bat98.   Batanin, M.A.: Monoidal globular categories as a natural environment for the theory of weak n-categories. Adv. Math. 136(1), 39–103 (1998)

Cis02.   Cisinski, D.-C.: Théories homotopiques dans les topos. J. Pure Appl. Algebra 174(1), 43–82 (2002); MR1924082 (2003i:18021)

Cis06.   Cisinski, D.-C.: Les préfaisceaux comme modèles des types d'homotopie. Astérisque, (308), xxiv+390 (2006); MR 2294028 (2007k:55002)

Con72.   Conduché, F.: Au sujet de l'existence d'adjoints à droite aux foncteurs "image réciproque" dans la catégorie des catégories. C. R. Acad. Sci. Paris Sér. A-B 275, A891–A894 (1972)

Gar08.   Garner, R.: 2-dimensional models of type theory. Mathematical Structures in Computer Science (2008) (to appear)

GG08.   Gambino, N., Garner, R.: The identity type weak factorisation system. Theoretical Computer Science 409(1), 94–109 (2008)

Gir64.   Giraud, J.: Méthode de la descente. Mémoires de la Soc. Math. de France 2 (1964)

GK09.   Gambino, N., Kock, J.: Polynomial functors and polynomial monads (to appear)

GvdB08.   Garner, R., van den Berg, B.: Types are weak ω-groupoids (2008) (submitted)

GvdB10.   Garner, R., van den Berg, B.: Topological and simplicial models of identity types (2010) (submitted)

Hir03.   Hirschhorn, P.S.: Model categories and their localizations. Mathematical Surveys and Monographs, vol. 99. American Mathematical Society, Providence (2003), MR 1944041 (2003j:18018)

Hov99.   Hovey, M.: Model categories. Mathematical Surveys and Monographs, vol. 63. American Mathematical Society, Providence (1999)

HS98.   Hofmann, M., Streicher, T.: The groupoid interpretation of type theory. In: Twenty-Five Years of Constructive Type Theory (Venice, 1995). Oxford Logic Guides, vol. 36, pp. 83–111. Oxford Univ. Press, New York (1998)

Kap10.   Kapulkin, C.: Homotopy theoretic models of type theory, Masters Thesis (University of Warsaw) (2010)

Lei04.   Leinster, T.: Higher operads, higher categories. London Mathematical Society Lecture Note Series, vol. 298. Cambridge University Press, Cambridge (2004)

Lum08.   Lumsdaine, P.L.: Weak ω-categories from intensional type theory, extended version (2008)

Lum10.   Lumsdaine, P.L.: Higher categories from type theories, Ph.D. thesis, CMU (2010)

Lur09.   Lurie, J.: Higher topos theory. Annals of Mathematics Studies, vol. 170. Princeton University Press, Princeton (2009)

ML72.   Martin-Löf, P.: An intuitionstic theory of types, Technical Report, University of Stockholm (1972)

MV99.   Morel, F., Voevodsky, V.: A^1-homotopy theory of schemes. Inst. Hautes Études Sci. Publ. Math. (1999) (90), 45–143 (2001); MR 1813224 (2002f:14029)

NPS90.   Nordström, B., Petersson, K., Smith, J.M.: Programming in Martin-Löf's type theory. Oxford University Press, Oxford (1990)

Qui67.   Quillen, D.: Homotopical algebra. Lecture Notes in Mathematics, vol. 43. Springer, Heidelberg (1967)

Rez02.   Rezk, C.: Every homotopy theory of simplicial algebras admits a proper
          model. Topology Appl. 119(1), 65–94 (2002); MR 1881711 (2003g:55033)
See84.   Seely, R.A.G.: Locally cartesian closed categories and type theory. Math.
          Proc. Cambridge Philos. Soc. 95(1), 33–48 (1984)
SU06.   Sørensen, M.H., Urzyczyn, P.: Lectures on Curry–Howard isomorphism.
          Studies in Logic and the Foundations of Mathematics, vol. 149. Eslevier,
          Amsterdam (2006)
Voe98.   Voevodsky, V.: $\mathbf{A}^1$-homotopy theory. In: Proceedings of the International
          Congress of Mathematicians, Berlin, vol. I, pp. 579–604 (electronic) (1998);
          MR 1648048 (99j:14018)
War08.   Warren, M.A.: Homotopy theoretic aspects of constructive type theory, Ph.D.
          thesis, CMU (2008)

A   Structural Rules of Martin-Löf Type Theory

The structural rules of Martin-Löf Type Theory are:

Variables   Substitution

$$\frac{\Gamma \vdash A \text{ type}}{\Gamma,\, x\colon A \vdash x\colon A} \text{ Vble} \qquad \frac{\Gamma \vdash a\colon A \quad \Gamma,\, x\colon A,\, \Delta \vdash \mathcal{I}}{\Gamma,\, \Delta[a/x] \vdash \mathcal{I}[a/x]} \text{ Subst}$$

Weakening   Exchange

$$\frac{\Gamma \vdash A \text{ type} \quad \Gamma \vdash \mathcal{I}}{\Gamma,\, x\colon A \vdash \mathcal{I}} \text{ Wkg} \qquad \frac{\Gamma,\, x\colon A,\, y\colon B,\, \Delta \vdash \mathcal{I}}{\Gamma,\, y\colon B,\, x\colon A,\, \Delta \vdash \mathcal{I}} \text{ Exch} \text{ if } x \text{ is not free in } B.$$

Definitional (Syntactic) Equality

$$\frac{\Gamma \vdash A \text{ type}}{\Gamma \vdash A = A \text{ type}} \qquad \frac{\Gamma \vdash A = B \text{ type}}{\Gamma \vdash B = A \text{ type}} \qquad \frac{\Gamma \vdash A = B \text{ type} \quad \Gamma \vdash B = C \text{ type}}{\Gamma \vdash A = C \text{ type}}$$

$$\frac{\Gamma \vdash a\colon A}{\Gamma \vdash a = a\colon A} \qquad \frac{\Gamma \vdash a = b\colon A}{\Gamma \vdash b = a\colon A} \qquad \frac{\Gamma \vdash a = b\colon A \quad \Gamma \vdash b = c\colon A}{\Gamma \vdash a = c\colon A}$$

Game Semantics and Uniqueness of Type Inhabitance in the Simply-Typed λ-Calculus

Pierre Bourreau and Sylvain Salvati

LaBRI - INRIA Sud-Ouest
351, Cours de la Libération
33405 Talence Cedex, France
{bourreau,salvati}@labri.fr

Abstract. The problem of characterizing sequents for which there is a unique proof in intuitionistic logic was first raised by Mints [Min77], initially studied in [BS82] and later in [Aot99]. We address this problem through game semantics and give a new and concise proof of [Aot99]. We also fully characterize a family of λ-terms for Aoto's theorem. The use of games also leads to a new characterization of principal typings for simply-typed λ-terms. These results show that game models can help proving strong structural properties in the simply-typed λ-calculus.

Keywords: games semantics, simply-typed λ-calculus, principal typing, coherence theorem, uniqueness of type inhabitance.

1 Introduction

Coherence theorems in category theory are used to ensure the equality of the composition of certain morphisms. In particular such conditions have been studied for cartesian closed categories - models of the simply-typed λ-calculus - in [BS82], a result which was later extended in [Aot99]. These results imply that a unique λ-term inhabits a given typing that verifies certain syntactic constraints. In [BS82], the exhibited typings are *balanced* while in[Aot99] they are *negatively non-duplicated*. A question that arises from these results, is that of the consequences of these constraints on types on their inhabitants. It can easily be observed that balanced typings are exactly inhabited by *affine λ-terms*. It was showed in [Kan07] that the family of *almost linear λ-terms* was included in that of the terms inhabiting negatively non-duplicated typings. One of the aims of this paper is to completely characterize the family of λ-terms that can be typed with negatively non-duplicated typings.

More precisely, we show that using game semantics leads to a concise proof of Aoto's theorem, and in general, gives an accurate method to address structural properties of simply-typed terms. The dialogic-game representation of proofs originates in [Lor59,Lor68,Bla92] and while game semantics has been widely used to study programming language semantics [AM99,HO00] and λ-calculus semantics [Hug00,GFH99,KNO02], to our knowledge it has never been used to address issues on proofs/terms structures, while it offers two main advantages:

L. Ong (Ed.): TLCA 2011, LNCS 6690, pp. 61–75, 2011.

first, it brings closer representations of typings and λ-terms which helps associating families of λ-terms with families of typings; second, it provides the analysis of proofs with a fine grained and natural access to the interplay of atomic types occurring in sequents. Our study leads indeed to a rather simple proof of Aoto's theorem, and to a syntactic characterization of the inhabitants of negatively non-duplicating types as *first-order copying λ-terms*, which extends both notions of linear and almost linear terms. These λ-terms could be depicted in Kanazawa's vocabulary as *almost affine*[1]. From a more general perspective, game semantics offers a simple way of investigating the relationship between typings and their inhabitants. As an example, we give a new characterization of the principal typings of β-normal terms.

In this paper, we will first recall basic notions on the simply-typed λ-calculus and introduce typing games and strategies; in the third section the correspondence between strategies and simply-typed terms will be presented; finally, we give a new characterization of principal typings, a concise proof of [Aot99] and a full characterization of the terms for [Aot99].

2 Preliminaries

2.1 Simply-Typed λ-Calculus

Let $\mathscr{A}$ be a countable set of atomic types. The set $\mathscr{T}(\mathscr{A})$ of simple types built upon $\mathscr{A}$ is the smallest set built as the closure of $\mathscr{A}$ under the right-associative connector $\rightarrow$. We call a *type substitution* σ an endomorphism of $\mathscr{T}(\mathscr{A})$ *i.e.* a function that verifies $\sigma(\alpha \rightarrow \beta) = \sigma(\alpha) \rightarrow \sigma(\beta)$, for α, β in $\mathscr{T}(\mathscr{A})$. Note that a type substitution is completely defined by the values it takes on $\mathscr{A}$. A *type relabelling* σ is a type substitution such that $\sigma(\mathscr{A})$ is included in $\mathscr{A}$. Finally, a *type renaming* denotes a bijective type substitution.

Let us consider a countable set of variables $\mathscr{V}$. The set Λ of λ-terms built on $\mathscr{V}$ is inductively defined with the following syntactic rules:

$$\Lambda ::= \mathscr{V} \mid \lambda\mathscr{V}.\Lambda \mid (\Lambda\Lambda)$$

We write λ-terms with the usual conventions, omitting sequences of λ's and unnecessary parentheses. For a term M, the notions of free variables (noted $FV(M)$), bound variables $(BV(M))$ and variables $(V(M) = FV(M) \cup BV(M))$ are defined as usual. We also take for granted the notions of α-conversion, β-reduction and η-conversion. A precise definition of all these notions can be found in [Bar84]. A context is a λ-term with a hole, built according to the following rules:

$$\Lambda_{[]} ::= [] \mid \lambda\mathscr{V}.\Lambda_{[]} \mid \Lambda_{[]}\Lambda \mid \Lambda\Lambda_{[]}$$

---

[1] This result was independently proved by Kanazawa in a yet unpublished work. Nevertheless, we believe the use of game semantics allows to express more directly the relation between syntactic properties of types and their inhabitants, giving a simpler proof.

The notation adopted for contexts will be $C[], C_1[], \ldots$ and grafting a term N in a context $C[]$ is noted $C[N]$. An *occurrence of a subterm* N in M is given by a context $C[]$ such that $M = C[N]$. We say that an occurrence of N in M is *characterized by* $C[]$ when $C[N] = M$. In general, we simply speak about an occurrence N of a subterm of M without mentionning the context that characterizes it. A term N is called a *subterm* of a term M when it has at least an occurrence in M, *i.e.* when there is a context $C[]$ such that $C[N] = M$.

A *typing environment* Γ is a finite subset of $\mathcal{V} \times \mathcal{T}(\mathscr{A})$ such that if (x, α) and (x, β) are in Γ, $\alpha = \beta$ holds. We write such an environment as a sequence of *type assignments* of the form $x : \alpha$. A *typing* $\langle \Gamma; \gamma \rangle$ is a pair made of a typing environment Γ and a type γ. A *typing judgement* of a term M is written $\Gamma \vdash M : \gamma$ and is said *valid* (we also say $\langle \Gamma; \gamma \rangle$ is a typing of M, or M is an inhabitant of $\langle \Gamma; \gamma \rangle$) if it can be obtained with the following inference system:

$$\frac{}{\Gamma, x : \alpha \vdash x : \alpha} \qquad \frac{\Gamma, x : \alpha \vdash M : \beta}{\Gamma \vdash \lambda x.M : \alpha \to \beta}$$

$$\frac{\Gamma \vdash M : \alpha \to \beta \qquad \Delta \vdash N : \alpha}{\Gamma \cup \Delta \vdash MN : \beta}$$

As a derivation of a typing judgement for a term M is always constructed by closely following the syntactic structure of M, for a given derivation $\mathcal{D}$ of the judgement $\Gamma \vdash M : \gamma$, there is a unique subderivation (where subderivation is taken in the obvious sense) $\mathcal{D}'$ of $\mathcal{D}$ that corresponds to an occurrence of a subterm N of M. Thus a derivation $\mathcal{D}$ of $\Gamma \vdash M : \gamma$ assigns to each occurrence N of a subterm of M a unique judgement of the form $\Delta \vdash N : \delta$. Given a derivation $\mathcal{D}$ of $\Gamma \vdash M : \gamma$, M is said in *η-long form relative to $\mathcal{D}$* when for each occurrence of any subterm N of M which is assigned a judgement of the form $\Delta \vdash N : \delta_1 \to \delta_2$ relative to $\mathcal{D}$, either N is of the form $\lambda x.N'$ or the context $C[]$ which characterizes this occurrence of N is of the form $C'[[]N']$. Terms in η-long form present several advantages (see [JG95]) and every simply-typed term M can be put in η-long form with respect to a derivation of a judgement [Hue76]. Furthermore, whenever M is in normal form, there is a unique way of deriving any typing judgement of M. In such a case, we say that to an occurrence of N in M is *associated the pair* $\langle \Delta; \delta \rangle$ *relative to* $\langle \Gamma; \gamma \rangle$ when the judgement associated to this occurrence of N relative to the unique derivation of $\Gamma \vdash M : \gamma$ is $\Delta \vdash N : \delta$. A term is said in *long normal form* when it is β-normal and η-long for some typing. A typing $\langle \Gamma; \gamma \rangle$ is a *principal typing* (or *most general typing*) of a term M, if $\Gamma \vdash M : \gamma$ is a valid judgement and if for every typing $\langle \Delta; \delta \rangle$ of M:

- $\Delta = \Delta_1 \cup \Delta_2$, where a type assignment $x : \alpha$ appears in Δ_1 if and only if x is a free variable of M
- there is a type substitution σ such that $\langle \Delta_1; \delta \rangle = \langle \Gamma; \gamma \rangle \cdot \sigma$.

An important result is that if a term M has a typing, then it has a principal typing which is unique up to renaming (see [Hin97]). This is the reason why from now on we will speak of the principal typing of a simply-typed term M. Given any typing $\langle \Gamma; \gamma \rangle$, polarity of types in $\langle \Gamma; \gamma \rangle$ is defined as follows:

- γ has a positive occurrence in $\langle \Gamma; \gamma \rangle$.
- for every type assignment $x : \delta \in \Gamma$, δ has a negative occurrence in $\langle \Gamma; \gamma \rangle$.
- if $\alpha \to \beta$ has a positive (*resp.* negative) occurrence in $\langle \Gamma; \gamma \rangle$, then α has a negative (*resp.* positive) occurrence and β a positive (*resp.* negative) one in this typing.

A typing $\langle \Gamma; \gamma \rangle$ is said *balanced* if each of its atomic types has at most one positive and at most one negative occurrence in it. It is said *negatively non-duplicating* if each of its atomic types has at most one negative occurrence in it.

Theorem 1. *[BS82] If a term M inhabits a balanced typing, then M is the unique inhabitant of this typing modulo $=_{\beta\eta}$.*
[Aot99] If a term M inhabits a negatively non-duplicating typing, then M is the unique inhabitant of this typing modulo $=_{\beta\eta}$.

2.2   Arenas and Typings

In what follows, arenas and games are presented as a restriction of Hyland-Ong games (HO-games) [HO00] and Nickau games [Nic94] as we do not use notions of question and answer. Following [KNO02], arenas associated to types are considered as ordered labelled tree; arenas associated to typings are considered as unordered labelled trees. Given a type γ, a move in the arena of γ is a finite sequence of natural numbers (we write $\mathbb{N}^*$ for the set of such sequences, ϵ for the empty sequence, and $s_1 \cdot s_2$ for the concatenation of such sequences). Elements of $\mathbb{N}$ will be written $i, j, i_1, \ldots$, and elements of $\mathbb{N}^*$ $s, s_1, \ldots$. Given a subset N of $\mathbb{N}^*$ and $i \in \mathbb{N}$, we denote by $i \cdot N$ the set $\{i \cdot s \mid s \in N\}$. Finally, given $s \in \mathbb{N}^*$, we write $|s|$ for the length of s defined as $|\epsilon| = 0$, $|i| = 1$ for $i \in \mathbb{N}$ and $|s_1 \cdot s_2| = |s_1| + |s_2|$.

Let us consider $\gamma = \gamma_1 \to \ldots \to \gamma_n \to a \in \mathscr{T}(\mathscr{A})$, where a is an atomic type. We inductively define the arena $A_\gamma = (M_\gamma, \tau_\gamma)$ (where M_γ is a finite set of *moves* and the *typing function* τ_γ is a function from M_γ to $\mathscr{A}$) from the arenas (M_{γ_i}, τ_i) associated to γ_i for every $i \in [1, n]$:

1. $M_\gamma = \{\epsilon\} \cup \bigcup_{i=1 \ldots n} i \cdot M_{\gamma_i}$,
2. $\tau_\gamma : M_\gamma \mapsto \mathscr{A}$ is defined by:
   (a) $\tau_\gamma(\epsilon) = a$
   (b) $\tau_\gamma(i \cdot s) = \tau_{\gamma_i}(s)$, for $i \in [1 \ldots n]$ and $s \in \mathbb{N}^*$

The arena associated to a type $\gamma \in \mathscr{T}(\mathscr{A})$ is therefore a finite prefix-closed subset of $\mathbb{N}^*$ whose elements are labelled with atomic types. The parent relation expresses the *enabling* relation [HO00]: given s_1 and s_2 in $\mathbb{N}^*$, s_1 enables s_2 (written $s_1 \vdash s_2$) when there is i in $\mathbb{N}$ such that $s_2 = s_1 \cdot i$. Given an arena (M, τ), we define the function $pl : M \mapsto \{O, P\}$ which associates moves to players (P stands for the proponent and O for the opponent), and $\overline{pl} : M \mapsto \{O, P\}$ its inverse function (*i.e.* $pl(m) = O$ iff $\overline{pl}(m) = P$ for every $m \in M$, and $\overline{\overline{pl}} = pl$), by $pl(\epsilon) = O$ and $pl(s_2) = \overline{pl}(s_1)$ for $s_1 \vdash s_2$. In the rest of the document, we

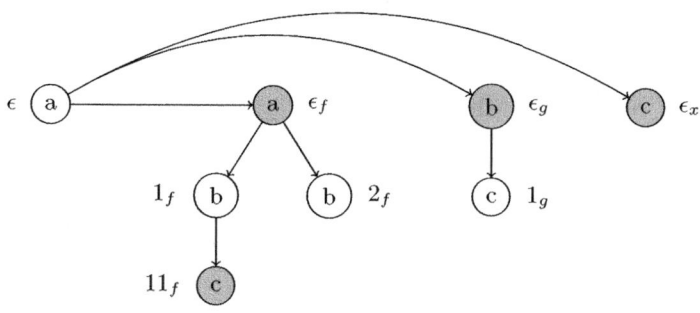

Fig. 1. Example of an arena

use the partition $M^O = \{s \in M \mid p\ell(s) = O\}$ and $M^P = \{s \in M \mid p\ell(s) = P\}$ of M which separates P-moves (*i.e.* negative atoms) and O-moves (*i.e.* positive atoms) in M. Remark the correspondence between, atomic types and moves, polarity and $p\ell$.

Given a type assignment $x : \gamma$, the associated arena $A_{x:\gamma} = (M_{x:\gamma}, \tau_{x:\gamma})$ is defined similarly to A_γ but moves in $M_{x:\gamma}$ are paired with the variable x, and written ϵ_x, i_x, s_x, ... for the pairs (ϵ, x), (i, x), (s, x)... respectively. In the rest of the document, we will use $m, n, m_1, \ldots$ for moves. For a variable x, a sequence s in $\mathbb{N}^*$ and $i \in \mathbb{N}$, given the move $m = s_x$, we write $i \cdot m$ (*resp.* $m \cdot i$) for the move $(i \cdot s)_x$ (*resp.* $(s \cdot i)_x$). For a set of moves M, the *preceding* relation $m_1 \prec m_2$ stands for $m_1, m_2 \in M$ iff there exist $i, j \in \mathbb{N}$ and $m \in M$ such that $m_1 = m \cdot i$, $m_2 = m \cdot j$ and $i < j$.

Let us consider a typing $\langle \Gamma; \gamma \rangle$, where $\Gamma = x_1 : \gamma_1, \ldots, x_n : \gamma_n$. Its associated arena $A = (M, \tau)$ is defined as:

1. $M = M_\gamma \cup \bigcup_{i \in [1,n]} M_{x_i:\gamma_i}$

2. $\tau(m) = \begin{cases} \tau_\gamma(m) & \text{if } m \in M_\gamma \\ \tau_{x_i:\gamma_i}(m) & \text{if } m \in M_{x_i:\gamma_i} \text{ for some } i \in [1, n] \end{cases}$

Moreover, the enabling relation is extended so that $m_1 \vdash m_2$ holds iff

- $m_2 = m_1 \cdot j$ for some j in $\mathbb{N}$ or
- $m_1 = \epsilon \in M_\gamma$ and $m_2 = \epsilon_{x_i}$ for some $i \in [1, n]$

The arena associated to a typing $\langle \Gamma; \gamma \rangle$ is therefore represented as a forest made of the labelled trees associated to the type γ and to the type assignments in Γ. The enabling relation gives an arborescent structure to this forest, but while the arena associated to a type defines an ordered tree (with the left-to-right relation given by $\prec$), the arborescent structure of an arena associated to a typing is not ordered (two subtrees which roots are ϵ_x, ϵ_y can be permuted). Note that the notation m_x for a move in the arena associated to some typing $\langle \Gamma; \gamma \rangle$ allows to differentiate this arena from the arena of $\langle \Gamma - \{x : \delta\}; \delta \rightarrow \gamma \rangle$. Figure 1 presents the arena for the sequent $\langle f : (c \rightarrow b) \rightarrow b \rightarrow a, g : c \rightarrow b, x : c; a \rangle$, where

O-moves are represented by white nodes, P-moves by dark nodes and oriented black edges represent the enabling relation.

2.3   Games and Strategies

In the rest of the document, we use $S, S_1, \ldots$ to denote finite sequences of moves $m_1.m_2 \ldots m_n$ of a given arena. We write $S_1 \sqsubseteq S_2$ when the sequence of moves S_1 is a prefix of the sequence of moves S_2. A finite sequence $S = m_1 \ldots m_n$ of moves in an arena A is said *justified* if to each occurrence of a move m in S (except the initial occurrence m_1), there is an occurrence of a move n which precedes m in S and which enables m (the occurrence of n is said to *justify* the occurrence of m). Formally, a justified sequence S is written $(m_1, 1, 0) \ldots (m_n, n, l)$ where (m, i, j) denotes the move in position i in S, justified by (m', j, k) where $j < i$. This notation will be simplified for the particular case of strategies we are interested in.

Given a finite sequence of moves S, the *P-view* $\ulcorner S \urcorner$ of S removes moves which are irrelevant for player P in S and is inductively defined as:

$$\ulcorner \epsilon \urcorner = \epsilon \qquad\qquad\qquad\qquad \text{for the initial move } \epsilon$$
$$\ulcorner S.(m, i, j) \urcorner = \ulcorner S \urcorner.(m, i, j) \qquad\qquad \text{for a } P\text{-move } m$$
$$\ulcorner S_1.(m_1, j, k).S_2.(m_2, i, j) \urcorner = \ulcorner S_1 \urcorner.(m_1, j, k).(m_2, i, j) \text{ for an } O\text{-move } m_2.$$

The O-view of a sequence S can be defined similarly, but is not needed for our purpose. We call a justified sequence $S = m_1 \ldots m_n$ a *legal position* if it verifies:

1. $m_1 = \epsilon$ (Initial move condition)
2. for $S = S_1.(m, i, j).(n, i+1, k).S_2$, $pl(m) = \overline{pl(n)}$ (Alternation)
3. $\ulcorner S \urcorner = S$ (P-view condition)

For a given arena A, we write L_A the set of its legal positions.

In general, a game is defined as an arena to which we associate a set of positions, in the present case, a set of prefix-closed legal positions. Because our study only requires the particular notion of games where available positions are precisely all legal positions in the given arena, we confuse the usual notion of game with that of an arena. Note that, given a legal position L, because L is P-views, an occurrence of an O-move $M \neq \epsilon$ in L is justified by the immediately preceding move (which is a P-move). The notation for occurrences of moves in a legal position S can therefore be simplified as follows:

- an occurrence of an O-move n in S will be denoted by (n, i) if it is in position $2i + 1$ in S in a left-to-right order;
- an occurrence of a P-move m in S will be written (m, i) if this occurrence is justified by a preceding occurrence (n, i) of the O-move n.

Remark that this notation confuses occurrences of a P-move justified by the same occurrence of an O-move (n, i). In the next section, we will see that these occurrences correspond to the same variable in the term associated to the strategy. When unnecessary, information about justification will be omitted from our notation and an occurrence of a move m will simply be noted m.

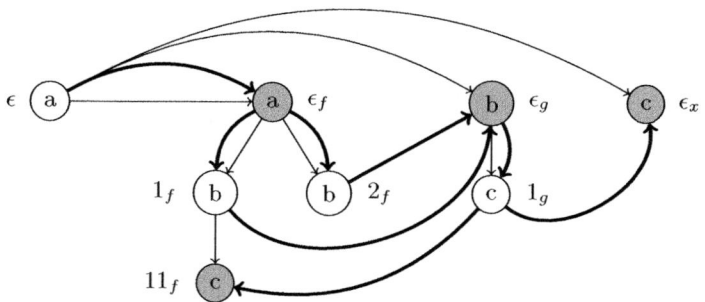

Fig. 2. Example of a typing strategy

Definition 1. *Let $A = (M, \tau)$ be an arena and $\Sigma \subseteq L_A$ a finite non-empty and prefix-closed set of legal positions. Σ is a $(P\text{-})$strategy (or* typing strategy*) if:*

1. *the length $|S|$ of any sequence $S \in \Sigma$ is even and different from 0.*
2. *if $S.m.n_1, S.m.n_2 \in \Sigma$, then $n_1 = n_2$*
3. *if $S.m.n \in \Sigma$ then $\tau(m) = \tau(n)$*

According to the second property of this definition, a strategy Σ is uniquely determined by the set $max(\Sigma) \subseteq \Sigma$ made of the sequences which are maximal for the partial order $\sqsubseteq$.

As an example, Figure 2 represents the arena for the sequent $\langle f : ((c \rightarrow b) \rightarrow b \rightarrow a), g : c \rightarrow b, x : c; a \rangle$. In this Figure, paths made of thick arrows starting from ϵ denote sequences of moves which are legal positions in the arena if the justification constraint is respected. An example of strategy Σ is given by $max(\Sigma) = \{(\epsilon, 0).(\epsilon_f, 0).(1_f, 1).(\epsilon_g, 0), (\epsilon, 0).(\epsilon_f, 0).(2_f, 1).(\epsilon_g, 0).(1_g, 2).(\epsilon_x, 0)\}$.

Definition 2. *Given an arena $A = (M, \tau)$, a typing strategy Σ on A is called a* winning strategy *if*

- *for all $S.m$ in $max(\Sigma)$, there is no move $n \in M^O$, such that $m \vdash n$.*
- *for every sequences $S.m_1 \in \Sigma$ and $S.m_1.n \in L_A$, there is a P-move m_2 such that $S.m_1.n.m_2 \in \Sigma$.*

The first condition ensures P wins when O can play no move while the second one ensures that P wins with this strategy independently from O's strategy. In the rest of the document, if not specified otherwise, only winning strategies will be considered.

Example 1. In the arena of Figure 2, there is a winning strategy Σ such that $max(\Sigma) = \{S_1, S_2\}$ with:

- $S_1 = (\epsilon, 0).(\epsilon_f, 0).(1_f, 1).(\epsilon_g, 0).(1_g, 2).(11_f, 1)$
- $S_2 = (\epsilon, 0).(\epsilon_f, 0).(2_f, 1).(\epsilon_g, 0).(1_g, 2).(\epsilon_x, 0)$

Remark that there is another typing strategy Σ' such that $max(\Sigma') = \{S_1', S_2\}$ where $S_1' = (\epsilon, 0).(\epsilon_f, 0).(1_f, 1).(\epsilon_g, 0).(1_g, 2).(\epsilon_x, 0)$.

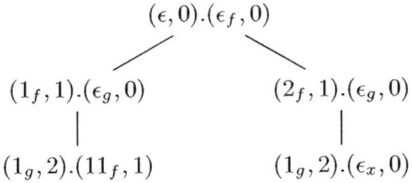

Fig. 3. Example of an arborescent reading

3   Strategies as Terms

3.1   Interpreting λ-Terms

Given an arena $A = (M, \tau)$, the prefix-closure property of a strategy Σ on A can be used to represent $max(\Sigma)$ as a tree denoted $\mathbb{T}_\Sigma$. We call this tree *the arborescent reading of* Σ and given a strategy Σ it is inductively defined on prefix-closed set of alternating sequences as $\mathbb{T}_\Sigma = m_1.m_2[\mathbb{T}_{\Sigma_1}, \ldots, \mathbb{T}_{\Sigma_p}]$ if:

- for every S in $max(\Sigma)$, $S = m_1.m_2.S'$ for some S',
- $\{n_1, \ldots, n_p\} = \{m' \mid m_2 \vdash m'\}$ and $n_i \prec n_j$ iff $i < j$,
- $\Sigma_i = \{n_i.S' \mid m_1.m_2.n_i.S' \in \Sigma\}$.

For example the strategy defined by $max(\Sigma) = \{S_1, S_2\}$ in Example 1 has the arborescent reading (also pictured on Figure 3)

$$\mathbb{T}_\Sigma = (\epsilon, 0).(\epsilon_f, 0)[(1_f, 1).(\epsilon_g, 0)[(1_g, 2).(11_f, 1)], (2_f, 1).(\epsilon_g, 0)[(1_g, 2).(\epsilon_x, 0)]]$$

Given an arena A, let us note $FV_A = \{((\epsilon_x, 0), x) \mid \text{ there is a variable } x, \epsilon_x \in M\}$

Definition 3. *Let the interpretation of a strategy* Σ *on an arena* $A = (M, \tau)$ *be* $[\![\Sigma]\!] = [\![\mathbb{T}_\Sigma, FV_A]\!]$ *which is inductively defined on* $\mathbb{T}_\Sigma$ *as:*

$$[\![(m, i).(n, j)[\mathbb{T}_{\Sigma_1}, \ldots, \mathbb{T}_{\Sigma_q}], V]\!] = \lambda x_1 \ldots x_p.x[\![\mathbb{T}_{\Sigma_1}, W]\!] \ldots [\![\mathbb{T}_{\Sigma_q}, W]\!]$$

where $m \in M^O$, $n \in M^P$, *the set* W *is equal to* $V \cup \{((m \cdot k, i), x_k) \mid m \cdot k \in M\}$, *the* x_k*'s being fresh variables, and* $((n, j), x)$ *is in* W.

The second rule of the interpretation associates variables to occurrences of P-moves. We say that an occurrence of a variable x is a *realization* of a P-move m (or of (m, i)) in $[\![\Sigma]\!]$ when $[\![\Sigma]\!] = C[xN_1 \ldots N_n] = C[[\![(m, i)[\mathbb{T}_{\Sigma_1}, \ldots, \mathbb{T}_{\Sigma_n}], V]\!]]$. Note that $((m, i), x)$ is in V. Moreover, if we suppose $[\![\Sigma]\!]$ respects the syntactic convention of Barendregt, *i.e.* if every variable in $[\![\Sigma]\!]$ is uniquely identified by its name, then it is easy to see that every occurrences of a variable x are realizations of the same (m, i). In this case, we simply say that *the variable* x *is a realization of the P-move* m (or of (m, i)). In the rest of the document, $[\![\Sigma]\!]$ is supposed to verify Barendregt's convention for every strategy Σ.

This interpretation of winning strategies as λ-terms allows us establishing a bijection between the set of strategies in the arena associated to the pair $\langle \Gamma; \gamma \rangle$ and the set of inhabitants of that pair modulo $\beta\eta$-conversion.

Lemma 1. *Given the arena A associated to the typing pair $\langle \Gamma; \gamma \rangle$, and a winning strategy Σ on A, $\Gamma \vdash [\![\Sigma]\!] : \gamma$ is a valid judgement.*

Proof. The proof is done by induction on the structure of $\mathbb{T}_\Sigma$.

In the course of the proof of this Lemma, it can be noted that $[\![\Sigma]\!]$ is β-normal and η-long with respect to $\langle \Gamma; \gamma \rangle$.

Lemma 2. *Given $\Gamma \vdash N : \gamma$, there is a winning strategy Σ on A such that $[\![\Sigma]\!] =_{\beta\eta} N$.*

Proof. Without loss of generality we suppose that N is in long normal form with respect to the typing pair $\langle \Gamma; \gamma \rangle$. We then construct Σ inductively on the structure of N.

To establish the bijection between terms and strategies, it suffices to complete the picture with the following Lemma.

Lemma 3. *Given an arena A, and two winning strategies Σ_1 and Σ_2 on A, $[\![\Sigma_1]\!] = [\![\Sigma_2]\!]$ iff $\Sigma_1 = \Sigma_2$.*

3.2 Game-Theoretic Characterization of Principal Typings

Given a simply-typed λ-term N, its principal typing enjoys two equivalent characterizations:

1. an intentional one, through the Hindey-Milner inference algorithm [DM82].
2. an extensional one, by specifying that every other typing of N can be found by substitution on the principal one.

In this section we propose a third characterization of principal typings of β-normal terms by means of game semantics.

Definition 4. *Given a typing arena $A = (M, \tau)$ and a typing strategy Σ on A, the binary relations $\lhd$ and $\lhd_\Sigma$ on $M^O \times M^P$ are defined as follows:*

1. *$n \lhd m$ iff $\tau(m) = \tau(n)$*
2. *$n \lhd_\Sigma m$ iff there is $S \in \Sigma$, such that $S = S'.n.m$.*

For a P-move m, we note $\lhd^m = \{ n \in M^O \mid n \lhd m \}$ the set of its *possible antecedents*, and $\lhd_\Sigma^m = \{ n \in M^O \mid n \lhd_\Sigma m \}$ the set of its antecedents in Σ. For any typing strategy Σ, $\lhd_\Sigma^m$ is a subset of $\lhd^m$. Remark that for any P-moves m_1, m_2, $\tau(m_1) = \tau(m_2)$ is equivalent to $\lhd^{m_1} = \lhd^{m_2}$.

Let us define $R_\Sigma \subseteq M^P \times M^P$ as $m_1 R_\Sigma m_2$ iff $\lhd_\Sigma^{m_1} \cap \lhd_\Sigma^{m_2} \neq \emptyset$. We note R_Σ^* the transitive closure of R_Σ; it is easy to verify that R_Σ^* is an equivalence relation. Then, all elements (m_1, m_2) in R_Σ, have at least one antecedent $n \in M^O$ in common in Σ; this implies $\tau(n) = \tau(m_1) = \tau(m_2)$. Thus, whenever $m_1 R_\Sigma^* m_2$, $\tau(m_1) = \tau(m_2)$.

Definition 5. *Let us consider a typing arena $A = (M, \tau)$ and a strategy Σ on A. Σ is a* covering *typing strategy if*

1. *Σ is a winning typing strategy. (winning condition)*
2. *For all moves $m = \epsilon_x \in M^P$, there is a sequence $S.m \in \Sigma$. (ϵ-completeness)*
3. *For all O-moves m, there is a P-move n and a sequence $S \in \Sigma$ (or S is empty), such that $S.m.n \in \Sigma$. (O-completeness)*
4. *For all m_1, m_2 in M^P, $\tau(m_1) = \tau(m_2)$ iff $m_1 R_{\Sigma}^* m_2$. (least typing constraint)*

Theorem 2. *Given a term N, a typing pair $\langle \Gamma; \gamma \rangle$ and the corresponding arena $A = (M, \tau)$. The two following properties are equivalent*

1. *the strategy Σ on A such that $\llbracket \Sigma \rrbracket =_{\beta\eta} N$ is a covering strategy*
2. *$\langle \Gamma; \gamma \rangle$ is the principal typing of the long normal form of N relative to $\langle \Gamma; \gamma \rangle$*

Proof. Without loss of generality we assume that N is in long normal form relatively to $\langle \Gamma; \gamma \rangle$. We prove the equivalence between the two following propositions:

1'. Σ is not a covering strategy on A
2'. $\langle \Gamma; \gamma \rangle$ is not the principal typing of N

The implication 2'. $\Rightarrow$ 1'. is proved using a disjunction of cases. First, we suppose that there is a type assignment $x : \alpha$ in Γ such that x has no occurrence in N; then Σ violates the condition of ϵ-completeness and thus is not a covering strategy. Secondly, we let $\langle \Gamma'; \gamma' \rangle$ be the principal typing of N and suppose there is type substitution σ such that $\langle \Gamma; \gamma \rangle = \langle \Gamma'; \gamma' \rangle \cdot \sigma$; we associate to a type substitution a transformation on the initial arena (a morphism of arena) and show that if σ is not a relabelling then Σ violates one of the condition in Definition 5. The implication 1'. $\Rightarrow$ 2'. is proved with similar arguments.

4   Expressing Structural Properties

4.1   First-Order Copying Terms

We are now about to define syntactically the family of λ-terms that correspond to strategies in arenas associated to negatively non-duplicating typing pairs. This definition requires some intricate relation between the variables that occur in a term. The main technical difficulty for defining this relation comes from α-conversion. Thus, we adopt Barendregt's naming convention of bound variables that consists in giving a different name to each variable introduced by each λ-abstraction. Under this convention, and given a normal λ-term N in η-long form for some typing $\langle \Gamma; \gamma \rangle$ and two of its variables x, y, the binary relation $x \mathtt{I}_{ij}^N y$ on $V(N) \times V(N)$ (where $i, j \in \mathbb{N}$) holds iff

$$N = C[x N_1 \ldots N_{i-1} (\lambda x_1 \ldots x_{j-1} y x_{j+1} \ldots x_n.N) N_{i+1} \ldots N_m]$$

Definition 6. *Let us consider a λ-term N in η-long form for some typing and which variables have distinct names. Given two variables x and y in $V(N)$, $x \approx_N y$ is verified iff:*

1. *$x = y$.*
2. *or there are two variables z_1, z_2 and i, j in $\mathbb{N}$ such that $z_1 I_{ij}^N x$, $z_2 I_{ij}^N y$, and $z_1 \approx_N z_2$*

The idea behind the relation $x \approx_N y$ on a term N is to express that the variables x and y are *recursively introduced* by the same variable.

Example 2. Given $N = \lambda f g.g(f(\lambda x_1 y_1.x_1(\lambda z_1.z_1)))(f(\lambda x_2 y_2.y_2(x_2(\lambda z_2.z_2)))))$, the relations $x_1 \approx_N x_2$, $y_1 \approx_N y_2$ and $z_1 \approx_N z_2$ hold.

This definition is naturally extended to subterms of N:

Definition 7. *Let us consider a term N in long normal form for some typing $\langle \Gamma; \gamma \rangle$, and two of its subterms N_1 and N_2; $N_1 \approx_N N_2$ is verified if:*

- *$N_1 = x_1$, $N_2 = x_2$, and $x_1 \approx_N x_2$*
- *$N_1 = \lambda x_1.P_1$, $N_2 = \lambda x_2.P_2$, $x_1 \approx_N x_2$ and $P_1 \approx_N P_2$*
- *$N_1 = x_1 P_1 \ldots P_n$, $N_2 = x_2 Q_1 \ldots Q_n$, $x_1 \approx_N x_2$ and $P_i \approx_N Q_i$ for every $i \in [1, n]$.*

Example 3. In the previous example, we have $x_1(\lambda z_1.z_1) \approx_N x_2(\lambda z_2.z_2)$. But we do not have $f(\lambda x_1 y_1.x_1(\lambda z_1.z_1)) \approx_N f(\lambda x_2 y_2.y_2(x_2(\lambda z_2.z_2)))$.

Remark that for a term $N = C_1[C_2[P_1][P_2]]$ in long normal form for some typing $\langle \Gamma; \gamma \rangle$ and where $P_1 \approx_N P_2$, the term $N' = C_1'[(\lambda x.C_2'[x][x])P]$ that is β-convertible to N is also an inhabitant of $\langle \Gamma; \gamma \rangle$. In the previous example, $(\lambda F f g.g(f(\lambda x_1 y_1.Fx_1))(f(\lambda x_2 y_2.y_2(Fx_2))))(\lambda x.x(\lambda z.z))$ is simply typable and has the same most general typing as N.

In the rest of the document, we focus on a particular case for this relation:

Definition 8. *Let us consider a term N in long normal form for its principal typing $\langle \Gamma; \gamma \rangle$. N is said first-order copying (written copy(1)) whenever two subterms N_1 and N_2 are assigned the same type α in $\langle \Gamma; \gamma \rangle$ iff $N_1 \approx_N N_2$.*

Example 4. The term in Example 2 is not copy(1) because $f(\lambda x_1 y_1.x_1(\lambda z_2.z_2))$ and $f(\lambda x_2 y_2.y_2(x_2(\lambda z_2.z_2)))$ have the same type in the principal typing of N but are not in relation with $\approx_N$. The term $N = f(h(\lambda x_1 y_1.y_1))(h(\lambda x_2 y_2.y_2))$ is copy(1) and can be expanded into $(\lambda z.fzz)(h(\lambda xy.y))$ preserving its principal typing. Note that z is of atomic type in any typing for which N is in η-long form.

The notion is called first-order copying because the maximal subterms of N which verify the relation $\approx_N$ are of atomic type in the principal typing of N just as in the last example above.

4.2   Proof Uniqueness for Copy(1)-Terms

We are now going to see that first-order copying are precisely the inhabitants of negatively non-duplicating typings. But we first prove that negatively non-duplicating typings have at most one inhabitant.

Lemma 4. *Let us consider a negatively non-duplicating typing* $\langle \Gamma; \gamma \rangle$ *and the associated arena* $A = (M, \tau)$. *There is at most one occurrence of each O-move* m *in every sequence of a winning strategy* Σ *on* A.

Proof. If we suppose there is a strategy Σ which contains a sequence S of the form $S_1.n.m_1.S_2.n.m_2$ where $n \in M^O$ then we can prove Σ must contain an infinite number of sequences which is absurd.

Theorem 3. *[Aot99] Let* $\langle \Gamma; \gamma \rangle$ *be a negatively non-duplicating typing. There is at most one inhabitant of* $\langle \Gamma; \gamma \rangle$.

Proof. Let us suppose that N is an inhabitant of $\langle \Gamma; \gamma \rangle$ and let $A = (M, \tau)$ be the arena $\langle \Gamma; \gamma \rangle$. Then we let Σ be the strategy on A such that $N = [\![\Sigma]\!]$. We recall that a P-move m corresponds to a negative occurrence of an atomic type $a = \tau(m)$. Suppose there is another strategy $\Sigma' \neq \Sigma$ in A; we show by induction on the length of a sequence S in Σ' that $S \in \Sigma$.

1. if $|S| = 2$, then $S = (\epsilon, 0).(m, 0)$ for some P-move m. Because the pair is negatively non-duplicating, every P-move $m' \neq m$ in M verifies $\tau(m') \neq \tau(m)$. Therefore S must also be in Σ.
2. if $|S| = 2p+2$, then S is of the form $S_1.(m, p).(n, i)$. By induction hypothesis, S_1 is in Σ. Moreover, if $S_1.(n, p).(m, i)$ is in Σ', there exists a P-move m' such that $S_1 = S_2.(m', j)$ (where $m' \vdash n$). But because Σ is a winning strategy there must exist a P-move m'' in M such that there is a sequence of the form $S_1.(n, p).(m'', k)$ in Σ. Again, the typing being negatively non-duplicating, m is the only P-move such that $\tau(m) = \tau(n)$ and therefore we must have $m = m''$. Moreover, according to Lemma 4, $k = j$.

We now turn to proving that negatively non-duplicating typings are inhabited by first-order copying λ-terms and that conversely the principal typing of first-order copying λ-terms is negatively non-duplicating. This correspondence is similar to the one between balanced typings and affine λ-terms [Bel76,Hir91].

Lemma 5. *Let us consider an arena* $A = (M, \tau)$, *a strategy* Σ *on it and the term* $N = [\![\Sigma]\!]$. *Two variables* x_1, x_2 *in* $V(N)$ *verify* $x_1 \approx_N x_2$ *iff there is* $m \in M^P$ *such that* x_1 *and* x_2 *are realizations of* m *in* $[\![\Sigma]\!]$.

Proof. First, the right implication is shown by induction on the definition of the $\approx_N$ relation on variables.

For the left implication, let us first define $J(S)$ for a sequence S in a strategy Σ as follows:

- $J((\epsilon, 0)) = 0.$
- $J(S_1.(m, i).S_2.(n, i)) = 1 + J(S_1.(m, i))$ if n is a P-move.
- $J(S.(m, i)) = J(S)$ if m is a O-move.

Let us suppose the two variables x_1 and x_2 are realizations of the same P-move m in $[\![\Sigma]\!]$, which is formally written as:

$$N = \begin{cases} C_1[\![[(m, i_1).[\mathbb{T}_1, \ldots, \mathbb{T}_n], V_1]\!] = C_1[x_1[\![\mathbb{T}_1, V_1]\!] \ldots [\![\mathbb{T}_n, V_1]\!]] \\ \\ C_2[\![[(m, i_2).[\mathbb{T}_1', \ldots, \mathbb{T}_n'], V_2]\!] = C_2[x_2[\![\mathbb{T}_1', V_2]\!] \ldots [\![\mathbb{T}_n', V_2]\!]] \end{cases}$$

Let us consider $S_1.(n, i_1)$ and $S_2.(n, i_2)$, the sequences in Σ for the respective occurrences of n in the equation above. We prove the rest of the lemma by induction on $p = max(J(S_1.n), J(S_2.n))$.

Theorem 4. *If a negatively non-duplicating typing $\langle \Gamma; \gamma \rangle$ types a term N then N is copy(1).*

Proof. Consider the strategy Σ on the arena A associated to $\langle \Gamma; \gamma \rangle$ such that $[\![\Sigma]\!] = N$. According to Theorem 3, Σ is the unique winning strategy in A. Suppose $N = C_1[N_1] = C_2[N_2]$, such that $N_1 = \lambda x_1 \ldots x_q.z_1 P_1 \ldots P_{p_1}$ and $N_2 = \lambda y_1 \ldots y_q.z_2 Q_1 \ldots Q_{p_2}$ are of the same type δ in $\langle \Gamma; \gamma \rangle$. We can write $[\![\Sigma]\!]$ as:

$$[\![\Sigma]\!] = \begin{cases} C_1[\![\mathbb{T}_1, V_1]\!] = C_1[\![[n_1.m_1.[\mathbb{T}_{1,1}, \ldots, \mathbb{T}_{1,p_1}], V_1]\!] = C_1[N_1] \\ C_2[\![\mathbb{T}_2, V_2]\!] = C_2[\![[n_2.m_2.[\mathbb{T}_{2,1}, \ldots, \mathbb{T}_{2,p_2}], V_2]\!] = C_2[N_2] \end{cases}$$

where n_1, n_2 are O-moves, and m_1, m_2 P-moves. Because N_1 and N_2 are of the same type, $\tau(m_1) = \tau(m_2)$ and because $\langle \Gamma; \gamma \rangle$ is negatively non-duplicating $m_1 = m_2 = m$. This implies $p_1 = p_2 = p$. By induction on the trees $\mathbb{T}_1$ and $\mathbb{T}_2$, we prove $N_1 \approx_N N_2$.

Theorem 5. *Given a first-order copying term N in long normal form for its principal typing $\langle \Gamma; \gamma \rangle$, $\langle \Gamma; \gamma \rangle$ is negatively non-duplicating.*

Proof. Given the arena $A = (M, \tau)$ associated to $\langle \Gamma; \gamma \rangle$, and the strategy Σ on A, such that $N = [\![\Sigma]\!]$, we know Σ is a covering strategy (Theorem 2). Suppose there exists $n_1 \neq n_2$ in M^P, such that $\tau(n_1) = \tau(n_2)$. By Theorem 2, this leads to $n_1 R_\Sigma^* n_2$, which implies the existence of two sequences $S_1.n_1$ and $S_2.n_2$ in Σ such that $N = C_i[\![[n_i[\mathbb{T}_{i1}, \ldots, \mathbb{T}_{im}], V_i]\!]] = C_i[N_i]$ for $i \in \{1, 2\}$. N_1 and N_2 are of the same atomic type $\tau(n_2) = \tau(n_1)$ then $N_1 = x_1 P_1 \ldots P_n$ and $N_2 = x_2 Q_1 \ldots Q_m$. According to Lemma 5, $x_1 \approx_N x_2$ does not hold which implies that $N_1 \approx_N N_2$ does not hold either: we obtain a contradiction with the hypothesis that N is first-order copying.

5   Conclusion

Games are widely used in logic for the study of programming language semantics [HO00], logic semantics [Bla92], and λ-calculi semantics [Hug00]. In this article, we show that game semantics brings a new point of view on the structural properties of the simply-typed λ-calculus. It was already known that the balanced types used in [BS82] are exactly the principal typings of affine λ-terms $i.e.$ terms M for which each subterm $\lambda x.M'$ in it has at most one free occurrence of x in M' and each free variable x has at most one occurrence in M [Bel76,Hir91]. Using games, we obtain a full correspondence between negatively non-duplicating typings and the newly introduced class of *first-order copying λ-terms*.

Even though the problems of coherence for the simply-typed λ-calculus seem quite theoretical, they have recently proved to be useful in designing parsing algorithms for grammars of simply-typed λ-terms [dG01,Mus01]. Determining a λ-term only with its typing properties allows indeed to partially characterize it up to $\beta\eta$-conversion. This idea has been used in [Kan07] to propose a very general parsing technique using [Aot99] coherence theorem for *almost linear* λ-terms and in [Sal10] to prove that parsing grammars of simply-typed λ-terms is decidable. Based on the result presented in this paper, we expect to extend Kanazawa's algorithm to first-order copying terms.

The question of extending coherence theorems to a wider class of terms remains an open-question. A family of typings has been given in [BD05] under the name of *deterministic typings*, with a method similar to the one given in this article, but it seems hard to give a syntactic characterization of the terms which inhabits such typings. Nevertheless, the use of game semantics gives an easy framework to study proof structures and while we focused on normal proofs, we shall try to extend our results to non-normal ones. For example, it would be interesting to investigate a generalization of the game-theoretical characterization of principal typings of β-normal terms to non β-normal ones will also be addressed.

References

AM99.   Abramsky, S., McCusker, G.: Game semantics. In: Schwichtenberg, H., Berger, U. (eds.) Computational Logic: Proceedings of the 1997 Marktoberdorf Summer School, pp. 1–56. Springer, Heidelberg (1999)

Aot99.   Aoto, T.: Uniqueness of normal proofs in implicational intuitionistic logic. Journal of Logic, Language and Information 8, 217–242 (1999)

Bar84.   Barendregt, H.: λ-calculus: its syntax and semantics. Elsevier Science Publishers Ltd., Amsterdam (1984)

BD05.   Broda, S., Damas, L.: On long normal inhabitants of a type. J. Log. Comput. 15(3), 353–390 (2005)

Bel76.   Belnap, N.: The two-property. Relevance Logic Newsletter, 173–180 (1976)

Bla92.   Blass, A.: A game semantics for linear logic. Annals of Pure and Applied Logic 56, 183–220 (1992)

BS82.   Babaev, A., Soloviev, S.: A coherence theorem for canonical morphism in cartesian closed categories. Journal of Soviet Mathematics 20, 2263–2279 (1982)

dG01.   de Groote, P.: Towards abstract categorial grammars. In: Proceedings of the Conference Association for Computational Linguistics, 39th Annual Meeting and 10th Conference of the European Chapter, pp. 148–155 (2001)

DM82.   Damas, L., Milner, R.: Principal type-schemes for functional programs. In: POPL 1982: Proceedings of the 9th ACM SIGPLAN-SIGACT Symposium on Principles of Programming Languages, pp. 207–212. ACM, New York (1982)

GFH99.   Di Gianantonio, P., Franco, G., Honsell, F.: Game semantics for untyped $\lambda\beta\eta$-calculus. In: Girard, J.-Y. (ed.) TLCA 1999. LNCS, vol. 1581, pp. 114–128. Springer, Heidelberg (1999)

Hin97.   Hindley, R.J.: Basic Simple Type Theory. Cambridge Press University, Cambridge (1997)

Hir91.   Hirokawa, S.: Balanced formulas, minimal formulas and their proofs. Technical report, Research Institute of Fundamental Information Science, Kyochu University (1991)

HO00.   Hyland, J.M.E., Ong, C.-H.L.: On full abstraction for PCF. In: Information and Computation, vol. 163(2), ch. 2, pp. 285–408. Elsevier Science B.V., Amsterdam (2000)

Hue76.   Huet, G.: Résolution d'équations dans les langages d'ordre 1,2,.,ω. PhD thesis, Université Paris 7 (1976)

Hug00.   Hughes, D.J.D.: Hypergame Semantics: Full Completeness for System F. PhD thesis, Oxford University (2000)

JG95.   Jay, C.B., Ghani, N.: The virtues of η-expansion. J. of Functional Programming 5(2), 135–154 (1995); Also appeared as tech. report ECS-LFCS-92-243

Kan07.   Kanazawa, M.: Parsing and generation as Datalog queries. In: Proceedings of the 45th Annual Meeting of the Association for Computational Linguistics, Prague, pp. 176–183. Association for Computational Linguistics (2007)

KNO02.   Ker, A.D., Nickau, H., Ong, L.: Innocent game models of untyped λ-calculus. Theoretical Computer Science 272(1-2), 247–292 (2002)

Lor59.   Lorenzen, P.: Ein dialogisches konstruktivitatskriterium. Infinitistic Methods, 193–200 (1959)

Lor68.   Lorenz, K.: Dialogspiele als semantische grundlage von logikkalkiilen. Arch. Math. Logik Grundlag 11, 32–55 (1968)

Min77.   Mints, G.E.: Closed categories and the theory of proofs. Journal of Mathematical Sciences 15, 45–62 (1977)

Mus01.   Muskens, R.: Lambda Grammars and the Syntax-Semantics Interface. In: van Rooy, R., Stokhof, M. (eds.) Proceedings of the Thirteenth Amsterdam Colloquium, Amsterdam, pp. 150–155 (2001)

Nic94.   Nickau, H.: Hereditarily sequential functionals. In: Nerode, A., Matiyasevich, Y. (eds.) LFCS 1994. LNCS, vol. 813, pp. 253–264. Springer, Heidelberg (1994)

Sal10.   Salvati, S.: On the membership problem for non-linear abstract categorial grammars. Journal of Logic, Language and Information 19(2), 163–183 (2010)

Orthogonality and Boolean Algebras for Deduction Modulo

Aloïs Brunel[1], Olivier Hermant[2], and Clément Houtmann[3]

[1] ENS de Lyon
Alois.Brunel@ens-lyon.org
[2] ISEP
Olivier.Hermant@isep.fr
[3] INRIA Saclay
Clement.Houtmann@inria.fr

Abstract. Originating from automated theorem proving, *deduction modulo* removes computational arguments from proofs by interleaving rewriting with the deduction process. From a proof-theoretic point of view, deduction modulo defines a generic notion of cut that applies to any first-order theory presented as a rewrite system. In such a setting, one can prove cut-elimination theorems that apply to many theories, provided they verify some generic criterion. *Pre-Heyting algebras* are a generalization of Heyting algebras which are used by Dowek to provide a semantic intuitionistic criterion called *superconsistency* for generic cut-elimination. This paper uses pre-Boolean algebras (generalizing Boolean algebras) and biorthogonality to prove a generic cut-elimination theorem for the classical sequent calculus modulo. It gives this way a novel application of reducibility candidates techniques, avoiding the use of proof-terms and simplifying the arguments.

1 Introduction

In the usual models of predicate logic (Boolean algebras, Heyting algebras, Kripke models), the interpretations of logically equivalent formulæ are always equal. In particular, valid formulæ are all interpreted by one unique truth value representing truthness. This is adequate for the study of purely logical systems but insufficient for the study of deduction modulo [DHK03]: indeed, in order to remove irrelevant computational arguments from proofs, deduction modulo interleaves rewriting with the deduction process and therefore defines a computational equivalence which is usually strictly weaker than logical equivalence and that appeals to a distinction at the semantic level too. For example, Euclid's algorithm can be specified in deduction modulo: in particular when $a < b$ and $b \bmod a \neq 0$, the gcd of a and b is equal to the gcd of $(b \bmod a)$ and a. Propositions "2 *is the gcd of* 4 *and* 6" and "2 *is the gcd of* 2 *and* 4" are then computationally and logically equivalent (because $2 = 6 \bmod 4$). These two propositions are also logically equivalent to Fermat's Last Theorem (all of them are valid), but they are not computationally equivalent to it. Indeed reducing

L. Ong (Ed.): TLCA 2011, LNCS 6690, pp. 76–90, 2011.

this theorem to a trivial assertion such as "2 *is the gcd of* 4 *and* 6" involves *proving* the theorem. Such a proof hardly qualifies as a *computation*.

Introduced by Dowek [Dow06], *pre-Heyting algebras* are a generalization of Heyting algebras which take into account such a distinction between computational and logical equivalences. Interestingly, they provide a semantic intuitionistic criterion called *superconsistency* for generic cut-elimination in deduction modulo. A theory is superconsistent if it has an interpretation in any pre-Heyting algebra. Since reducibility candidates in deduction modulo [DW03] are a remarkable example of a pre-Heyting algebra, any superconsistent theory can be interpreted in this algebra and consequently verifies the generic notion of cut-elimination provided by deduction modulo. Therefore pre-Heyting algebras are adequate for deduction modulo in intuitionistic logic.

In this paper, we propose a similar notion of model for deduction modulo in *classical logic* that we call *pre-Boolean algebras*. We show that these models lead to a classical version of superconsistency which implies cut-elimination in classical sequent calculus modulo. Our approach significantly differs on two points from the original use of reducibility candidates in deduction modulo [DW03]. First, we do not use original Girard's reducibility candidates [Gir72] or Tait's saturated sets [Tai75], but rather orthogonality which easily adapts to classical sequent calculi: This technique has first been introduced to define reducibility candidates for proofnets and to prove strong normalization of a fragment of linear logic [Gir87] and has since been used many times for various linear logic fragments [Oka99,Gim09] but also for the classical version of system F_ω [LM08] and is the basis of Krivine's classical realizability [Kri09]. Second, we only prove cut-elimination instead of normalization, hence our proof is considerably simplified. Our technique is related to the proofs of cut-elimination for linear logic that use phase semantics [Oka02,Abr91,CT06], but whereas those cut-elimination models can be seen as projections of typed reducibility candidates models [Oka99], ours is crucially designed in a untyped fashion: superconsistency forecloses the degree of freedom to choose the interpretation of atomic formulæ, and the truth values must be forced to contain all the axioms, in order to be able to conclude.

This paper is organized as follows: Deduction modulo, impersonated by a classical sequent calculus, is presented in Section 2. In Section 3, we define pre-Boolean algebras, our generalization of Boolean algebra which acknowledge the distinction between computational and logical equivalences. Section 4 introduces orthogonality for classical deduction modulo using sets of pointed sequents, which allows us to construct a pre-Boolean algebra of sequents and prove adequacy (*i.e.* cut-elimination) in Section 5. Finally in Section 6, we extract a Boolean algebra from the pre-Boolean algebra of sequents presented in Section 5.

2   Classical Sequent Calculus Modulo

We suppose given a signature containing a set of variables $(x, y, z \ldots)$, a set of function symbols and a set of predicate symbols. Each function symbol and each predicate symbol has a fixed arity. Terms $(t, u, v \ldots)$ and atomic formulæ $(a, b, c \ldots)$ are constructed as usual. Formulæ $(A, B, C \ldots)$ are constructed from atomic formulæ, negated atomic formulæ ($\bar{a}, \bar{b}, \bar{c} \ldots$), conjunctions ($\wedge$), disjunctions ($\vee$), universal quantification ($\forall$) and existential quantification ($\exists$).

$$A, B ::= a \mid \bar{a} \mid \top \mid \bot \mid A \wedge B \mid A \vee B \mid \forall x.A \mid \exists x.A$$

Negation is the involutive function $(.)^{\perp}$ recursively defined as

$$
\begin{array}{llll}
a^{\perp} = \bar{a} & \bot^{\perp} = \top & (A \wedge B)^{\perp} = A^{\perp} \vee B^{\perp} & (\forall x.A)^{\perp} = \exists x.A^{\perp} \\
\bar{a}^{\perp} = a & \top^{\perp} = \bot & (A \vee B)^{\perp} = A^{\perp} \wedge B^{\perp} & (\exists x.A)^{\perp} = \forall x.A^{\perp}
\end{array}
$$

Capture avoiding substitutions are denoted $[t/x]$. Sequents are finite multisets of formulæ (denoted $\vdash A_1, A_2 \ldots$). If $\equiv$ is a congruence relation on formulæ, the (one-sided) sequent calculus LK modulo $\equiv$ is described in Figure 1.

$$\frac{}{\vdash A, A^{\perp}} \text{ (Axiom)} \qquad \frac{\vdash A, \Delta_1 \qquad \vdash A^{\perp}, \Delta_2}{\vdash \Delta_1, \Delta_2} \text{ (Cut)} \qquad \frac{\vdash A, \Delta \qquad A \equiv B}{\vdash B, \Delta} \text{ (Conv)}$$

$$\frac{\vdash A, A, \Delta}{\vdash A, \Delta} \text{ (Contr)} \qquad \frac{\vdash \Delta}{\vdash A, \Delta} \text{ (Weak)} \qquad \frac{}{\vdash \top} \text{ ($\top$)} \qquad \text{(no rule for } \bot)$$

$$\frac{\vdash A, \Delta_1 \qquad \vdash B, \Delta_2}{\vdash A \wedge B, \Delta_1, \Delta_2} \text{ ($\wedge$)} \qquad \frac{\vdash A, B, \Delta}{\vdash A \vee B, \Delta} \text{ ($\vee$)}$$

$$\frac{\vdash A[t/x], \Delta}{\vdash \exists x.A, \Delta} \text{ ($\exists$)} \qquad \frac{\vdash A, \Delta \qquad x \text{ fresh in } \Delta}{\vdash \forall x.A, \Delta} \text{ ($\forall$)}$$

Fig. 1. Sequent calculus LK modulo $\equiv$

3   A Generalized Semantics

Definition 1 (pre-Boolean algebra). *A pre-Boolean algebra is a structure* $\langle \mathcal{B}, \top, \bot, \wedge, \vee, (.)^{\perp}, \forall, \exists \rangle$ *where $\mathcal{B}$ is a set, $\top$ and $\bot$ are elements of $\mathcal{B}$, $\wedge$ and $\vee$ are functions from $\mathcal{B} \times \mathcal{B}$ to $\mathcal{B}$, $(.)^{\perp}$ is a function from $\mathcal{B}$ to $\mathcal{B}$, and $\forall$ and $\exists$ are (partial) functions from $\wp(\mathcal{B})$ to $\mathcal{B}$. Finally, we require the function $(.)^{\perp}$ to be involutive.*

A pre-Boolean algebra is said full if $\forall$ and $\exists$ are total functions. It is ordered if it can be equipped with an order relation $\sqsubseteq$ such that $\wedge, \vee, \forall, \exists$ are monotonous, $(.)^{\perp}$ is antimonotonous. If greatest lower bounds for $\sqsubseteq$ are defined for any $X \subseteq \mathcal{B}$, the pre-Boolean algebra is said complete.

The order relation of the above Definition 1 shall not be confused with the order relation of Boolean algebras. The terminology pre-Boolean algebra has been chosen because it is possible to equip our structure with an additional *pre*-order relation $\leq$ together with axioms on $\leq$ that are a weakening to pre-orders of the definition of a distributive complemented lattice: $\wedge$ and $\vee$ build a[1]

---

[1] And no more *the.*

binary greatest lower bound and a least upper bound, respectively, and they obey distributivity laws for the pre-order; $\top$ and $\bot$ are a greatest element and a least element, respectively; $(.)^{\bot}$ is a complement operation; $\forall$ and $\exists$ build arbitrary greatest lower bounds and lowest upper bounds, respectively. Of course, when $\leq$ is an order, the notion of pre-Boolean algebras boils down to Boolean algebras.

The presentation of pre-Boolean algebras with a pre-order has the advantage to make a straightforward analogy with Boolean algebras, as Dowek did when generalizing Heyting algebras to pre-Heyting algebras (also known as Truth Values algebras) [Dow06]. Loosening $\leq$ into a pre-order follows the intention to separate computational and logical equivalences: to the first correspond strict equality and to the second the equivalence relation generated by $\leq \cup \geq$. Notice that the pre-order relation is necessary to show that soundness and completeness still hold for this extended notion of model as it is the case for pre-Heyting algebras and intuitionistic natural deduction modulo [Dow06].

Except for the restriction on the involutivity of $(.)^{\bot}$, this matches exactly the pre-Boolean algebra definition of [Dow10], if we define, as usual, $a \Rightarrow b$ as $a^{\bot} \vee b$ (conversely, define $a^{\bot}$ as $a \Rightarrow \bot$).

Since in this paper, the pre-order $\leq$ does not play any role, we chose to get rid of it, just as Cousineau does [Cou09]. Indeed, the main concern is the computational equivalence, and not the logical one.

In doing this, we do not get more structures: any algebra can always be equipped with the trivial pre-order, for which $a \leq b$ whatever a and b are, and all the axioms for $\leq$ are obviously satisfied. In fact, we even have a strictly smaller class of models, by imposing the involutivity of $(.)^{\bot}$, leading to a potentially larger class of super-consistent theories.

Interpretations in pre-Boolean algebra are partial functions defined as usual. Note that our instances of pre-Boolean algebras ensure the totality of interpretations.

Definition 2 (Interpretation). *Let $\langle \mathcal{B}, \leq, \top, \bot, \wedge, \vee, (.)^{\bot}, \forall, \exists \rangle$ be a pre-Boolean algebra and $(.)^*$ be a function from n-ary atomic predicates P to functions in $M^n \rightarrow \mathcal{B}$ and from n-ary function symbols to functions in $M^n \rightarrow M$, for some chosen domain M. Let ϕ be a valuation assigning to each variable a value in M. If C is a formula and t is a term, then their respective interpretations C^*_ϕ and t^*_ϕ are defined inductively as:*

$$f(t_1, \ldots, t_n)^*_\phi = f^*((t_1)^*_\phi, \ldots, (t_n)^*_\phi) \qquad x^*_\phi = \phi(x)$$
$$P(t_1, \ldots, t_n)^*_\phi = P^*((t_1)^*_\phi, \ldots, (t_n)^*_\phi)$$

$$(\top)^* = \top \quad (A \wedge B)^*_\phi = A^*_\phi \wedge B^*_\phi \quad (\forall x.A)^*_\phi = \forall\{ (A)^*_{\phi+(d/x)} \mid d \in M \}$$
$$(\bot)^* = \bot \quad (A \vee B)^*_\phi = A^*_\phi \vee B^*_\phi \quad (\exists x.A)^*_\phi = ((\forall x.(A^{\bot}))^*_\phi)^{\bot}$$

where $\phi + (d/x)$ is the valuation assigning d to x and $\phi(y)$ to any $y \neq x$.

Lemma 1 (Substitution). *For any formula A, terms t and u, and valuation ϕ, $(u[t/x])^*_\phi = u^*_{\phi+(t^*_\phi/x)}$ and $(A[t/x])^*_\phi = A^*_{\phi+(t^*_\phi/x)}$.*

Proof. By structural induction on u (resp. A).

Definition 3 (Model interpretation). *Let $\equiv$ be a congruence on terms and formulæ. An interpretation $(.)^*$ is said to be a model interpretation for $\equiv$ if and only if for any valuation ϕ, any terms $t \equiv u$ and formulæ $A \equiv B$, $t_\phi^* = u_\phi^*$ and $A_\phi^* = B_\phi^*$.*

The usual definition of consistency states that a theory is consistent if it can be interpreted in a non-trivial model (*i.e.* where $\perp \neq \top$). In particular, a congruence $\equiv$ is consistent if there exists a model interpretation $(.)^*$ for $\equiv$ in *some non-trivial* model, *i.e.* in some non-trivial pre-Boolean algebra. Such a definition is modified to define *superconsistency* [Dow06] as follows.

Definition 4 (Superconsistency). *A congruence $\equiv$ is superconsistent if for all full, ordered and complete pre-Boolean algebra D, an interpretation can be found for $\equiv$ in D.*

4   Behaviours

We dedicate the following sections to a proof that superconsistency is a criterion which entails cut-elimination in our one-sided classical sequent calculus: if $\equiv$ is superconsistent, then cut-elimination holds in LK modulo $\equiv$. To establish such a cut-elimination result, we use orthogonality to design a pre-Boolean algebra of pointed sequents and demonstrate adequacy which in turn implies cut-elimination. The technique used here to prove cut-elimination differs from Dowek and Werner's approach [DW03] because we do not prove strong normalization but cut-elimination (*i.e.* admissibility of the Cut rule) and by the use of orthogonality. However the philosophy remains: in the process of proving cut-elimination, we demonstrate that a pre-Boolean algebra is constructed. Therefore we finally obtain a superconsistency criterion, based on our definition of pre-Boolean algebra, for cut-elimination in our classical sequent calculus modulo.

The notion of orthogonality that we will use in Section 5 relies on sets of *pointed sequents*. These are usual sequents where one formula is distinguished.

Definition 5 (Pointed Sequents). *We define pointed sequents as sequents of the form $\vdash \Delta$ where exactly one formula A of Δ is distinguished[2]: a pointed sequent is a pair consisting of a non-empty sequent and a pointer to some specific formula of the sequent. We denote this formula by A°. The set of pointed sequents is the set of sequents of the form $\vdash A^\circ, \Delta$ and is noted P°. The set of usual sequents is the set of sequents of the form $\vdash \Delta$ with no distinguished formula and is noted P. Pointed sequents are represented by letters $t, u, s \dots$ Moreover, the subset of P° which contains exactly all the sequents whose distinguished formula is A is denoted by $P^\circ(A)$. If $X \subseteq P^\circ$ we pose $X(A) = X \cap P^\circ(A)$.*

Pointed sequents are meant to interact through *cuts*, and therefore define *orthogonality*.

---

[2] In other words, the formula A is in the *stoup*.

Definition 6 (Cut). *If $t = (\vdash A^\circ, \Delta_1)$ and $u = (\vdash B^\circ, \Delta_2)$ are pointed sequents with $B \equiv A^\perp$, then the sequent $t \star u$ is defined by $t \star u = (\vdash \Delta_1, \Delta_2)$. Obviously $t \star u \in P$. Notice that if $B \not\equiv A^\perp$, $t \star u$ is undefined.*

We denote by Ax the set of all axioms, that is the sequents $\vdash A^\perp, A$ for every A. Ax° is the set of pointed axioms.

Definition 7 (Orthogonal). *In what follows, we pose*

$$\perp\!\!\!\perp = \{ \vdash \Delta \mid \vdash \Delta \text{ has a cut-free proof in LK modulo } \equiv \}.$$

We will write $\perp\!\!\!\perp^\circ$ for the set of pointed sequents which have a cut-free proof. If $X \subseteq P^\circ$, then we define the orthogonal of X as

$$X^\perp = \bigcup_B \{ u \in P^\circ(B) \mid \forall t \in X(C) \text{ with } B \equiv C^\perp, t \star u \in \perp\!\!\!\perp \}$$

Lemma 2. *The usual properties on orthogonality hold:*

$$X \subseteq X^{\perp\perp}, \qquad X \subseteq Y \text{ implies } Y^\perp \subseteq X^\perp, \qquad X^{\perp\perp\perp} = X^\perp.$$

Definition 8 (Behaviour). *A set of sequents X is said to be a behaviour when $X^{\perp\perp} = X$.*

Lemma 3. *Behaviours are always stable by conversion of the distinguished formula through $\equiv$. In other words, if X is a behaviour, if $\vdash A^\circ, \Delta$ is in X and if $A \equiv B$, then $\vdash B^\circ, \Delta$ is in X.*

Proof. Let us prove first that any orthogonal $X^\perp$ is stable by conversion: *if* $(\vdash A^\circ, \Delta) \in X^\perp$ and $A \equiv B$, then $(\vdash B^\circ, \Delta) \in X^\perp$. Let us assume that $(\vdash C^\circ, \Delta') \in X$ with $C \equiv B^\perp$. Then $C \equiv A^\perp$ (since $B^\perp \equiv A^\perp$) and since $(\vdash A^\circ, \Delta) \in X^\perp$, there exists a cut-free proof of $\vdash \Delta, \Delta'$. We just proved that $(\vdash B^\circ, \Delta) \in X^\perp$.

Now, any behaviour $X = X^{\perp\perp}$ is the orthogonal of $X^\perp$ and therefore is stable by conversion through $\equiv$. □

Lemma 4. *The set of behaviours is closed under unrestricted intersection.*

Proof. If $\mathcal{S}$ is a set of behaviours, then we show that

$$\left(\bigcap_{X \in \mathcal{S}} X \right)^{\perp\perp} \subseteq \bigcap_{X \in \mathcal{S}} X.$$

Let us take an element $t \in (\bigcap \mathcal{S})^{\perp\perp}$. Let X be an element of $\mathcal{S}$. Let $u \in X^\perp$. Because $\bigcap \mathcal{S} \subseteq X$, we have $X^\perp \subseteq (\bigcap \mathcal{S})^\perp$. Hence $u \in (\bigcap \mathcal{S})^\perp$ and so $t \star u \in \perp\!\!\!\perp$. That means $t \in X^{\perp\perp}$, but X is a behaviour, so $t \in X$. This is true for every $X \in \mathcal{S}$ so finally $t \in \bigcap_{X \in \mathcal{S}} X$. □

Definition 9 (Behaviours Operations). *if X and Y are behaviours and $\mathcal{S}$ is a set of behaviours, then $X \wedge Y$ and $\forall \mathcal{S}$ are respectively defined as $X \wedge Y = ((X.Y) \cup Ax^\circ)^{\perp\perp}$ where $X.Y$ is*

$$\{ \vdash (A \wedge B)^\circ, \Delta_A, \Delta_B \mid (\vdash A^\circ, \Delta_A) \in X \text{ and } (\vdash B^\circ, \Delta_B) \in Y \}$$

and

$$\forall \mathcal{S} = (\{ \vdash (\forall x A)^\circ, \Delta \mid \text{ for any } t \in \mathcal{T}, X \in \mathcal{S}, (\vdash (A[t/x])^\circ, \Delta) \in X \} \cup Ax^\circ)^{\perp\perp}$$

where $\mathcal{T}$ is the set of open terms of the language.

By definition and by Lemma 2, $X^\perp$, $X \wedge Y$ and $\forall \mathcal{S}$ are always behaviours.

5   The Pre-boolean Algebra of Sequents

The next step towards cut-elimination is the construction of a pre-Boolean algebra whose elements are behaviours. The base set of our algebra is

$$D = \{ X \mid Ax^\circ \subseteq X \subseteq \mathbb{\perp}^\circ \text{ and } X = X^{\perp\perp} \}$$

Let us construct a pre-Boolean algebra from D using operators $(.)^\perp$, $\wedge$ and $\forall$.

Lemma 5. *If $S \subseteq D$ then $\bigcap S$ is the greatest lower bound of S in D (for the inclusion order $\subseteq$).*

Proof. Since the base set D is closed under unrestricted intersection (Lemma 4), $\bigcap S \in D$. Now if $C \in D$ is a lower bound of S, then $C \subseteq \bigcap S$. Hence $\bigcap S$ is the greatest lower bound of S in D.   □

Lemma 6. *For all $X \in D$, $X^\perp \in D$.*

Proof. Let us notice that $(Ax^\circ)^\perp = \mathbb{\perp}^\circ$. Then $Ax^\circ \subseteq X \subseteq \mathbb{\perp}^\circ$ (since $X \in D$) and Lemma 2 imply $Ax^\circ \subseteq (Ax^\circ)^{\perp\perp} = (\mathbb{\perp}^\circ)^\perp \subseteq X^\perp \subseteq (Ax^\circ)^\perp = \mathbb{\perp}^\circ$.   □

Lemma 7. *If $X, Y \in D$, then for every C, $(\vdash (C^\perp)^\circ, C) \in (X.Y \cup Ax^\circ)^\perp$.*

Proof. We prove equivalently that for all $(\vdash C^\circ, \Delta) \in (X.Y \cup Ax^\circ)$, the sequent $((\vdash (C^\perp)^\circ, C) \star (\vdash C^\circ, \Delta)) = (\vdash C, \Delta)$ has a cut-free proof.

- If $(\vdash C^\circ, \Delta) \in Ax^\circ$, then $\Delta = C^\perp$. Therefore $(\vdash (C^\perp)^\circ, C) \star (\vdash C^\perp, C^\circ) = (\vdash C^\perp, C)$ has obviously a cut-free proof.
- If $(\vdash C^\circ, \Delta) \in X.Y$, then $C = A \wedge B$, $\Delta = \Delta_1, \Delta_2$ and both $\vdash A, \Delta_1$ and $\vdash B, \Delta_2$ have cut-free proofs. By application of the $(\wedge)$ rule, $\vdash A \wedge B, \Delta_1, \Delta_2$ has a cut-free proof.   □

Theorem 1. *D is stable under $(.)^\perp$, $\wedge$ and $\forall$.*

Proof. First, Lemma 6 implies stability under $(.)^\perp$.
    Let us prove stability under $\wedge$: let us assume $X, Y \in D$ and prove $X \wedge Y \in D$.

- $X \wedge Y$ is a behaviour by definition.
- $Ax^\circ \subseteq X \wedge Y$ since $Ax^\circ \subseteq (X.Y \cup Ax^\circ) \subseteq (X.Y \cup Ax^\circ)^{\perp\perp} = X \wedge Y$.
- Now, let us prove that $X \wedge Y \subseteq \perp\!\!\!\perp^\circ$. We take $(\vdash C^\circ, \Delta) \in X \wedge Y$ and we show that it has a cut-free proof. First, we can notice that $(\vdash (C^\perp)^\circ, C) \in (X.Y \cup Ax^\circ)^\perp$ (Lemma 7). Hence, $(\vdash C^\circ, \Delta) \star (\vdash (C^\perp)^\circ, C) = (\vdash C, \Delta) \in \perp\!\!\!\perp$ and so $(\vdash C, \Delta)$ has a cut-free proof: $(\vdash C^\circ, \Delta) \in \perp\!\!\!\perp^\circ$.

Finally let us prove stability under $\forall$: let us assume that $\mathcal{S}$ is a subset of D and prove that $\forall\mathcal{S} \in D$.

- $\forall\mathcal{S}$ is a behaviour by definition.
- The definition of $\forall\mathcal{S}$ shows that it is the biorthogonal $X^{\perp\perp}$ of a set X containing Ax°. Therefore $Ax^\circ \subseteq X \subseteq X^{\perp\perp} = \forall\mathcal{S}$.
- Finally to prove $\forall\mathcal{S} \subseteq \perp\!\!\!\perp^\circ$, it suffices to show that

$$(\{ \vdash (\forall x A)^\circ, \Delta \mid \text{ for any } t \in \mathcal{T}, X \in \mathcal{S}, (\vdash (A[t/x])^\circ, \Delta) \in X \} \cup Ax^\circ) \subseteq \perp\!\!\!\perp^\circ$$

because $\perp\!\!\!\perp^\circ$ is a behaviour. $Ax^\circ \subseteq \perp\!\!\!\perp^\circ$ obviously. Now, we assume that for any $t \in \mathcal{T}$ and any $X \in \mathcal{S}, (\vdash (D[t/x])^\circ, \Gamma) \in X$. Let us prove that $(\vdash (\forall x.D)^\circ, \Gamma) \in \perp\!\!\!\perp^\circ$. It suffices to take a fresh variable $y \in \mathcal{T}$: then $(\vdash (D[y/x]^\circ, \Gamma))$ is cut-free and by the $\forall$ rule, we obtain that $(\vdash \forall x.D, \Gamma)$ is cut-free too. $\qquad\square$

Remark 1. We need to inject the axioms in the construction of the operation $X \wedge Y$, because otherwise, we could not prove the stability of D with respect to this operation. Indeed, to show that $X \wedge Y \in D$ given $X, Y \in D$ we need to prove equivalently that all the sequents in $(X.Y)^\perp$ are cut-free, which is generally not true because $X.Y$ does not contain all the axioms.

Theorem 2. *The structure $\langle D, \leq, \top, \bot, \wedge, \vee, (.)^\perp, \forall, \exists \rangle$, where*

- $\leq$ *be the trivial pre-order on D,*
- $\top$ *is $\perp\!\!\!\perp^\circ$ and $\bot$ is $\perp\!\!\!\perp^{\circ\perp}$,*
- *the operators $\wedge, (.)^\perp, \forall$ are those defined in Definition 9 and 7*
- *and the operators $\vee$ and $\exists$ are the respective boolean dual of $\wedge$ and $\forall$, i.e.*
  $X \vee Y = (X^\perp \wedge Y^\perp)^\perp$ *and* $\exists S = (\forall S^\perp)^\perp$ *where* $S^\perp = \{ X^\perp \mid X \in S \}$,

is a pre-Boolean algebra.

Proof. Since we chose a trivial pre-order, there is nothing to check but the stability of D under all the operators, that holds by the above lemmata. $\qquad\square$

Remark 2. It should be noted that the pre-Boolean algebra D we have defined is not a monoid when equipped with the operation $\star$ (since when t and u are pointed sequents, $t \star u$ is not a pointed sequent) and so cannot be seen as a classical phase space.

Finally we can state our main result.

Theorem 3 (Adequacy). *Let $\equiv$ be a congruence on terms and formulæ and $(.)^*$ be a model interpretation for $\equiv$ in D. Let $\vdash C_1, ..., C_k$ be a provable sequent in LK modulo $\equiv$, let σ be a substitution whose domain does not contain any bounded variable in $C_1, ..., C_k$, ϕ be a valuation and let $(\vdash (\sigma C_1^\perp)^\circ, \Delta_1) \in ((C_1)_\phi^*)^\perp, ..., (\vdash (\sigma C_k^\perp)^\circ, \Delta_k) \in ((C_k)_\phi^*)^\perp$. Then $\vdash \Delta_1, ..., \Delta_k \in \bot\!\bot$.*

Proof. The proof is done by induction on the last rule of the proof of $\vdash C_1, \ldots, C_k$.

Axiom. For simplicity we suppose that the axiom is performed on C_1 and $C_2 = C_1^\perp$. Therefore $(C_2)_\phi^* = ((C_1)_\phi^*)^\perp$ and since $(\vdash (\sigma C_1^\perp)^\circ, \Delta_1) \in ((C_1)_\phi^*)^\perp = (C_2)_\phi^*$ and $(\vdash (\sigma C_2^\perp)^\circ, \Delta_2) \in ((C_2)_\phi^*)^\perp$, then $\vdash \Delta_1, \Delta_2 \in \bot\!\bot$.

Conjunction. We assume that the derivation is

$$\frac{\vdash A_1, C_2, \ldots, C_k \qquad \vdash A_2, C_{k+1}, \ldots, C_n}{\vdash \underbrace{A_1 \wedge A_2}_{C_1}, C_2, \ldots, C_k, C_{k+1}, \ldots, C_n} (\wedge)$$

Let us assume $(\vdash (\sigma C_i^\perp)^\circ, \Delta_i) \in ((C_i)_\phi^*)^\perp$ for all $i > 1$. By induction hypothesis, $\vdash \sigma A_1^\circ, \Delta_2, \ldots, \Delta_k$ is in $((A_1)_\phi^*)^{\perp\perp} = (A_1)_\phi^*$ and $\vdash \sigma A_2^\circ, \Delta_{k+1}, \ldots, \Delta_n$ is in $((A_2)_\phi^*)^{\perp\perp} = (A_2)_\phi^*$. Therefore $\vdash \sigma C_1^\circ, \Delta_2, \ldots, \Delta_n$ is in

$$(A_1)_\phi^*.(A_2)_\phi^* \subseteq ((A_1)_\phi^*.(A_2)_\phi^*)^{\perp\perp} \subseteq ((A_1)_\phi^*.(A_2)_\phi^* \cup Ax^\circ)^{\perp\perp}$$
$$= (C_1)_\phi^* = ((C_1)_\phi^*)^{\perp\perp}.$$

Then $\vdash \Delta_1, \ldots, \Delta_n \in \bot\!\bot$.

Disjunction. We assume that the derivation is

$$\frac{\vdash A_1, A_2, C_2, \ldots, C_k}{\vdash \underbrace{A_1 \vee A_2}_{C_1}, C_2, \ldots, C_k} (\vee)$$

Let us assume that $(\vdash (\sigma C_i^\perp)^\circ, \Delta_i)$ is in $((C_i)_\phi^*)^\perp$ for all $i > 1$ and let us prove that for all $(\vdash (\sigma C_1^\perp)^\circ, \Delta_1) \in ((C_1)_\phi^*)^\perp$, the sequent $\vdash \Delta_1, \ldots, \Delta_k$ is in $\bot\!\bot$. It is equivalent to prove that $\vdash \sigma C_1^\circ, \Delta_2, \ldots, \Delta_k$ is in

$$((C_1)_\phi^*)^{\perp\perp} = ((A_1)_\phi^* \vee (A_2)_\phi^*)^{\perp\perp} = (((A_1)_\phi^*)^\perp \wedge ((A_2)_\phi^*)^\perp)^{\perp\perp\perp}$$
$$= (((A_1)_\phi^*)^\perp \wedge ((A_2)_\phi^*)^\perp)^\perp = (((A_1)_\phi^*)^\perp.((A_2)_\phi^*)^\perp \cup Ax^\circ)^{\perp\perp\perp}$$
$$= (((A_1)_\phi^*)^\perp.((A_2)_\phi^*)^\perp \cup Ax^\circ)^\perp$$

or equivalently to prove that for all sequent $(\vdash (\sigma C_1^\perp)^\circ, \Delta_1)$ in the set $((A_1)_\phi^*)^\perp.((A_2)_\phi^*)^\perp \cup Ax^\circ$, the sequent $\vdash \Delta_1, \ldots, \Delta_k$ is in $\bot\!\bot$.
- if $(\vdash (\sigma C_1^\perp)^\circ, \Delta_1) \in Ax^\circ$, then $\Delta_1 = \sigma C_1$. Since $\vdash (\sigma A_1^\perp)^\circ, \sigma A_1$ and $\vdash (\sigma A_2^\perp)^\circ, \sigma A_2$ are respectively in $((A_1)_\phi^*)^\perp$ and $((A_2)_\phi^*)^\perp$, then by induction hypothesis, $\vdash \sigma A_1, \sigma A_2, \Delta_2, \ldots, \Delta_k \in \bot\!\bot$. Using $(\vee_R)$,

$$\vdash A_1 \vee A_2, \Delta_2, \ldots, \Delta_k \quad = \quad \vdash \Delta_1, \ldots, \Delta_k \quad \in \bot\!\bot.$$

– if $(\vdash (\sigma C_1^\perp)^\circ, \Delta_1) \in ((A_1)_\phi^*)^\perp.((A_2)_\phi^*)^\perp$, then there exist sequents with $\vdash (\sigma A_1^\perp)^\circ, \Delta_a$ and $\vdash (\sigma A_2^\perp)^\circ, \Delta_b$ respectively in $((A_1)_\phi^*)^\perp$ and $((A_2)_\phi^*)^\perp$ such that Δ_1 is Δ_a, Δ_b. Then by induction hypothesis,

$$\vdash \Delta_a, \Delta_b, \Delta_2, \ldots, \Delta_k \quad = \quad \vdash \Delta_1, \Delta_2, \ldots, \Delta_k \quad \in \perp\!\!\!\perp .$$

Universal quantifier. We assume that the derivation is

$$\dfrac{\vdash A, C_2, \ldots, C_k \qquad x \text{ is fresh in each } C_i}{\vdash \underbrace{\forall x.A}_{C_1}, C_2, \ldots, C_k} \ (\forall)$$

Let us assume that $(\vdash (\sigma C_i^\perp)^\circ, \Delta_i)$ is in $((C_i)_\phi^*)^\perp$ for all $i > 1$ and that the sequent $(\vdash ((\sigma \forall x.A)^\perp)^\circ, \Gamma)$ is in $((\forall x A)_\phi^*)^\perp$. We now want to prove that the sequent $(\vdash \Delta_2, \ldots, \Delta_k, \Gamma)$ is in $\perp\!\!\!\perp$. It is sufficient to prove that the sequent $(\vdash \Delta_2, \ldots, \Delta_k, (\sigma(\forall x.A))^\circ)$ is in $(\forall x.A)_\phi^*$. By noticing that σ only substitutes variables that are free in $\forall x.A$, we get that $\sigma(\forall x.A) = \forall x.(\sigma A)$. It remains to prove that if $t \in \mathcal{T}$ and $d \in M$, then $(\vdash (\sigma A[t/x])^\circ, \Gamma) \in (A_{\phi+[d/x]}^*)$. But, because x is fresh in C_i,

$$(\vdash ((\sigma + [t/x])C_i^\perp)^\circ, \Delta_i) = (\vdash (\sigma C_i^\perp)^\circ, \Delta_i)$$

for each $i > 1$. Again since x is fresh in each C_i, it is easy to see that $((C_i)_\phi^*)^\perp = ((C_i)_{\phi+[d/x]}^*)^\perp$ for each $i > 1$. Hence the induction hypothesis applies to $(\sigma + [t/x])$ and $(\phi + [d/x])$. We then know that

$$\vdash \Delta_1, \ldots, \Delta_k, ((\sigma + [t/x])A)^\circ \in A_{\phi+[d/x]}^*$$

which is what we wanted.

Existential quantifier. We assume that the derivation is

$$\dfrac{\vdash A[t/x], C_2, \ldots, C_k}{\vdash \underbrace{\exists x.A}_{C_1}, C_2, \ldots, C_k} \ (\exists)$$

Let us assume that $(\vdash (\sigma C_i^\perp)^\circ, \Delta_i)$ is in $((C_i)_\phi^*)^\perp$ for all $i > 1$ and that $(\vdash ((\sigma \exists x.A)^\perp)^\circ, \Gamma) \in ((\exists x.A)_\phi^*)^\perp = (\forall x.A^\perp)_\phi^*$. Since x is not in the domain of σ (because x is bounded in $\exists x.A$), and by definition of $(.)^\perp$, we have $(\sigma \exists x.A)^\perp = \forall x.(\sigma A)^\perp$. Hence we know that $(\vdash (\forall x.(\sigma A)^\perp)^\circ, \Gamma) \in \forall \{ (A^\perp)_{\phi+[d/x]}^* \mid \forall d \in M \}$. In particular, we have $(\vdash (\sigma A[t/x])^\perp, \Gamma) \in (A^\perp)_{\phi+[t_\phi^*/x]}^*$. By Lemma 1, $(A^\perp)_{\phi+[t_\phi^*/x]}^* = (A[t/x]^\perp)_\phi^*$, so we can apply the induction hypothesis and finally obtain that $\vdash \Delta_1, \ldots, \Delta_k, \Gamma \in \perp\!\!\!\perp$.

Cut. The derivation is

$$\dfrac{\vdash A, C_1, \ldots, C_p \qquad \vdash A^\perp, C_{p+1}, \ldots, C_k}{\vdash C_1, \ldots, C_k} \ (\text{Cut})$$

for some $1 \leqslant p \leqslant k$. Let us suppose that $(\vdash (\sigma C_i^\perp)^\circ, \Delta_i)$ is in $((C_i)_\phi^*)^\perp$ for all i. Then by induction hypothesis,

– if $\vdash (\sigma A^\perp)^\circ, \Delta$ is in $(A_\phi^*)^\perp$, then $\vdash \Delta, \Delta_1, \ldots, \Delta_p \in \bot\!\!\!\bot$
– and if $\vdash \sigma A^\circ, \Delta$ is in A_ϕ^*, then $\vdash \Delta, \Delta_{p+1}, \ldots, \Delta_k \in \bot\!\!\!\bot$.

Therefore $\vdash \sigma A^\circ, \Delta_1, \ldots, \Delta_p$ is in A_ϕ^* and $\vdash (\sigma A^\perp)^\circ, \Delta_{p+1}, \ldots, \Delta_k$ is in $(A^\perp)_\phi^* = (A_\phi^*)^\perp$. Then $\vdash \Delta_1, \ldots, \Delta_k, \Gamma_1, \ldots, \Gamma_n \in \bot\!\!\!\bot$.

Weakening. We assume that the derivation is

$$\frac{\vdash\quad C_2, \ldots, C_k}{\vdash C_1, C_2 \ldots, C_k} \text{(Weak)}$$

Let us suppose that $(\vdash (\sigma C_i^\perp)^\circ, \Delta_i)$ is in $((C_i)_\phi^*)^\perp$ for all $i > 1$. Then by induction hypothesis, $\vdash \Delta_2, \ldots, \Delta_k \in \bot\!\!\!\bot$. Therefore by weakening in cut-free proofs, $\vdash \Delta_1, \ldots, \Delta_k \in \bot\!\!\!\bot$.

Contraction. We assume that the derivation is

$$\frac{\vdash C_1, C_1, \ldots, C_k}{\vdash\quad C_1, \ldots, C_k} \text{(Contr)}$$

Let us suppose that $(\vdash (\sigma C_i^\perp)^\circ, \Delta_i)$ is in $((C_i)_\phi^*)^\perp$ for all $i > 0$. Then by induction hypothesis, $\vdash \Delta_1, \Delta_1, \ldots, \Delta_k \in \bot\!\!\!\bot$. Therefore by contraction in cut-free proofs, $\vdash \Delta_1, \ldots, \Delta_k \in \bot\!\!\!\bot$.

Conversion. We assume that the derivation is

$$\frac{\vdash A, C_2, \ldots, C_k \qquad A \equiv C_1}{\vdash C_1, C_2, \ldots, C_k} (\equiv)$$

and since $A \equiv C_1$, we know that $A_\phi^* = (C_1)_\phi^*$ and $(A_\phi^*)^\perp = ((C_1)_\phi^*)^\perp$. Let us suppose that $\vdash (\sigma C_i^\perp)^\circ, \Delta_i$ is in $((C_i)_\phi^*)^\perp$ for all $i > 1$. Then since $\sigma C_1^\perp \equiv \sigma A^\perp$, the sequent $\vdash \sigma A^\perp, \Delta$ is also in $((C_i)_\phi^*)^\perp = (A_\phi^*)^\perp$. Finally by induction hypothesis, $\vdash \Delta_1, \ldots, \Delta_k \in \bot\!\!\!\bot$. $\square$

Cut-elimination is a corollary of our adequacy result.

Corollary 1 (Superconsistency implies cut-elimination). *If $\equiv$ is a superconsistent theory, then cut-elimination holds for LK modulo $\equiv$, i.e. any sequent $\vdash \Delta$ derivable in LK modulo $\equiv$ has a cut-free proof in LK modulo $\equiv$.*

Proof. Superconsistency of $\equiv$ implies that there exists a model interpretation $(.)^*$ for $\equiv$ in the pre-Boolean algebra of sequents D (corresponding to $\equiv$). Let $\vdash C_1, \ldots, C_k$ be some provable sequent in LK modulo $\equiv$. Let us remark that $(\vdash (C_i^\perp)^\circ, C_i) \in ((C_i)_\phi^*)^\perp$ for each C_i (where ϕ is the empty valuation). Then by our adequacy result (Theorem 3), $(\vdash C_1, \ldots, C_k) \in \bot\!\!\!\bot$. In other words, this sequent has a cut-free proof in LK modulo $\equiv$. $\square$

Remark 3. To prove cut-elimination, we crucially rely on the fact that for each formula A and whatever the model interpretation $(.)^*$ given by the superconsistency is, A_ϕ^* contains all the axioms of the form $\vdash B^\perp, B^\circ$, including $\vdash A^\perp, A^\circ$. Otherwise, the interpretation of an atom A which is given by the superconsistency criterion would not necessarily contain all the provable sequents of the form $\vdash \Gamma, A^\circ$.

This property does not hold in usual phase semantics based cut-elimination models: in these settings, the elements of A_ϕ^* represent sequents of the form $\vdash \Gamma, A$. Whereas phase semantics cut-elimination models can be seen as projections of typed reducibility candidates models [Oka99], our model can be seen as the projection of a untyped reducibility candidates model.

6   An Underlying Boolean Algebra

We exhibit a (non-trivial) Boolean algebra, similar but simpler than the one of [DH07], extracted from the pre-Boolean algebra of sequents of section 5.

Definition 10 (Context Extraction). *Let A be a formula, we define $\lfloor A \rfloor$ to be the set of contexts $\Gamma = A_1, \cdots, A_n$ such that for any valuation ϕ, substitution σ, and any sequence of contexts Δ_i such that $\vdash \Delta_i, ((\sigma A_i)^\perp)^\circ \in (A_i)_\phi^{*\perp}$, $\vdash \Delta_1, \cdots, \Delta_n, (\sigma A)^\circ \in A_\phi^*$.*

Equivalently, one may impose that for any context Δ such that $\vdash \Delta, ((\sigma A)^\perp)^\circ \in A_\phi^{*\perp}$, we have $\vdash \Delta_1, \cdots, \Delta_n, \Delta \in \perp\!\!\!\perp$.

Definition 11 (Boolean algebra). *We define $\langle \mathcal{B}, \leq, \top, \perp, \wedge, \vee, .^\perp, \forall, \exists \rangle$ as follows. $\mathcal{B}$ is the set containing $\lfloor A \rfloor$ for any A. The order is inclusion, and the operations are*

$$\top = \lfloor \top \rfloor \qquad \lfloor A \rfloor \wedge \lfloor B \rfloor = \lfloor A \wedge B \rfloor \qquad \lfloor A \rfloor^\perp = \lfloor A^\perp \rfloor$$
$$\perp = \lfloor \perp \rfloor \qquad \lfloor A \rfloor \vee \lfloor B \rfloor = \lfloor A \vee B \rfloor$$

$\forall$ *and* $\exists$ *are defined only on sets of the form* $\{ \lfloor A[t/x] \rfloor \mid t \in \mathcal{T} \}$, *where* $\mathcal{T}$ *is the set of equivalence classes modulo $\equiv$ of open terms:*

$$\forall \{ \lfloor A[t/x] \rfloor \mid t \in \mathcal{T} \} = \lfloor \forall x A \rfloor \qquad \exists \{ \lfloor A[t/x] \rfloor \mid t \in \mathcal{T} \} = \lfloor \exists x A \rfloor$$

Notice that $\lfloor A[t/x] \rfloor$ for $t \in \mathcal{T}$ does not depend of the chosen representative of the equivalence class t since as soon as $t_1 \equiv t_2$, $\lfloor A[t_1/x] \rfloor = \lfloor A[t_2/x] \rfloor$.

Lemma 8. *Let A and B be two formulæ. Then:*

- $A^\perp \in \lfloor A \rfloor$
- *if $A_1, ..., A_n \in \lfloor A \rfloor$ then $\vdash A_1, ..., A_n, A$ has a cut-free proof.*
- $\vdash A^\perp, B$ *has a cut-free proof if, and only if, $\lfloor A \rfloor \subseteq \lfloor B \rfloor$*

Proof.

- $\vdash \Delta, (\sigma A)^{\perp\perp\circ} \in (A^\perp)_\phi^{*\perp}$ and $\vdash \Delta, \sigma A^\circ \in A_\phi^*$ are the same statement.
- $\vdash (A_i^\perp)^\circ, A_i \in A_{i\,\phi}^{*\perp}$ for each A_i and ϕ. By Definition 10, $\vdash A_1, ..., A_n, A \in \perp\!\!\!\perp$.
- the if part follows from the two previous points: $A^\perp \in \lfloor A \rfloor \subseteq \lfloor B \rfloor$ and therefore $\vdash A^\perp, B$ has a proof. For the only if part, let $A_1, ..., A_n \in \lfloor A \rfloor$, σ be a substitution and ϕ be a valuation. Let Δ_i such that $\vdash \Delta_i, ((\sigma A_i)^\perp)^\circ \in (A_i)_\phi^{*\perp}$. By hypothesis $\vdash \Delta_1, ..., \Delta_n, (\sigma A)^\circ \in A_\phi^*$, so Theorem 3 applied to the proof of $\vdash A^\perp, B$ implies that $\vdash \Delta_1, ..., \Delta_n, (\sigma B)^\circ \in B_\phi^*$. Therefore $A_1, ..., A_n \in \lfloor B \rfloor$.

Proposition 1. $\langle \mathcal{B}, \leq, \top, \bot, \wedge, \vee, .^{\bot}, \forall, \exists \rangle$ *is a boolean algebra, and* $\lfloor . \rfloor$ *is a model interpretation in this algebra, where the domain for terms is* $\mathcal{T}$.

Proof. This proposition is a consequence of the adequacy Theorem 3. Let us check the points of Definition 11:

1. $\lfloor A \rfloor \wedge \lfloor B \rfloor$ is the greatest lower bound of $\lfloor A \rfloor$ and $\lfloor B \rfloor$.
   - $\lfloor A \wedge B \rfloor \subseteq \lfloor A \rfloor$: by Lemma 8 since $\vdash (A \wedge B)^{\bot}, A$ has a two-step proof.
   - $\lfloor A \wedge B \rfloor \subseteq \lfloor B \rfloor$: by Lemma 8 since $\vdash (A \wedge B)^{\bot}, B$ has a two-step proof.
   - $\lfloor C \rfloor \subseteq \lfloor A \rfloor$ and $\lfloor C \rfloor \subseteq \lfloor B \rfloor$ implies $\lfloor C \rfloor \subseteq \lfloor A \wedge B \rfloor$: by hypothesis and Lemma 8, $C^{\bot} \in \lfloor C \rfloor \subseteq \lfloor A \rfloor \cap \lfloor B \rfloor$, and we have two proofs of $\vdash C^{\bot}, A$ and $\vdash C^{\bot}, B$. We combine them to form a proof of $\vdash C^{\bot}, A \wedge B$ and conclude by Lemma 8.
2. $\lfloor A \rfloor \vee \lfloor B \rfloor$ is the least upper bound of $\lfloor A \rfloor$ and $\lfloor B \rfloor$.
   - $\lfloor A \rfloor \subseteq \lfloor A \vee B \rfloor$: by Lemma 8 since $\vdash A^{\bot}, A \vee B$ has a two-step proof.
   - $\lfloor B \rfloor \subseteq \lfloor A \vee B \rfloor$: by Lemma 8 since $\vdash B^{\bot}, A \vee B$ has a two-step proof.
   - $\lfloor A \rfloor \subseteq \lfloor C \rfloor$ and $\lfloor B \rfloor \subseteq \lfloor C \rfloor$ implies $\lfloor A \vee B \rfloor \subseteq \lfloor C \rfloor$: by hypothesis and Lemma 8, $A^{\bot} \in \lfloor A \rfloor \subseteq \lfloor C \rfloor$ and $\vdash A^{\bot}, C$ has a proof. By a similar argument $\vdash B^{\bot}, C$ has also a proof. We combine them to form a proof of $\vdash (A \vee B)^{\bot}, C$ and conclude by Lemma 8.
3. properties of greatest and least elements.
   - $\lfloor C \rfloor \subseteq \lfloor \top \rfloor$: by Lemma 8 since $\vdash C^{\bot}, \top$ has a two-step proof.
   - $\lfloor \bot \rfloor \subseteq \lfloor C \rfloor$: by Lemma 8 since $\vdash \bot^{\bot}, C$ has a two-step proof.
4. distributivity of $\wedge$ and $\vee$ follow from the same laws in the logic, through Lemma 8: if two formulæ A and B are equivalent then $\lfloor A \rfloor = \lfloor B \rfloor$.
5. $\lfloor A^{\bot} \rfloor$ is a complement of $\lfloor A \rfloor$.
   - $\lfloor \top \rfloor \subseteq \lfloor A^{\bot} \vee A \rfloor$: by Lemma 8 since $\vdash \top^{\bot}, A^{\bot} \vee A$ has a two-step proof.
   - $\lfloor A^{\bot} \wedge A \rfloor \subseteq \lfloor \bot \rfloor$: by Lemma 8 since $\vdash (A^{\bot} \wedge A)^{\bot}, \bot$ has a two-step proof.
6. idempotency of $(.)^{\bot}$: $\lfloor A \rfloor^{\bot\bot} = \lfloor A^{\bot\bot} \rfloor = \lfloor A \rfloor$.

Lastly, we check that the operators $\forall$ and $\exists$ define a greatest lower bound and a least upper bound, respectively. Notice that although the order is set inclusion, those operators are *not* set intersection and union[3]: $\mathcal{B}$ is neither complete, nor closed under arbitrary union and intersection and it misses many sets.

- $\lfloor \forall x A \rfloor \subseteq \bigcap \{ \lfloor A[t/x] \rfloor \mid t \in \mathcal{T} \}$: by Lemma 8 since for any $t, \vdash (\forall x A)^{\bot}, A[t/x]$ has a one-step proof.
- $\lfloor C \rfloor \subseteq \bigcap \{ \lfloor A[t/x] \rfloor \mid t \in \mathcal{T} \}$ implies $C \subseteq \lfloor \forall x A \rfloor$: assume without loss of generality that x does not appear freely in C. By hypothesis and Lemma 8, $C^{\bot} \in \lfloor C \rfloor \subseteq \bigcap \{ \lfloor A[t/x] \rfloor \mid t \in \mathcal{T} \} \subseteq \lfloor A[x/x] \rfloor$ and $\vdash C^{\bot}, A$ has a proof. Adding a $(\forall)$ rule yields a proof of $\vdash C^{\bot}, \forall x A$. We conclude by Lemma 8.
- $\bigcup \{ \lfloor A[t/x] \rfloor \mid t \in \mathcal{T} \} \subseteq \lfloor \exists x A \rfloor$: by Lemma 8 since for any $t, \vdash A[t/x]^{\bot}, \exists x A$ has a one-step proof.

---

[3] So, for instance, the greatest lower bound is allowed to be smaller than set intersection.

- $\bigcup\{ \lfloor A[t/x] \rfloor \mid t \in \mathcal{T} \} \subseteq \lfloor C \rfloor$ implies $\lfloor \exists x A \rfloor \subseteq \lfloor C \rfloor$: assume without loss of generality that x does not appear freely in C. By hypothesis and Lemma 8, $A^\perp \in \lfloor A[x/x] \rfloor \subseteq C$ and $\vdash A^\perp, C$ has a proof. Adding a $(\forall)$ rule yields a proof of $\vdash (\exists x A)^\perp, C$. We conclude by Lemma 8.

Definition 10 ensures that $\lfloor . \rfloor$ is an interpretation (Definition 2), provided that terms are interpreted by their equivalence class modulo $\equiv$. Lastly, if $A \equiv B$ then they are logically equivalent and by Lemma 8 $\lfloor A \rfloor = \lfloor B \rfloor$.

A direct proof of Proposition 1, bypassing Theorem 3, is possible. In this option, each of its case uses the same arguments than the corresponding case of Theorem 3. Such a proof would be made easier by considering the definition of [Dow10] for pre-Boolean algebra where one has conditions on $\Rightarrow$ rather than distributivity laws.

The benefits of a direct proof would be an alternative proof of the cut-elimination theorem, as it is done in [DH07], through the usual soundness theorem with respect to Boolean Algebras and strong completeness with respect to the particular Boolean Algebra we presented here.

7 Conclusion

We have generalized Boolean algebras into pre-Boolean algebras, a notion of model for classical logic which acknowledges the distinction between computational and logical equivalences. We also have demonstrated how super-consistency —a semantic criterion for generic cut-elimination in (intuitionistic) deduction modulo— adapts to classical logic: We have proposed a classical version of superconsistency based on pre-Boolean algebras. Using orthogonality, we have constructed a pre-Boolean algebra of sequents which allows to prove that our classical superconsistency criterion implies cut-elimination in classical sequent calculus modulo. In the last section, we have explained how a non-trivial Boolean algebra of contexts can be extracted from the pre-Boolean algebra of sequents, therefore relating our orthogonality cut-elimination proof with the usual semantics of classical logic (*i.e.* Boolean algebras). Finally we have proved that the same cut-elimination result can be obtained from this particular Boolean algebra, without going through the proof of adequacy for the pre-Boolean algebra.

Let us notice that any pre-Boolean algebra is also a pre-Heyting algebra. Therefore a theory which is superconsistent on pre-Heyting algebras is automatically superconsistent on pre-Boolean algebras. (The converse does not hold in general and pre-Heyting algebras are not always pre-Boolean algebras.) Dowek has proved [Dow06] that several theories of interest are superconsistent on pre-Heyting algebras: arithmetic, simple type theory, the theories defined by a confluent, terminating and quantifier free rewrite system, the theories defined by a confluent, terminating and positive rewrite system and the theories defined by a positive rewrite system such that each atomic formula has at most one one-step reduct. We automatically obtain that these theories are also superconsistent on pre-Boolean algebras, and therefore that cut-elimination holds in classical sequent calculus modulo these theories.

Using pre-Boolean algebras is not the unique way of connecting the superconsistency criterion with classical logic. For instance, one can use double-negation translations and prove that superconsistency (on pre-Heyting algebras) of a theory implies superconsistency (still on pre-Heyting algebras) of its double-negation translation which in turn implies cut-elimination in classical logic, using [DW03]. Superconsistency of double-negated theories on pre-Heyting algebras and superconsistency on pre-Boolean algebras remain to be compared. Both are implied by superconsistency on pre-Heyting algebras, and in both cases, no counterexample of the inverse has been found yet.

References

Abr91. Abrusci, V.M.: Phase semantics and sequent calculus for pure noncommutative classical linear propositional logic. Journal of Symbolic Logic 56(4), 1403–1451 (1991)

Cou09. Cousineau, D.: Complete reducibility candidates. In: PSTT 2009 (2009)

CT06.   Ciabattoni, A., Terui, K.: Towards a semantic characterization of cut-elimination. Studia Logica 82(1), 95–119 (2006)

DH07.   Dowek, G., Hermant, O.: A simple proof that super-consistency implies cut elimination. In: Baader, F. (ed.) RTA 2007. LNCS, vol. 4533, pp. 93–106. Springer, Heidelberg (2007)

DHK03. Dowek, G., Hardin, T., Kirchner, C.: Theorem proving modulo. Journal of Automated Reasoning 31(1), 33–72 (2003)

Dow06. Dowek, G.: Truth values algebras and proof normalization. In: Altenkirch, T., McBride, C. (eds.) TYPES 2006. LNCS, vol. 4502, pp. 110–124. Springer, Heidelberg (2007)

Dow10. Dowek, G.: Fondements des systèmes de preuve. Course notes (2010)

DW03.   Dowek, G., Werner, B.: Proof normalization modulo. Journal of Symbolic Logic 68(4), 1289–1316 (2003)

Gim09. Gimenez, S.: Programmer, Calculer et Raisonner avec les Réseaux de la Logique Linéaire. PhD thesis, Université Paris 7 (2009)

Gir72. Girard, J.-Y.: Interprétation Fonctionnelle et Élimination des Coupures de l'Arithmétique dOrdre Supérieur. PhD thesis, Université Paris 7 (1972)

Gir87. Girard, J.-Y.: Linear logic. Theor. Comput. Sci. 50, 1–102 (1987)

Kri09. Krivine, J.-L.: Realizability in classical logic. Panoramas et synthèses 27, 197–229 (2009)

LM08.   Lengrand, S., Miquel, A.: Classical F [omega], orthogonality and symmetric candidates. Annals of Pure and Applied Logic 153(1-3), 3–20 (2008)

Oka99. Okada, M.: Phase semantic cut-elimination and normalization proofs of first- and higher-order linear logic. Theoretical Computer Science 227(1-2), 333–396 (1999)

Oka02. Okada, M.: A uniform semantic proof for cut-elimination and completeness of various first and higher order logics. Theoretical Computer Science 281(1-2), 471–498 (2002)

Tai75. Tait, W.W.: A realizability interpretation of the theory of species. In: Parikh, R.J. (ed.) Logic Colloquium, pp. 240–251. Springer, Heidelberg (1975)

The Biequivalence
of Locally Cartesian Closed Categories
and Martin-Löf Type Theories

Pierre Clairambault and Peter Dybjer

University of Bath and Chalmers University of Technology

Abstract. Seely's paper *Locally cartesian closed categories and type theory* contains a well-known result in categorical type theory: that the category of locally cartesian closed categories is equivalent to the category of Martin-Löf type theories with Π, Σ, and extensional identity types. However, Seely's proof relies on the problematic assumption that substitution in types can be interpreted by pullbacks. Here we prove a corrected version of Seely's theorem: that the Bénabou-Hofmann interpretation of Martin-Löf type theory in locally cartesian closed categories yields a biequivalence of 2-categories. To facilitate the technical development we employ categories with families as a substitute for syntactic Martin-Löf type theories. As a second result we prove that if we remove Π-types the resulting categories with families are biequivalent to left exact categories.

1 Introduction

It is "well-known" that locally cartesian closed categories (lcccs) are equivalent to Martin-Löf's intuitionistic type theory [9,10]. But how *known* is it really? Seely's original proof [13] contains a flaw, and the papers by Curien [3] and Hofmann [5] who address this flaw only show that Martin-Löf type theory can be interpreted in locally cartesian closed categories, but not that this interpretation is an equivalence of categories provided the type theory has Π, Σ, and extensional identity types. Here we complete the work and fully rectify Seely's result except that we do not prove an equivalence of categories but a *biequivalence* of 2-categories. In fact, a significant part of the endeavour has been to find an appropriate formulation of the result, and in particular to find a suitable notion analogous to Seely's "interpretation of Martin-Löf theories".

Categories with families and democracy. Seely turns a given Martin-Löf theory into a category where the objects are *closed* types and the morphisms from type A to type B are functions of type $A \to B$. Such categories are the objects of Seely's "category of Martin-Löf theories".

Instead of syntactic Martin-Löf theories we shall employ *categories with families (cwfs)* [4]. A cwf is a pair $(\mathbb{C}, T)$ where $\mathbb{C}$ is the category of contexts and explicit substitutions, and $T : \mathbb{C}^{op} \to \mathbf{Fam}$ is a functor, where $T(\Gamma)$ represents

L. Ong (Ed.): TLCA 2011, LNCS 6690, pp. 91–106, 2011.

the family of sets of terms indexed by types in context Γ and $T(\gamma)$ performs the substitution of γ in types and terms. Cwf is an appropriate substitute for syntax for dependent types: its definition unfolds to a variable-free calculus of explicit substitutions [4], which is like Martin-Löf's [11,14] except that variables are encoded by projections. One advantage of this approach compared to Seely's is that we get a natural definition of morphism of cwfs, which preserves the structure of cwfs up to isomorphism. In contrast Seely's notion of "interpretation of Martin-Löf theories" is defined indirectly via the construction of an lccc associated with a Martin-Löf theory, and basically amounts to a functor preserving structure between the corresponding lcccs, rather than directly as something which preserves all the "structure" of Martin-Löf theories.

To prove our biequivalences we require that our cwfs are *democratic*. This means that each context is *represented* by a type. Our results require us to build local cartesian closed structure in the category of contexts. To this end we use available constructions on types and terms, and by democracy such constructions can be moved back and forth between types and contexts. Since Seely works with closed types only he has no need for democracy.

The coherence problem. Seely interprets type substitution in Martin-Löf theories as pullbacks in lcccs. However, this is problematic, since type substitution is already defined by induction on the structure of types, and thus fixed by the interpretation of the other constructs of type theory. It is not clear that the pullbacks can be chosen to coincide with this interpretation.

In the paper *Substitution up to isomorphism* [3] Curien describes the fundamental nature of this problem. He sets out

> ... to solve a difficulty arising from a mismatch between syntax and semantics: in locally cartesian closed categories, substitution is modelled by pullbacks (more generally pseudo-functors), that is, only up to isomorphism, unless split fibrational hypotheses are imposed. ... but not all semantics do satisfy them, and in particular not the general description of the interpretation in an arbitrary locally cartesian closed category. In the general case, we have to show that the isomorphisms between types arising from substitution are *coherent* in a sense familiar to category theorists.

To solve the problem Curien introduces a calculus with explicit substitutions for Martin-Löf type theory, with special terms witnessing applications of the type equality rule. In this calculus type equality can be interpreted as isomorphism in lcccs. The remaining coherence problem is to show that Curien's calculus is equivalent to the usual formulation of Martin-Löf type theory, and Curien proves this result by cut-elimination.

Somewhat later, Hofmann [5] gave an alternative solution based on a technique which had been used by Bénabou [1] for constructing a *split* fibration from an arbitrary fibration. In this way Hofmann constructed a model of Martin-Löf type theory with Π-types, Σ-types, and (extensional) identity types from a locally cartesian closed category. Hofmann used categories with attributes (cwa) in the sense of Cartmell [2] as his notion of model. In fact, cwas and cwfs are closely related: the notion of cwf arises by reformulating the axioms of cwas to make the connection with the usual syntax of dependent type theory more transparent. Both cwas and cwfs are split notions of model of Martin-Löf type theory, hence the relevance of Bénabou's construction.

However, Seely wanted to prove an equivalence of categories. Hofmann conjectured [5]:

> We have now constructed a cwa over $\mathcal{C}$ which can be shown to be equivalent to $\mathcal{C}$ in some suitable 2-categorical sense.

Here we spell out and prove this result, and thus fully rectify Seely's theorem. It should be apparent from what follows that this is not a trivial exercise. In our setting the result is a biequivalence analogous to Bénabou's (much simpler) result: that the 2-category of fibrations (with non-strict morphisms) is biequivalent to the 2-category of split fibrations (with non-strict morphisms).

While carrying out the proof we noticed that if we remove Π-types the resulting 2-category of cwfs is biequivalent to the 2-category of left exact (or finitely complete) categories. We present this result in parallel with the main result.

Plan of the paper. An equivalence of categories consists of a pair of functors which are inverses up to natural isomorphism. Biequivalence is the appropriate notion of equivalence for bicategories [8]. Instead of functors we have *pseudofunctors* which only preserve identity and composition up to isomorphism. Instead of natural isomorphisms we have *pseudonatural transformations* which are inverses up to *invertible modification*.

A 2-category is a strict bicategory, and the remainder of the paper consists of constructing two biequivalences of 2-categories. In Section 2 we introduce cwfs and show how to turn a cwf into an indexed category. In Section 3 we define the 2-categories $\mathbf{CwF}_{\mathrm{dem}}^{\mathrm{Iext}\,\Sigma}$ of democratic cwfs which support extensional identity types and Σ-types and $\mathbf{CwF}_{\mathrm{dem}}^{\mathrm{Iext}\,\Sigma\Pi}$ which also support Π-types. We also define the notions of pseudo cwf-morphism and pseudo cwf-transformation. In Section 4 we define the 2-categories $\mathbf{FL}$ of left exact categories and $\mathbf{LCC}$ of locally cartesian closed categories. We show that there are forgetful 2-functors $U : \mathbf{CwF}_{\mathrm{dem}}^{\mathrm{Iext}\,\Sigma} \to \mathbf{FL}$ and $U : \mathbf{CwF}_{\mathrm{dem}}^{\mathrm{Iext}\,\Sigma\Pi} \to \mathbf{LCC}$. In section 5 we construct the pseudofunctors $H : \mathbf{FL} \to \mathbf{CwF}_{\mathrm{dem}}^{\mathrm{Iext}\,\Sigma}$ and $H : \mathbf{LCC} \to \mathbf{CwF}_{\mathrm{dem}}^{\mathrm{Iext}\,\Sigma\Pi}$ based on the Bénabou-Hofmann construction. In section 6 we prove that H and U give rise to the biequivalences of $\mathbf{FL}$ and $\mathbf{CwF}_{\mathrm{dem}}^{\mathrm{Iext}\,\Sigma}$ and of $\mathbf{LCC}$ and $\mathbf{CwF}_{\mathrm{dem}}^{\mathrm{Iext}\,\Sigma\Pi}$.

An appendix containing the full proof of the biequivalences can be found at http://www.cse.chalmers.se/~peterd/papers/categorytypetheory.html/.

2   Categories with Families

2.1   Definition

Definition 1. *Let* **Fam** *be the category of families of sets defined as follows. An object is a pair (A, B) where A is a set and $B(x)$ is a family of sets indexed by*

$x \in A$. A morphism with source (A, B) and target (A', B') is a pair consisting of a function $f : A \rightarrow A'$ and a family of functions $g(x) : B(x) \rightarrow B'(f(x))$ indexed by $x \in A$.

Note that **Fam** is equivalent to the arrow category $\mathbf{Set}^{\rightarrow}$.

Definition 2. *A **category with families (cwf)** consists of the following data:*

- *A base category $\mathbb{C}$. Its objects represent* contexts *and its morphisms represent* substitutions. *The identity map is denoted by* id $: \Gamma \rightarrow \Gamma$ *and the composition of maps $\gamma : \Delta \rightarrow \Gamma$ and $\delta : \Xi \rightarrow \Delta : \Xi \rightarrow \Gamma$ is denoted by $\gamma \circ \delta$ or more briefly by $\gamma\delta : \Xi \rightarrow \Gamma$.*
- *A functor $T : \mathbb{C}^{op} \rightarrow \mathbf{Fam}$. $T(\Gamma)$ is a pair, where the first component represents the set $\mathrm{Type}(\Gamma)$ of types in context Γ, and the second component represents the type-indexed family $(\Gamma \vdash A)_{A \in \mathrm{Type}(\Gamma)}$ of sets of terms in context Γ. We write $a : \Gamma \vdash A$ for a term $a \in \Gamma \vdash A$. Moreover, if γ is a morphism in $\mathbb{C}$, then $T(\gamma)$ is a pair consisting of the type substitution function $A \mapsto A[\gamma]$ and the type-indexed family of term substitution functions $a \mapsto a[\gamma]$.*
- *A terminal object $[]$ of $\mathbb{C}$ which represents the* empty context *and a terminal map $\langle\rangle : \Delta \rightarrow []$ which represents the* empty substitution.
- *A context comprehension which to an object Γ in $\mathbb{C}$ and a type $A \in \mathrm{Type}(\Gamma)$ associates an object $\Gamma \cdot A$ of $\mathbb{C}$, a morphism $\mathrm{p}_A : \Gamma \cdot A \rightarrow \Gamma$ of $\mathbb{C}$ and a term $\mathrm{q} \in \Gamma \cdot A \vdash A[\mathrm{p}]$ such the following universal property holds: for each object Δ in $\mathbb{C}$, morphism $\gamma : \Delta \rightarrow \Gamma$, and term $a \in \Delta \vdash A[\gamma]$, there is a unique morphism $\theta = \langle\gamma, a\rangle : \Delta \rightarrow \Gamma \cdot A$, such that $\mathrm{p}_A \circ \theta = \gamma$ and $\mathrm{q}[\theta] = a$. (We remark that a related notion of comprehension for hyperdoctrines was introduced by Lawvere [7].)*

The definition of cwf can be presented as a system of axioms and inference rules for a variable-free generalized algebraic formulation of the most basic rules of dependent type theory [4]. The correspondence with standard syntax is explained by Hofmann [6] and the equivalence is proved in detail by Mimram [12]. The easiest way to understand this correspondence might be as a translation between the standard lambda calculus based syntax of dependent type theory and the language of cwf-combinators. In one direction the key idea is to translate a variable (de Bruijn number) to a projection of the form $\mathrm{q}[\mathrm{p}^n]$. In the converse direction, recall that the cwf-combinators yield a calculus of explicit substitutions whereas substitution is a meta-operation in usual lambda calculus. When we translate cwf-combinators to lambda terms, we execute the explicit substitutions, using the equations for substitution in types and terms as rewrite rules. The equivalence proof is similar to the proof of the equivalence of cartesian closed categories and the simply typed lambda calculus.

We shall now define what it means that a cwf supports extra structure corresponding to the rules for the various type formers of Martin-Löf type theory.

Definition 3. *A cwf supports (extensional) identity types provided the following conditions hold:*

Form. If $A \in \mathrm{Type}(\Gamma)$ and $a, a' : \Gamma \vdash A$, there is $\mathrm{I}_A(a, a') \in \mathrm{Type}(\Gamma)$;

Intro. If $a : \Gamma \vdash A$, there is $\mathrm{r}_{A,a} : \Gamma \vdash \mathrm{I}_A(a, a)$;

Elim. If $c : \Gamma \vdash \mathrm{I}_A(a, a')$ then $a = a'$ and $c = \mathrm{r}_{A,a}$.

Moreover, we have stability under substitution: if $\delta : \Delta \to \Gamma$ then

$$\mathrm{I}_A(a, a')[\delta] = \mathrm{I}_{A[\delta]}(a[\delta], a'[\delta])$$
$$\mathrm{r}_{A,a}[\delta] = \mathrm{r}_{A[\delta],a[\delta]}$$

Definition 4. *A cwf supports Σ-types iff the following conditions hold:*

Form. If $A \in \mathrm{Type}(\Gamma)$ and $B \in \mathrm{Type}(\Gamma \cdot A)$, there is $\Sigma(A, B) \in \mathrm{Type}(\Gamma)$,

Intro. If $a : \Gamma \vdash A$ and $b : \Gamma \vdash B[\langle \mathrm{id}, a \rangle]$, there is $\mathrm{pair}(a, b) : \Gamma \vdash \Sigma(A, B)$,

Elim. If $a : \Gamma \vdash \Sigma(A, B)$, there are $\pi_1(a) : \Gamma \vdash A$ and $\pi_2(a) : \Gamma \vdash B[\langle \mathrm{id}, \pi_1(a) \rangle]$
  such that

$$\pi_1(\mathrm{pair}(a, b)) = a$$
$$\pi_2(\mathrm{pair}(a, b)) = b$$
$$\mathrm{pair}(\pi_1(c), \pi_2(c)) = c$$

Moreover, we have stability under substitution:

$$\Sigma(A, B)[\delta] = \Sigma(A[\delta], B[\langle \delta \circ \mathrm{p}, \mathrm{q} \rangle])$$
$$\mathrm{pair}(a, b)[\delta] = \mathrm{pair}(a[\delta], b[\delta])$$
$$\pi_1(c)[\delta] = \pi_1(c[\delta])$$
$$\pi_2(c)[\delta] = \pi_2(c[\delta])$$

Note that in a cwf which supports extensional identity types and Σ-types surjective pairing, $\mathrm{pair}(\pi_1(c), \pi_2(c)) = c$, follows from the other conditions [10].

Definition 5. *A cwf supports Π-types iff the following conditions hold:*

Form. If $A \in \mathrm{Type}(\Gamma)$ and $B \in \mathrm{Type}(\Gamma \cdot A)$, there is $\Pi(A, B) \in \mathrm{Type}(\Gamma)$.

Intro. If $b : \Gamma \cdot A \vdash B$, there is $\lambda(b) : \Gamma \vdash \Pi(A, B)$.

Elim. If $c : \Gamma \vdash \Pi(A, B)$ and $a : \Gamma \vdash A$ then there is a term $\mathrm{ap}(c, a) : \Gamma \vdash B[\langle \mathrm{id}, a \rangle]$ *such that*

$$\mathrm{ap}(\lambda(b), a) = b[\langle \mathrm{id}, a \rangle] : \Gamma \vdash B[\langle \mathrm{id}, a \rangle]$$
$$c = \lambda(\mathrm{ap}(c[\mathrm{p}], \mathrm{q})) : \Gamma \vdash \Pi(A, B)$$

Moreover, we have stability under substitution:

$$\Pi(A, B)[\gamma] = \Pi(A[\gamma], B[\langle \gamma \circ \mathrm{p}, \mathrm{q} \rangle])$$
$$\lambda(b)[\gamma] = \lambda(b[\langle \gamma \circ \mathrm{p}, \mathrm{q} \rangle])$$
$$\mathrm{ap}(c, a)[\gamma] = \mathrm{ap}(c[\gamma], a[\gamma])$$

Definition 6. *A cwf $(\mathbb{C}, T)$ is democratic iff for each object Γ of $\mathbb{C}$ there is $\overline{\Gamma} \in \mathrm{Type}([])$ and an isomorphism $\Gamma \cong_{\gamma_\Gamma} []\cdot\overline{\Gamma}$. Each substitution $\delta : \Delta \to \Gamma$ can then be represented by the term $\overline{\delta} = q[\gamma_\Gamma \delta \gamma_\Delta^{-1}] : []\cdot\overline{\Delta} \vdash \overline{\Gamma}[\mathrm{p}]$.*

Democracy does not correspond to a rule of Martin-Löf type theory. However, a cwf generated inductively by the standard rules of Martin-Löf type theory with a one element type N_1 and Σ-types is democratic, since we can associate N_1 to the empty context and the closed type $\Sigma x_1 : A_1. \cdots . \Sigma x_n : A_n$ to a context $x_1 : A_1, \ldots, x_n : A_n$ by induction on n.

2.2   The Indexed Category of Types in Context

We shall now define the indexed category associated with a cwf. This will play a crucial role and in particular introduce the notion of *isomorphism* of types.

Proposition 7 (The Context-Indexed Category of Types). *If* $(\mathbb{C}, T)$ *is a cwf, then we can define a functor* $\boldsymbol{T} : \mathbb{C}^{op} \to \mathbf{Cat}$ *as follows:*

- *The objects of* $\boldsymbol{T}(\Gamma)$ *are types in* $\mathrm{Type}(\Gamma)$*. If* $A, B \in \mathrm{Type}(\Gamma)$*, then a morphism in* $\boldsymbol{T}(\Gamma)(A, B)$ *is a morphism* $\delta : \Gamma{\cdot}A \to \Gamma{\cdot}B$ *in* $\mathbb{C}$ *such that* $\mathrm{p}\delta = \mathrm{p}$*.*
- *If* $\gamma : \Delta \to \Gamma$ *in* $\mathbb{C}$*, then* $\boldsymbol{T}(\gamma) : \mathrm{Type}(\Gamma) \to \mathrm{Type}(\Delta)$ *maps an object* $A \in \mathrm{Type}(\Gamma)$ *to* $A[\gamma]$ *and a morphism* $\delta : \Gamma{\cdot}A \to \Gamma{\cdot}B$ *to* $\langle \mathrm{p}, \mathrm{q}[\delta\langle \gamma \circ \mathrm{p}, \mathrm{q}\rangle]\rangle :$ $\Delta{\cdot}A[\gamma] \to \Delta{\cdot}B[\gamma]$*.*

We write $A \cong_\theta B$ if $\theta : A \to B$ is an isomorphism in $\boldsymbol{T}(\Gamma)$. If $a : \Gamma \vdash A$, we write $\{\theta\}(a) = \mathrm{q}[\theta\langle id, a\rangle] : \Gamma \vdash B$ for the *coercion* of a to type B and $a =_\theta b$ if $a = \{\theta\}(b)$. Moreover, we get an alternative formulation of democracy.

Proposition 8. $(\mathbb{C}, T)$ *is democratic iff the functor from* $\boldsymbol{T}([])$ *to* $\mathbb{C}$*, which maps a closed type* A *to the context* $[]{\cdot}A$*, is an equivalence of categories.*

Seely's category **ML** of Martin-Löf theories [13] is essentially the category of categories $\boldsymbol{T}([])$ of closed types.

Fibres, slices and lcccs. Seely's interpretation of type theory in lcccs relies on the idea that a type $A \in \mathrm{Type}(\Gamma)$ can be interpreted as its *display map*, that is, a morphism with codomain Γ. For instance, the type $\mathtt{list}(n)$ of lists of length $n : \mathtt{nat}$ would be mapped to the function $l : \mathtt{list} \to \mathtt{nat}$ which to each list associates its length. Hence, types and terms in context Γ are interpreted in the *slice category* $\mathbb{C}/\Gamma$, since terms are interpreted as global sections. Syntactic types are connected with types-as-display-maps by the following result, an analogue of which was one of the cornerstones of Seely's paper.

Proposition 9. *If* $(\mathbb{C}, T)$ *is democratic and supports extensional identity and* Σ*-types, then* $\boldsymbol{T}(\Gamma)$ *and* $\mathbb{C}/\Gamma$ *are equivalent categories for all* Γ*.*

Proof. To each object (type) A in $\boldsymbol{T}(\Gamma)$ we associate the object p_A in $\mathbb{C}/\Gamma$. A morphism from A to B in $\boldsymbol{T}(\Gamma)$ is by definition a morphism from p_A to p_B in $\mathbb{C}/\Gamma$.

Conversely, to each object (morphism) $\delta : \Delta \to \Gamma$ of $\mathbb{C}/\Gamma$ we associate a type in $\mathrm{Type}(\Gamma)$. This is the inverse image $x : \Gamma \vdash \mathrm{Inv}(\delta)(x)$ which is defined type-theoretically by

$$\mathrm{Inv}(\delta)(x) = \Sigma y : \overline{\Delta}.\mathrm{I}_{\overline{\Gamma}}(\overline{x}, \overline{\delta}(y))$$

written in ordinary notation. In cwf combinator notation it becomes

$$\text{Inv}(\delta) = \Sigma(\overline{\Delta}[\langle\rangle], I_{\overline{\Gamma}[\langle\rangle]}(q[\gamma_\Gamma p], \overline{\delta}[\langle\langle\rangle, q\rangle]) \in \text{Type}(\Gamma)$$

These associations yield an equivalence of categories since $p_{\text{Inv}(\delta)}$ and δ are isomorphic in $\mathbb{C}/\Gamma$.

It is easy to see that $\boldsymbol{T}(\Gamma)$ has binary products if the cwf supports Σ-types and exponentials if it supports Π-types. Simply define $A \times B = \Sigma(A, B[p])$ and $B^A = \Pi(A, B[p])$. Hence by Proposition 9 it follows that $\mathbb{C}/\Gamma$ has products and $\mathbb{C}$ has finite limits in any democratic cwf which supports extensional identity types and Σ-types. If it supports Π-types too, then $\mathbb{C}/\Gamma$ is cartesian closed and $\mathbb{C}$ is locally cartesian closed.

3 The 2-Category of Categories with Families

3.1 Pseudo Cwf-Morphisms

A notion of *strict cwf-morphism* between cwfs $(\mathbb{C}, T)$ and $(\mathbb{C}', T')$ was defined by Dybjer [4]. It is a pair (F, σ), where $F : \mathbb{C} \to \mathbb{C}'$ is a functor and $\sigma : T \xrightarrow{\bullet} T'F$ is a natural transformation of family-valued functors, such that terminal objects and context comprehension are preserved on the nose. Here we need a weak version where the terminal object, context comprehension, and substitution of types and terms of a cwf are only preserved up to isomorphism. The pseudonatural transformations needed to prove our biequivalences will be families of cwf-morphisms which do not preserve cwf-structure on the nose.

The definition of pseudo cwf-morphism will be analogous to that of *strict* cwf-morphism, but cwf-structure will only be preserved up to coherent isomorphism.

Definition 10. *A* **pseudo cwf-morphism** *from* $(\mathbb{C}, T)$ *to* $(\mathbb{C}', T')$ *is a pair* (F, σ) *where:*

- *$F : \mathbb{C} \to \mathbb{C}'$ is a functor,*
- *For each context Γ in $\mathbb{C}$, σ_Γ is a **Fam**-morphism from $T\Gamma$ to $T'F\Gamma$. We will write $\sigma_\Gamma(A) : \text{Type}'(F\Gamma)$ for the type component and $\sigma_\Gamma^A(a) : F\Gamma \vdash \sigma_\Gamma(A)$ for the term component of this morphism.*

The following preservation properties must be satisfied:

- *Substitution is preserved: For each context $\delta : \Delta \to \Gamma$ in $\mathbb{C}$ and $A \in \text{Type}(\Gamma)$, there is an isomorphism of types $\theta_{A,\delta} : \sigma_\Gamma(A)[F\delta] \to \sigma_\Delta(A[\delta])$ such that substitution on terms is also preserved, that is, $\sigma_\Delta^{A[\gamma]}(a[\gamma]) =_{\theta_{A,\gamma}} \sigma_\Gamma^A(a)[F\gamma]$.*
- *The terminal object is preserved: $F[]$ is terminal.*
- *Context comprehension is preserved: $F(\Gamma \cdot A)$ with the projections $F(p_A)$ and $\{\theta_{A,p}^{-1}\}(\sigma_{\Gamma A}^{A[p]}(q_A))$ is a context comprehension of $F\Gamma$ and $\sigma_\Gamma(A)$. Note that the universal property on context comprehensions provides a unique isomorphism $\rho_{\Gamma,A} : F(\Gamma \cdot A) \to F\Gamma \cdot \sigma_\Gamma(A)$ which preserves projections.*

These data must satisfy naturality and coherence laws which amount to the fact that if we extend σ_Γ to a functor $\boldsymbol{\sigma}_\Gamma : \boldsymbol{T}(\Gamma) \rightarrow \boldsymbol{T}'F(\Gamma)$, then $\boldsymbol{\sigma}$ is a pseudo natural transformation from $\boldsymbol{T}$ to $\boldsymbol{T}'F$. This functor is defined by $\boldsymbol{\sigma}_\Gamma(A) = \sigma_\Gamma(A)$ on an object A and $\boldsymbol{\sigma}_\Gamma(f) = \rho_{\Gamma,B}F(f)\rho_{\Gamma,A}^{-1}$ on a morphism $f : A \rightarrow B$.

A consequence of this definition is that all cwf structure is preserved.

Proposition 11. *Let (F, σ) be a pseudo cwf-morphism from $(\mathbb{C}, T)$ to $(\mathbb{C}', T')$.*

(1) Then substitution extension is preserved: for all $\delta : \Delta \rightarrow \Gamma$ in $\mathbb{C}$ and $a : \Delta \vdash A[\delta]$, we have $F(\langle \delta, a \rangle) = \rho_{\Gamma,A}^{-1}\langle F\delta, \{\theta_{A,\delta}^{-1}\}(\sigma_\Delta^{A[\delta]}(a)) \rangle$.

(2) Redundancy terms/sections: for all $a \in \Gamma \vdash A$, $\sigma_\Gamma^A(a) = \mathrm{q}[\rho_{\Gamma,A}F(\langle \mathrm{id}, a \rangle)]$.

If $(F, \sigma) : (\mathbb{C}_0, T_0) \rightarrow (\mathbb{C}_1, T_1)$ and $(G, \tau) : (\mathbb{C}_1, T_1) \rightarrow (\mathbb{C}_2, T_2)$ are two pseudo cwf-morphisms, we define their composition $(G, \tau)(F, \sigma)$ as $(GF, \tau\sigma)$ where:

$$(\tau\sigma)_\Gamma(A) = \tau_{F\Gamma}(\sigma_\Gamma(A))$$
$$(\tau\sigma)_\Gamma^A(a) = \tau_{F\Gamma}^{\sigma_\Gamma(A)}(\sigma_\Gamma^A(a))$$

The families θ^{GF} and ρ^{GF} are obtained from θ^F, θ^G and ρ^F and ρ^G in the obvious way. The fact that these data satisfy the necessary coherence and naturality conditions basically amounts to the stability of pseudonatural transformation under composition. There is of course an identity pseudo cwf-morphism whose components are all identities, which is obviously neutral for composition. So, there is a category of cwfs and pseudo cwf-morphisms.

Since the isomorphism $(\Gamma \cdot A) \cdot B \cong \Gamma \cdot \Sigma(A, B)$ holds in an arbitrary cwf which supports Σ-types, it follows that pseudo cwf-morphisms automatically preserve Σ-types, since they preserve context comprehension. However, if cwfs support other structure, we need to define what it means that cwf-morphisms preserve this extra structure up to isomorphism.

Definition 12. *Let (F, σ) be a pseudo cwf-morphism between cwfs $(\mathbb{C}, T)$ and $(\mathbb{C}', T')$ which support identity types, Π-types, and democracy, respectively.*

- *(F, σ) preserves identity types provided $\sigma_\Gamma(\mathrm{I}_A(a, a')) \cong \mathrm{I}_{\sigma_\Gamma(A)}(\sigma_\Gamma^A(a), \sigma_\Gamma^A(a))$;*
- *(F, σ) preserves Π-types provided $\sigma_\Gamma(\Pi(A, B)) \cong \Pi(\sigma_\Gamma(A), \sigma_{\Gamma A}(B)[\rho_{\Gamma,A}^{-1}])$;*
- *(F, σ) preserves democracy provided $\sigma_{[]}(\overline{\Gamma}) \cong_{d_\Gamma} \overline{F\Gamma}[\langle \rangle]$, and the following diagram commutes:*

$$
\begin{array}{ccc}
F\Gamma & \xrightarrow{\;\; F\gamma_\Gamma \;\;} & F([] \cdot \overline{\Gamma}) \\
{\scriptstyle \gamma_{F\gamma}} \downarrow & & \downarrow {\scriptstyle \rho_{[],\overline{\Gamma}}} \\
[] \cdot \overline{F\Gamma} \xleftarrow{\langle\langle\rangle,\mathrm{q}\rangle} F[] \cdot \overline{F\Gamma}[\langle\rangle] & \xleftarrow{\;\; d_\Gamma \;\;} & F[] \cdot \sigma_{[]}(\overline{\Gamma})
\end{array}
$$

These preservation properties are all stable under composition and thus yield several different 2-categories of structure-preserving pseudo cwf-morphisms.

3.2 Pseudo Cwf-Transformations

Definition 13 (Pseudo cwf-transformation). *Let (F, σ) and (G, τ) be two cwf-morphisms from $(\mathbb{C}, T)$ to $(\mathbb{C}', T')$. A pseudo cwf-transformation from (F, σ) to (G, τ) is a pair (ϕ, ψ) where $\phi : F \xrightarrow{\bullet} G$ is a natural transformation, and for each Γ in $\mathbb{C}$ and $A \in \text{Type}(\Gamma)$, a morphism $\psi_{\Gamma, A} : \sigma_\Gamma(A) \to \tau_\Gamma(A)[\phi_\Gamma]$ in $T'(F\Gamma)$, natural in A and such that the following diagram commutes:*

$$
\begin{array}{ccc}
\sigma_\Gamma(A)[F\delta] & \xrightarrow{\; T'(F\delta)(\psi_{\Gamma,A}) \;} & \tau_\Gamma(A)[\phi_\Gamma F(\delta)] \\
\Big\downarrow{\scriptstyle \theta_{A,\delta}} & & \Big\downarrow{\scriptstyle T'(\phi_\Delta)(\theta'_{A,\delta})} \\
\sigma_\Delta(A[\delta]) & \xrightarrow[\; \psi_{\Delta, A[\delta]} \;]{} & \tau_\Delta(A[\delta])[\phi_\Delta]
\end{array}
$$

where θ and θ' are the isomorphisms witnessing preservation of substitution in types in the definition of pseudo cwf-morphism.

Pseudo cwf-transformations can be composed both vertically (denoted by (ϕ', ψ') (ϕ, ψ)) and horizontally (denoted by $(\phi', \psi') \star (\phi, \psi)$), and these compositions are associative and satisfy the interchange law. Note that just as coherence and naturality laws for pseudo cwf-morphisms ensure that they give rise to pseudonatural transformations (hence morphisms of indexed categories) σ to τ, this definition exactly amounts to the fact that pseudo cwf-transformations between (F, σ) and (F, τ) correspond to modifications from σ to τ.

3.3 2-Categories of Cwfs with Extra Structure

Definition 14. *Let $\mathbf{CwF}_{\text{dem}}^{\text{I}_{\text{ext}} \Sigma}$ be the 2-category of small democratic categories with families which support extensional identity types and Σ-types. The 1-cells are cwf-morphisms preserving democracy and extensional identity types (and Σ-types automatically) and the 2-cells are pseudo cwf-transformations.*

Moreover, let $\mathbf{CwF}_{\text{dem}}^{\text{I}_{\text{ext}} \Sigma \Pi}$ be the sub-2-category of $\mathbf{CwF}_{\text{dem}}^{\text{I}_{\text{ext}} \Sigma}$ where also Π-types are supported and preserved.

4 Forgetting Types and Terms

Definition 15. *Let $\mathbf{FL}$ be the 2-category of small categories with finite limits (left exact categories). The 1-cells are functors preserving finite limits (up to isomorphism) and the 2-cells are natural transformations.*

Let $\mathbf{LCC}$ be the 2-category of small locally cartesian closed categories. The 1-cells are functors preserving local cartesian closed structure (up to isomorphism), and the 2-cells are natural transformations.

$\mathbf{FL}$ is a sub(2-)category of the 2-category of categories: we do not provide a choice of finite limits. Similarly, $\mathbf{LCC}$ is a sub(2-)category of $\mathbf{FL}$. The first component of our biequivalences will be *forgetful* 2-functors.

Proposition 16. *The forgetful 2-functors*

$$U : \mathbf{CwF}_{\text{dem}}^{\text{I}_{\text{ext}}\Sigma} \to \mathbf{FL}$$
$$U : \mathbf{CwF}_{\text{dem}}^{\text{I}_{\text{ext}}\Sigma\Pi} \to \mathbf{LCC}$$

defined as follows on 0-, 1-, and 2-cells

$$U(\mathbb{C}, T) = \mathbb{C}$$
$$U(F, \sigma) = F$$
$$U(\phi, \psi) = \phi$$

are well-defined.

Proof. By definition U is a 2-functor from $\mathbf{CwF}$ to $\mathbf{Cat}$, it remains to prove that it sends a cwf in $\mathbf{CwF}_{\text{dem}}^{\text{I}_{\text{ext}}\Sigma}$ to $\mathbf{FL}$ and a cwf in $\mathbf{CwF}_{\text{dem}}^{\text{I}_{\text{ext}}\Sigma\Pi}$ to $\mathbf{LCC}$, along with the corresponding properties for 1-cells and 2-cells.

For 0-cells we already proved as corollaries of Proposition 9 that if $(\mathbb{C}, T)$ supports Σ-types, identity types and democracy, then $\mathbb{C}$ has finite limits; and if $(\mathbb{C}, T)$ also supports Π-types, then $\mathbb{C}$ is an lccc.

For 1-cells we need to prove that if (F, σ) preserves identity types and democracy, then F preserves finite limits; and if (F, σ) also preserves Π-types then F preserves local exponentiation. Since finite limits and local exponentiation in $\mathbb{C}$ and $\mathbb{C}'$ can be defined by the inverse image construction, these two statements boil down to the fact that if (F, σ) preserves identity types and democracy then inverse images are preserved. Indeed we have an isomorphism $F(\Gamma \cdot \text{Inv}(\delta)) \cong F\Gamma \cdot \text{Inv}(F\delta)$. This can be proved by long but mostly direct calculations involving all components and coherence laws of pseudo cwf-morphisms.

There is nothing to prove for 2-cells.

5   Rebuilding Types and Terms

Now, we turn to the reverse construction. We use the Bénabou-Hofmann construction to build a cwf from any finitely complete category, then generalize this operation to functors and natural transformations, and show that this gives rise to a pseudofunctor.

Proposition 17. *There are pseudofunctors*

$$H : \mathbf{FL} \to \mathbf{CwF}_{\text{dem}}^{\text{I}_{\text{ext}}\Sigma}$$
$$H : \mathbf{LCC} \to \mathbf{CwF}_{\text{dem}}^{\text{I}_{\text{ext}}\Sigma\Pi}$$

defined by

$$H\mathbb{C} = (\mathbb{C}, T_{\mathbb{C}})$$
$$HF = (F, \sigma_F)$$
$$H\phi = (\phi, \psi_\phi)$$

on 0-cells, 1-cells, and 2-cells, respectively, and where $T_{\mathbb{C}}, \sigma_F$, and ψ_ϕ are defined in the following three subsections.

Proof. The remainder of this Section contains the proof. We will in turn show the action on 0-cells, 1-cells, 2-cells, and then prove pseudofunctoriality of H.

5.1   Action on 0-Cells

As explained before, it is usual (going back to Cartmell [2]) to represent a type-in-context $A \in \mathrm{Type}(\Gamma)$ in a category as a *display map* [15], that is, as an object p_A in $\mathbb{C}/\Gamma$. A term $\Gamma \vdash A$ is then represented as a section of the display map for A, that is, a morphism a such that $\mathrm{p}_A \circ a = \mathrm{id}_\Gamma$. Substitution in types is then represented by pullback. This is essentially the technique used by Seely for interpreting Martin-Löf type theory in lcccs. However, as we already mentioned, it leads to a coherence problem.

To solve this problem Hofmann [5] used a construction due to Bénabou [1], which from any fibration builds an equivalent *split* fibration. Hofmann used it to build a category with attributes (cwa) [2] from a locally cartesian closed category. He then showed that this cwa supports Π, Σ, and extensional identity types. This technique essentially amounts to associating to a type A, not only a display map, but a whole family of display maps, one for each substitution instance $A[\delta]$. In other words, we choose a pullback square for every possible substitution and this choice is split, hence solving the coherence problem. As we shall explain below this family takes the form of a functor, and we refer to it as a *functorial family.*

Here we reformulate Hofmann's construction using cwfs. See Dybjer [4] for the correspondence between cwfs and cwas.

Lemma 18. *Let $\mathbb{C}$ be a category with terminal object. Then we can build a democratic cwf $(\mathbb{C}, T_\mathbb{C})$ which supports Σ-types. If $\mathbb{C}$ has finite limits, then $(\mathbb{C}, T_\mathbb{C})$ also supports extensional identity types. If $\mathbb{C}$ is locally cartesian closed, then $(\mathbb{C}, T_\mathbb{C})$ also supports Π-types.*

Proof. We only show the definition of types and terms in $T_\mathbb{C}(\Gamma)$. This construction is essentially the same as Hofmann's [5].

A *type* in $\mathrm{Type}_\mathbb{C}(\Gamma)$ is a *functorial family*, that is, a functor $\overrightarrow{A} : \mathbb{C}/\Gamma \to \mathbb{C}^\rightarrow$ such that $\mathrm{cod} \circ \overrightarrow{A} = \mathrm{dom}$ and if $\Omega \xrightarrow{\alpha} \Delta$ is a morphism in $\mathbb{C}/\Gamma$, then $\overrightarrow{A}(\alpha)$ is a pullback square:

$$
\begin{array}{ccc}
\delta\alpha \overset{\nwarrow}{} \;\Gamma\; \overset{\nearrow}{} \delta \\
\end{array}
$$

$$
\begin{array}{ccc}
 & \overrightarrow{A}(\delta,\alpha) & \\
\overrightarrow{A}(\delta\alpha) \downarrow & & \downarrow \overrightarrow{A}(\delta) \\
\Omega & \xrightarrow{\;\alpha\;} & \Delta
\end{array}
$$

Following Hofmann, we denote the upper arrow of the square by $\overrightarrow{A}(\delta, \alpha)$.

A *term* $a : \Gamma \vdash \overrightarrow{A}$ is a section of $\overrightarrow{A}(\mathrm{id}_\Gamma)$, that is, a morphism $a : \Gamma \to \Gamma \cdot \overrightarrow{A}$ such that $\overrightarrow{A}(\mathrm{id}_\Gamma)a = \mathrm{id}_\Gamma$, where we have defined context extension by $\Gamma \cdot \overrightarrow{A} = \mathrm{dom}(\overrightarrow{A}(\mathrm{id}_\Gamma))$. Interpreting types as functorial families makes it easy to define substitution in types. Substitution in terms is obtained by exploiting the universal property of pullback squares, yielding a functor $T_\mathbb{C} : \mathbb{C}^{op} \to \mathbf{Fam}$.

Note that $(\mathbb{C}, T_{\mathbb{C}})$ is a *democratic* cwf since to any context Γ we can associate a functorial family $\widehat{\langle\rangle} : \mathbb{C}/[] \to C^{\to}$, where $\langle\rangle : \Gamma \to []$ is the terminal projection. The isomorphism $\gamma_\Gamma : \Gamma \to [] \cdot \widehat{\langle\rangle}$ is just id_Γ.

5.2   Action on 1-Cells

Suppose that $\mathbb{C}$ and $\mathbb{C}'$ have finite limits and that $F : \mathbb{C} \to \mathbb{C}'$ preserves them. As described in the previous section, $\mathbb{C}$ and $\mathbb{C}'$ give rise to cwfs $(\mathbb{C}, T_{\mathbb{C}})$ and $(\mathbb{C}', T_{\mathbb{C}'})$. In order to extend F to a pseudo cwf-morphism, we need to define, for each object Γ in $\mathbb{C}$, a **Fam**-morphism $(\sigma_F)_\Gamma : T_{\mathbb{C}}(\Gamma) \to T_{\mathbb{C}'} F(\Gamma)$. Unfortunately, unless F is full, it does not seem possible to embed faithfully a functorial family $\overrightarrow{A} : \mathbb{C}/\Gamma \to \mathbb{C}^{\to}$ into a functorial family over $F\Gamma$ in $\mathbb{C}'$. However, there is such an embedding for display maps (just apply F) from which we will freely regenerate a functorial family from the obtained display map.

The "hat" construction. As remarked by Hofmann, any morphism $f : \Delta \to \Gamma$ in a category $\mathbb{C}$ with a (not necessarily split) choice of finite limits generates a functorial family $\widehat{f} : \mathbb{C}/\Gamma \to \mathbb{C}^{\to}$. If $\delta : \Delta \to \Gamma$ then $\widehat{f}(\delta) = \delta^*(f)$, where $\delta^*(f)$ is obtained by taking the pullback of f along δ (δ^* is known as the *pullback functor*):

$$
\begin{array}{ccc}
 & \longrightarrow & \\
\delta^*(f)\downarrow & & \downarrow f \\
\Delta & \xrightarrow{\ \delta\ } & \Gamma
\end{array}
$$

Note that we can always choose pullbacks such that $\widehat{f}(\mathrm{id}_\Gamma) = \mathrm{id}_\Gamma^*(f) = f$. If $\Omega \xrightarrow{\ \alpha\ } \Delta$ is a morphism in $\mathbb{C}/\Gamma$, we define $\widehat{f}(\alpha)$ as the left square in the
$$\delta\alpha \searrow \ \ \Gamma \ \ \swarrow \delta$$
following diagram:

$$
\begin{array}{ccccc}
 & \xrightarrow{\ \widehat{f}(\delta,\alpha)\ } & & \longrightarrow & \\
\widehat{f}(\delta\alpha)\downarrow & & \downarrow \widehat{f}(\delta) & & \downarrow f \\
\Delta' & \xrightarrow{\ \alpha\ } & \Delta & \xrightarrow{\ \delta\ } & \Gamma
\end{array}
$$

This is a pullback, since both the outer square and the right square are pullbacks.

Translation of types. The hat construction can be used to extend F to types:

$$
\sigma_F(\overrightarrow{A}) = \widehat{F(\overrightarrow{A}(\mathrm{id}))}
$$

Note that $F(\Gamma \cdot \overrightarrow{A}) = F(\mathrm{dom}(\overrightarrow{A}(\mathrm{id}))) = \mathrm{dom}(F(\overrightarrow{A}(\mathrm{id}))) = \mathrm{dom}(\sigma_\Gamma(\overrightarrow{A})(\mathrm{id})) = F\Gamma \cdot \sigma_\Gamma(\overrightarrow{A})$, so context comprehension is preserved on the nose. However, substitution on types is *not* preserved on the nose. Hence we have to define a coherent family of isomorphisms $\theta_{\overrightarrow{A},\delta}$.

Completion of cwf-morphisms. Fortunately, whenever F preserves finite limits there is a canonical way to generate all the remaining data.

Lemma 19 (Generation of isomorphisms). *Let $(\mathbb{C}, T)$ and $(\mathbb{C}', T')$ be two cwfs, $F : \mathbb{C} \to \mathbb{C}'$ a functor preserving finite limits, $\sigma_\Gamma : \mathrm{Type}(\Gamma) \to \mathrm{Type}'(F\Gamma)$ a family of functions, and $\rho_{\Gamma,A} : F(\Gamma{\cdot}A) \to F\Gamma{\cdot}\sigma_\Gamma(A)$ a family of isomorphisms such that $\mathrm{p}\rho_{\Gamma,A} = F\mathrm{p}$. Then there exists an unique choice of functions σ_Γ^A on terms and of isomorphisms $\theta_{A,\delta}$ such that (F, σ) is a pseudo cwf-morphism.*

Proof. By item *(2)* of Proposition 11, the unique way to extend σ to terms is to set $\sigma_\Gamma^A(a) = \mathrm{q}[\rho_{\Gamma,A}F(\langle \mathrm{id}, a\rangle)]$. To generate θ, we use the two squares below:

$$F\Delta{\cdot}\sigma_\Gamma(A)[F\delta] \xrightarrow{\langle (F\delta)\mathrm{pq}\rangle} F\Gamma{\cdot}\sigma_\Gamma(A) \quad F\Delta{\cdot}\sigma_\Delta(A[\delta]) \xrightarrow{\rho_{\Gamma,A}F(\langle \delta\mathrm{p},\mathrm{q}\rangle)\rho_{\Delta,A[\delta]}^{-1}} F\Gamma{\cdot}\sigma_\Gamma(A)$$

$$\mathrm{p}\downarrow \qquad\qquad \downarrow\mathrm{p} \qquad\qquad \mathrm{p}\downarrow \qquad\qquad\qquad \downarrow\mathrm{p}$$

$$F\Delta \xrightarrow{\quad F\delta \quad} F\Gamma \qquad\qquad F\Delta \xrightarrow{\quad F\delta \quad} F\Gamma$$

The first square is a substitution pullback. The second is a pullback because F preserves finite limits and $\rho_{\Gamma,A}$ and $\rho_{\Delta,A[\delta]}$ are isomorphisms. The isomorphism $\theta_{A,\delta}$ is defined as the unique mediating morphism from the first to the second. It follows from the universal property of pullbacks that the family θ satisfies the necessary naturality and coherence conditions. There is no other choice for $\theta_{A,\delta}$, because if (F, σ) is a pseudo cwf-morphism with families of isomorphisms θ and ρ, then $\rho_{\Gamma,A}F(\langle \delta\mathrm{p},\mathrm{q}\rangle)\rho_{\Delta,A[\delta]}^{-1}\theta_{A,\delta} = \langle (F\delta)\mathrm{p},\mathrm{q}\rangle$. Hence if F preserves finite limits, $\theta_{A,\delta}$ must coincide with the mediating morphism.

Preservation of additional structure. As a pseudo cwf-morphism, (F, σ_F) automatically preserves Σ-types. Since the democratic structure of $(\mathbb{C}, T_{\mathbb{C}})$ and $(\mathbb{C}', T_{\mathbb{C}'})$ is trivial it is clear that it is preserved by (F, σ_F). To prove that it also preserves type constructors, we use the following proposition.

Proposition 20. *Let (F, σ) be a pseudo cwf-morphism between $(\mathbb{C}, T)$ and $(\mathbb{C}', T')$ supporting Σ-types and democracy. Then:*

- *If $(\mathbb{C}, T)$ and $(\mathbb{C}', T')$ both support identity types, then (F, σ) preserves identity types provided F preserves finite limits.*
- *If $(\mathbb{C}, T)$ and $(\mathbb{C}', T')$ both support Π-types, then (F, σ) preserves Π-types provided F preserves local exponentiation.*

Proof. For the first part it remains to prove that if F preserves finite limits, then (F, σ) preserves identity types. Since $a, a' \in \Gamma \vdash A$, $\mathrm{p}_{\mathrm{I}_A(a,a')} : \Gamma{\cdot}\mathrm{I}_A(a, a') \to \Gamma$ is an equalizer of $\langle \mathrm{id}, a\rangle$ and $\langle \mathrm{id}, a'\rangle$ and F preserves equalizers, it follows that $F(\mathrm{p}_{\mathrm{I}_A(a,a')})$ is an equalizer of $\langle \mathrm{id}, \sigma_\Gamma^A(a)\rangle$ and $\langle \mathrm{id}, \sigma_\Gamma^A(a')\rangle$, and by uniqueness of equalizers it is isomorphic to $\mathrm{I}_{\sigma_\Gamma(A)}(\sigma_\Gamma^A(a), \sigma_\Gamma^A(a'))$.

The proof of preservation of Π-types exploits in a similar way the uniqueness (up to iso) of "Π-objects" of $A \in \mathrm{Type}(\Gamma)$ and $B \in \mathrm{Type}(\Gamma{\cdot}A)$.

5.3   Action on 2-Cells

Similarly to the case of 1-cells, under some conditions a natural transformation $\phi : F \overset{\bullet}{\to} G$ where (F, σ) and (G, τ) are pseudo cwf-morphisms can be completed to a pseudo cwf-transformation (ϕ, ψ_ϕ), as stated below.

Lemma 21 (Completion of pseudo cwf-transformations). *Suppose (F, σ) and (G, τ) are pseudo cwf-morphisms from $(\mathbb{C}, T)$ to $(\mathbb{C}', T)$ such that F and G preserve finite limits and $\phi : F \overset{\bullet}{\to} G$ is a natural transformation, then there exists a family of morphisms $(\psi_\phi)_{\Gamma, A} : \sigma_\Gamma(A) \to \tau_\Gamma(A)[\phi_\Gamma]$ such that (ϕ, ψ_ϕ) is a pseudo cwf-transformation from (F, σ) to (G, τ).*

Proof. We set $\psi_{\Gamma, A} = \langle \mathsf{p}, \mathsf{q}[\rho'_{\Gamma, A} \phi_{\Gamma \cdot A} \rho_{\Gamma, A}^{-1}] \rangle : F\Gamma \cdot \sigma_\Gamma A \to F\Gamma \cdot \tau_\Gamma(A)[\phi_\Gamma]$. To check the coherence law, we apply the universal property of a well-chosen pullback square (exploiting the fact that G preserves finite limits).

This completion operation on 2-cells commutes with units and both notions of composition, as will be crucial to prove pseudofunctoriality of H:

Lemma 22. *If $\phi : F \overset{\bullet}{\to} G$ and $\phi' : G \overset{\bullet}{\to} H$, then*

$$(\phi', \psi_{\phi'})(\phi, \psi_\phi) = (\phi'\phi, \psi_{\phi'\phi})$$
$$(\phi, \psi_\phi) \star 1 = (\phi \star 1, \psi_{\phi \star 1})$$
$$1 \star (\phi, \psi_\phi) = (1 \star \phi, \psi_{1 \star \phi})$$
$$(\phi', \psi_{\phi'}) \star (\phi, \psi_\phi) = (\phi' \star \phi, \psi_{\phi' \star \phi})$$

whenever these expressions typecheck.

Proof. Direct calculations.

5.4   Pseudofunctoriality of H

Note that H is *not* a functor, because for any $F : \mathbb{C} \to \mathbb{D}$ with finite limits and functorial family $\overrightarrow{A}$ over Γ (in $\mathbb{C}$), $\sigma_\Gamma(\overrightarrow{A})$ forgets all information on $\overrightarrow{A}$ except its display map $\overrightarrow{A}(\mathrm{id})$, and later extends $F(\overrightarrow{A}(\mathrm{id}))$ to an independent functorial family. However if $F : \mathbb{C} \to \mathbb{D}$ and $G : \mathbb{D} \to \mathbb{E}$ preserve finite limits, the two pseudo cwf-morphisms $(G, \sigma^G) \circ (F, \sigma^F) = (GF, \sigma^G \sigma^F)$ and (GF, σ^{GF}) are related by the pseudo cwf-transformation $(1_{GF}, \psi_{1_{GF}})$, which is obviously an isomorphism. The coherence laws only involve vertical and horizontal compositions of units and pseudo cwf-transformations obtained by completion, hence they are easy consequences of Lemma 22.

6   The Biequivalences

Theorem 23. *We have the following biequivalences of 2-categories.*

$$\mathbf{FL} \overset{H}{\underset{U}{\rightleftarrows}} \mathbf{CwF}^{\mathrm{Iext}\,\Sigma}_{\mathrm{dem}} \qquad\qquad \mathbf{LCC} \overset{H}{\underset{U}{\rightleftarrows}} \mathbf{CwF}^{\mathrm{Iext}\,\Sigma\Pi}_{\mathrm{dem}}$$

Proof. Since $UH = \mathrm{Id}$ (the identity 2-functor) it suffices to construct pseudo-natural transformations of pseudofunctors:

$$\mathrm{Id} \underset{\epsilon}{\overset{\eta}{\rightleftarrows}} HU$$

which are inverse up to invertible modifications. Since $HU(\mathbb{C},T) = (\mathbb{C},T^{\mathbb{C}})$, these pseudonatural transformations are families of equivalences of cwfs:

$$(\mathbb{C},T) \underset{\epsilon_{(\mathbb{C},T)}}{\overset{\eta_{(\mathbb{C},T)}}{\rightleftarrows}} (\mathbb{C},T^{\mathbb{C}})$$

which satisfy the required conditions for pseudonatural transformations.

Construction of $\eta_{(\mathbb{C},T)}$. Using Lemma 19, we just need to define a base functor, which will be $\mathrm{Id}_{\mathbb{C}}$, and a family σ_{Γ}^{η} which translates types (in the sense of T) to functorial families. This is easy, since types in the cwf $(\mathbb{C},T)$ come equipped with a chosen behaviour under substitution. Given $A \in \mathrm{Type}(\Gamma)$, we define:

$$\sigma_{\Gamma}^{\eta}(A)(\delta) = \mathrm{p}_{A[\delta]}$$
$$\sigma_{\Gamma}^{\eta}(A)(\delta,\gamma) = \langle \gamma\mathrm{p}, \mathrm{q}\rangle$$

For each pseudo cwf-morphism (F,σ), the pseudonaturality square relates two pseudo cwf-morphisms whose base functor is F. Hence, the necessary invertible pseudo cwf-transformation is obtained using Lemma 21 from the identity natural transformation on F. The coherence conditions are straightforward consequences of Lemma 22.

Construction of $\epsilon_{(\mathbb{C},T)}$. As for η, the base functor for $\epsilon_{(\mathbb{C},T)}$ is $\mathrm{Id}_{\mathbb{C}}$. Using Lemma 19 again we need, for each context Γ, a function $\sigma_{\Gamma}^{\epsilon}$ which given a functorial family $\overrightarrow{A}$ over Γ will build a syntactic type $\sigma_{\Gamma}^{\epsilon}(\overrightarrow{A}) \in \mathrm{Type}(\Gamma)$. In other terms, we need to find a syntactic representative of an arbitrary display map, that is, an arbitrary morphism in $\mathbb{C}$. We use the inverse image:

$$\sigma_{\Gamma}^{\epsilon}(\overrightarrow{A}) = \mathrm{Inv}(\overrightarrow{A}(\mathrm{id})) \in \mathrm{Type}(\Gamma)$$

The family ϵ is pseudonatural for the same reason as η above.

Invertible modifications. For each cwf $(\mathbb{C},T)$, we need to define invertible pseudo cwf-transformations $m_{(\mathbb{C},T)} : (\epsilon\eta)_{(\mathbb{C},T)} \to \mathrm{id}_{(\mathbb{C},T)}$ and $m'_{(\mathbb{C},T)} : (\eta\epsilon)_{(\mathbb{C},T)} \to \mathrm{id}_{(\mathbb{C},T)}$. As pseudo cwf-transformations between pseudo cwf-morphisms with the same base functor, their first component will be the identity natural transformation, and the second will be generated by Lemma 21. The coherence law for modifications is a consequence of Lemma 22.

Acknowledgement. We are grateful to the anonymous reviewers for several useful remarks which have helped us improve the paper. We would also like to acknowledge the support of the (UK) EPSRC grant RC-CM1025 for the first author and of the (Swedish) Vetenskapsrådet grant "Types for Proofs and Programs" for the second author.

References

1. Bénabou, J.: Fibred categories and the foundation of naive category theory. Journal of Symbolic Logic 50, 10–37 (1985)
2. Cartmell, J.: Generalized algebraic theories and contextual categories. Annals of Pure and Applied Logic 32, 209–243 (1986)
3. Curien, P.-L.: Substitution up to isomorphism. Fundamenta Informaticae 19(1,2), 51–86 (1993)
4. Dybjer, P.: Internal type theory. In: Berardi, S., Coppo, M. (eds.) TYPES 1995. LNCS, vol. 1158, pp. 120–134. Springer, Heidelberg (1996)
5. Hofmann, M.: On the interpretation of type theory in locally cartesian closed categories. In: Pacholski, L., Tiuryn, J. (eds.) CSL 1994. LNCS, vol. 933.Springer, Heidelberg (1995)
6. Hofmann, M.: Syntax and semantics of dependent types. In: Pitts, A., Dybjer, P. (eds.) Semantics and Logics of Computation.Cambridge University Press, Cambridge (1996)
7. Lawvere, F.W.: Equality in hyperdoctrines and comprehension schema as an adjoint functor. In: Heller, A. (ed.) Applications of Categorical Algebra, Proceedings of Symposia in Pure Mathematics. AMS, Providence (1970)
8. Leinster, T.: Basic bicategories. arXiv:math/9810017v1 (1999)
9. Martin-Löf, P.: Constructive mathematics and computer programming. In: Logic, Methodology and Philosophy of Science, VI, 1979, pp. 153–175. North-Holland, Amsterdam (1982)
10. Martin-Löf, P.: Intuitionistic Type Theory. Bibliopolis (1984)
11. Martin-Löf, P.: Substitution calculus. Notes from a lecture given in Göteborg (November 1992)
12. Mimram, S.: Decidability of equality in categories with families. Report, Magistère d'Informatique et Modelisation, École Normale Superieure de Lyon (2004), http://www.pps.jussieu.fr/~smimram/
13. Seely, R.: Locally cartesian closed categories and type theory. Math. Proc. Cambridge Philos. Soc. 95(1), 33–48 (1984)
14. Tasistro, A.: Formulation of Martin-Löf's theory of types with explicit substitutions. Technical report, Department of Computer Sciences, Chalmers University of Technology and University of Göteborg, Licentiate Thesis (1993)
15. Taylor, P.: Practical Foundations of Mathematics. Cambridge University Press, Cambridge (1999)

Realizability Proof for Normalization of Full Differential Linear Logic

Stéphane Gimenez

Laboratoire PPS, Université Paris-Diderot – Paris 7

Abstract. Realizability methods allowed to prove normalization results on many typed calculi. *Girard* adapted these methods to systems of nets and managed to prove normalization of second order *Linear Logic* [4]. Our contribution is to provide an extension of this proof that embrace *Full Differential Linear Logic* (a logic that can describe both single-use resources and inexhaustible resources). Anchored within the realizability framework our proof is modular enough so that further extensions (to second order, to additive constructs or to any other independent feature that can be dealt with using realizability) come for free.

Keywords: Linear Logic, Proof Nets, Differential Linear Logic, Differential Interaction Nets, Realizability, Weak Normalization.

Introduction

It happens that the three differential constructs of *Differential Linear Logic* (also abbreviated as *DiLL*), namely *co-weakening, co-contraction* and *co-dereliction,* and the original *promotion* construct from *Linear Logic* (*LL*) can be used altogether in a system that has been called *Full Differential Linear Logic* (*Full-DiLL*). In particular, this system embeds *Differential λ-calculus*, which was the first avatar of the differential paradigm introduced by *Ehrhard* and *Regnier* [1,3]. A similar system has been studied within an intuitionist setting by *Tranquilli* [9], and a combinatorial proof of its weak normalization has already been provided by *Pagani* [7]. We provide in this paper a new proof of its weak normalization, using a reducibility technique which is of great importance because it is the only known method to prove such a result in presence of second order quantifiers.

Plan. A brief presentation of *Full-DiLL* is given in Section 1. Realizability tools used by *Girard* are introduced in Section 2 and the proof he provided for weak normalization of *LL* is recalled in Section 3. Weak normalization of *Full-DiLL* will be addressed in Section 4.

1 Full Differential Linear Logic

1.1 Syntax

In this paper, we will depict proofs of *LL* and *Full-DiLL* graphically using the *interaction nets* formalism [5], which is well suited to multiplicative constructions [6], and is the usual way to represent differential constructions, see [2].

L. Ong (Ed.): TLCA 2011, LNCS 6690, pp. 107–122, 2011.

Our study will be restricted to *multiplicative* and *exponential* fragments of these logics (whose constructions are recalled bellow), because an extension of our result to *additive* and *second order* fragments can be extracted from *Girard's* proof [4] without any significant changes, and it would otherwise be cumbersome.

$$\overline{\vdash A^{\perp}, A}\ axiom$$

$$\overline{\vdash 1}\ unit$$

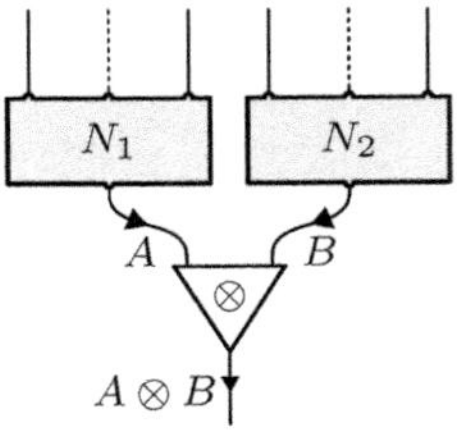

$$\frac{\vdash \Gamma_1, A \quad \vdash B, \Gamma_2}{\vdash \Gamma_1, A \otimes B, \Gamma_2}\ tensor$$

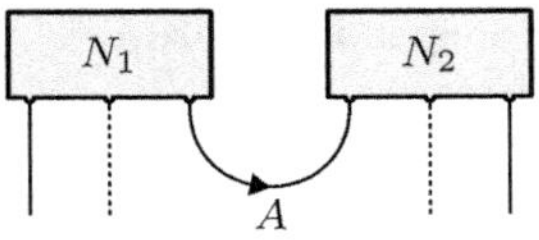

$$\frac{\vdash \Gamma_1, A \quad \vdash A^{\perp}, \Gamma_2}{\vdash \Gamma_1, \Gamma_2}\ cut$$

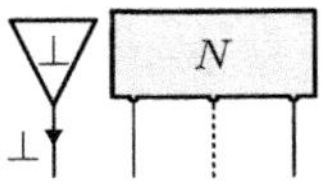

$$\frac{\vdash \Gamma}{\vdash \perp, \Gamma}\ co\text{-}unit$$

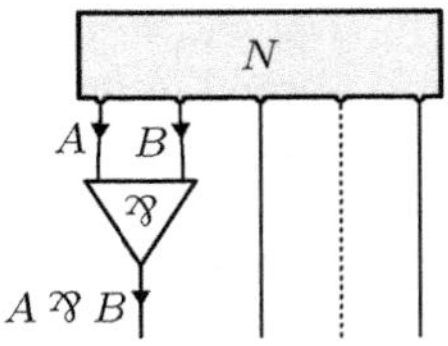

$$\frac{\vdash A, B, \Gamma}{\vdash A \,⅋\, B, \Gamma}\ par$$

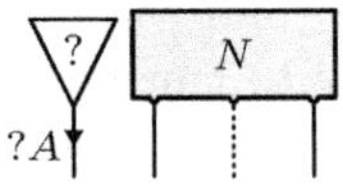

$$\frac{\vdash \Gamma}{\vdash ?A, \Gamma}\ weakening$$

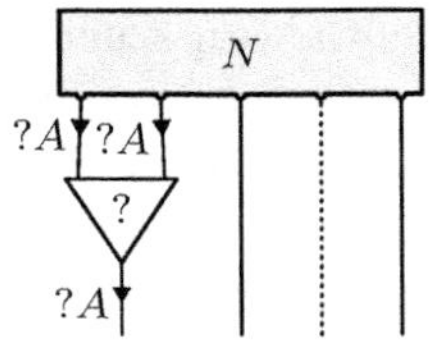

$$\frac{\vdash ?A, ?A, \Gamma}{\vdash ?A, \Gamma}\ contraction$$

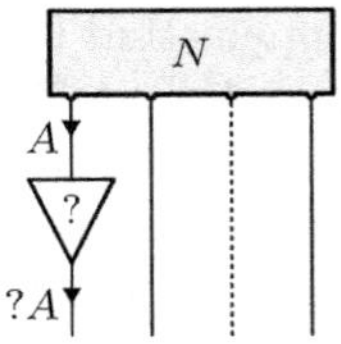

$$\frac{\vdash A, \Gamma}{\vdash ?A, \Gamma}\ dereliction$$

Specifically in the case of *Full-DiLL*, we add the three differential constructs which are represented like this:

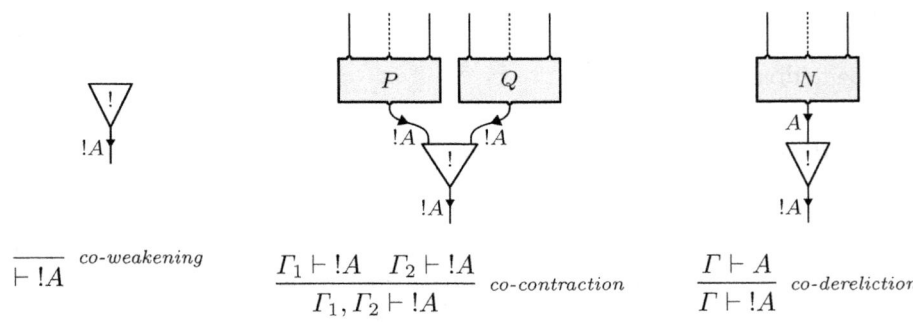

$$\overline{\vdash !A}\ co\text{-}weakening$$

$$\frac{\Gamma_1 \vdash !A \quad \Gamma_2 \vdash !A}{\Gamma_1, \Gamma_2 \vdash !A}\ co\text{-}contraction$$

$$\frac{\Gamma \vdash A}{\Gamma \vdash !A}\ co\text{-}dereliction$$

Intuitively, links labeled with exponential modalities like $!A$ can be thought as communication channels between resource producers and resource consumers. A *co-weakening* provides an empty bag of resources, a *co-dereliction* provides a bag with a single resource and *co-contractions* allow to merge such bags. As for *promotions* (bags providing inexhaustible resources), they will be displayed as *boxes*, similarly to what is done in the standard *Proof Net* formalism.

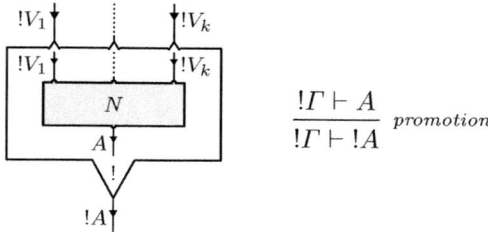

$$\frac{!\Gamma \vdash A}{!\Gamma \vdash !A} \ promotion$$

Informally, a (simple) net is just a set of *nodes* and *free ports* linked by *wires* such that some typing constraints are satisfied. In their graphical representation, wires are labeled with the type of the destination port (according to the chosen orientation) which has to be orthogonal to the type of the source port. A net is said to have *interface* $\vdash A_1, \ldots, A_n$ when its free ports have types $A_1, \ldots, A_n$. Sequent rules can then be thought as valid steps to build nets. In fact, despite the syntactical freedom suggested by their graphical representations, most of the nets we consider will be valid in this sense.

Definition 1 (Sequentializable nets). *A net that is the direct translation of a proof from usual linear logic will be called an* LL net. *If the use of formal sums and sequent rules for* co-weakening, co-dereliction, *and* co-contraction *is also allowed, the net will be called a* Full-DiLL net.

Inside a differential setting, we should indeed precise that all nets we consider are generalized to formal sums of simple nets that share the same interface. This generalization is needed to define their dynamics. Such sums distribute over any context but *promotion* which is the only construction that shall not be considered linear: a box is a special node which is itself parametrized by a formal sum of simple nets, and this sum will not distribute outside the scope of the box (unless it is open).

1.2   Dynamics

Written in a net formalism, this full system can be endowed with a dynamic similar to *cut-elimination*, by means of a reduction relation $\longrightarrow$ that is contextual, preserves validity, and combines:

- reduction rules of standard *Proof Nets*, see [4].
- reduction rules of *Differential Interaction Nets*, see [2].
- and the following set of three rules, which tells how *promotions* interact with differential constructions.

A *co-weakening* in front of a *box input* enters the box.

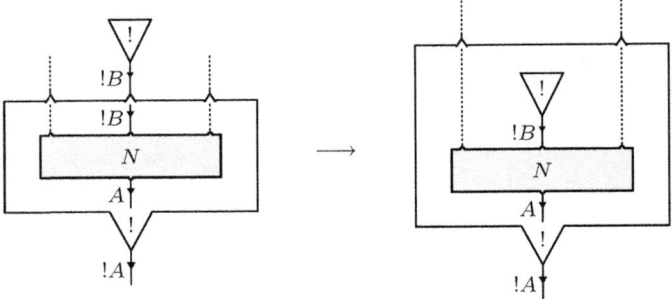

A *co-contraction* in front of a *box input* also enters the box.

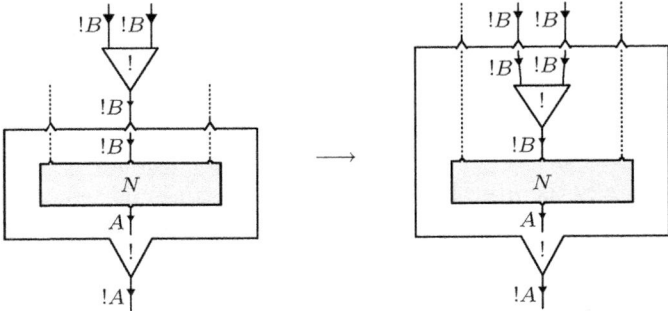

When a *co-dereliction* encounters a *box input*, their interaction is called *chain rule* (the reason being that somehow it plays the same role as the one with the same name which is more widely known in mathematics):

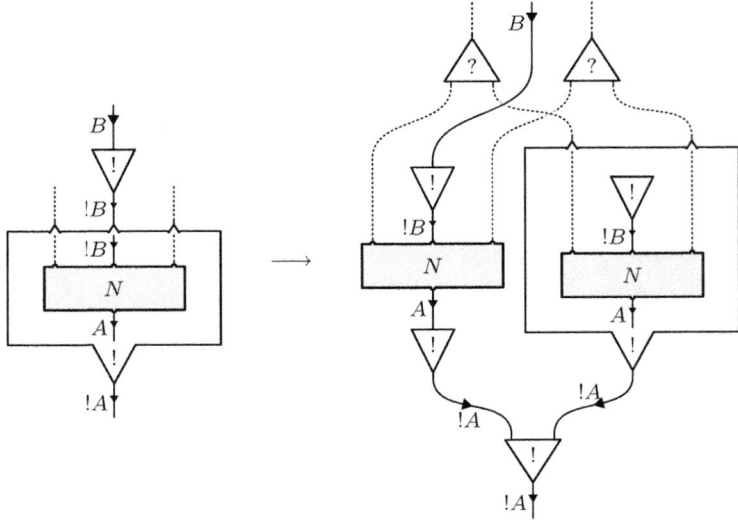

2   Realizability Tools

In this section, we recall what was Girard's approach to realizability, rephrasing it slightly to match with interaction nets syntax.

2.1   Observable and Duality

Definition 2 (Pointed net). *A pointed net is a net with a distinguished port (represented graphically by a dot).*

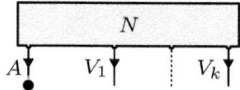

Such a net is said to have interface $\vdash A\,;V_1,\ldots,V_k$*. And* $\mathcal{N}_A$ *denote the set of pointed nets whose distinguished port is of type* A*. We will also use* $\mathcal{N}_A^\dagger$ *to denote the subset of* $\mathcal{N}_A$ *whose elements can be promoted, that is, nets with interface* $\vdash A\,;?V_1,\ldots,?V_k$*.*

A pointed axiom of type A will be denoted by $Axiom_A$:

Also, when $P \in \mathcal{N}_A$ and $Q \in \mathcal{N}_{A^\perp}$, we use $Cut_A(P,Q)$ to denote the cut between nets P and Q along their distinguished ports:

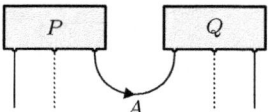

Let $\mathcal{O}$ be a chosen set of nets called *observable*, we define the binary relation $\star$ on pointed nets such that $P \star Q$ holds if and only if $P \in \mathcal{N}_A$ and $Q \in \mathcal{N}_{A^\perp}$ and we have $Cut_A(P,Q) \in \mathcal{O}$.

Definition 3 (Duality). *If* $\mathcal{P}$ *is a set of pointed nets of type* A*, its* dual *is defined as:*

$$\mathcal{P}^\star := \{Q \mid \forall P \in \mathcal{P},\ P \star Q\}$$

Property 1. For any set of pointed nets $\mathcal{P} \subseteq \mathcal{N}_A$ and $\mathcal{Q} \subseteq \mathcal{N}_A$, the following properties hold:

$$\mathcal{P} \subseteq \mathcal{Q} \implies \mathcal{Q}^\star \subseteq \mathcal{P}^\star \qquad \mathcal{P} \subseteq \mathcal{P}^{\star\star} \qquad \mathcal{P}^{\star\star\star} = \mathcal{P}^\star$$

Proof. The two first assertions may be deduced without difficulties from the definition of duality. As for the third one, $\mathcal{P}^{\star\star\star} \subseteq \mathcal{P}^\star$ is obtained substituting $\mathcal{P}$ by $\mathcal{P}^\star$ in the second one, and $\mathcal{P}^{\star\star\star} \supseteq \mathcal{P}^\star$ is obtained substituting $\mathcal{Q}$ by $\mathcal{P}^{\star\star}$ in the first one.

Let $\mathcal{O}_A$ be the set of all $N \in \mathcal{N}_A$ such that the underlying non pointed net belongs to $\mathcal{O}$.

Property 2

$$\mathcal{O}_A \;=\; \left\{\, Axiom_{A^\perp} \,\right\}^{\star} \;=\; \mathcal{O}_A{}^{\star\star}$$

Proof. This is true because we are in the interaction nets formalism which comes with the property $Cut_A(N, Axiom_{A^\perp}) = N$ for all $N \in \mathcal{N}_A$. Using this property, proof of the first stated equality is straightforward. The second equality follows since $\left\{\, Axiom_{A^\perp} \,\right\}^{\star\star\star} = \left\{\, Axiom_{A^\perp} \,\right\}^{\star}$.

2.2   Behaviors in *LL*

Definition 4 (Behavior). *A* behavior *at type A is any set* $\mathcal{P}$ *of pointed nets of type A which is equal to its double dual:*

$$\mathcal{P} \;=\; \mathcal{P}^{\star\star}$$

Let us chose arbitrary behaviors $[\![\alpha]\!]$ for every base type α of linear logic. Take for example $[\![\alpha]\!] = \mathcal{O}_\alpha$. From those, it is possible to build behaviors inductively for every type. Let us explain how it is done for multiplicative and exponential layers of original linear logic.

Multiplicatives. The behavior $[\![1]\!] \subseteq \mathcal{N}_1$ is defined as the double dual of the single construction associated with the corresponding logical connector:

The behavior $[\![\bot]\!] \subseteq \mathcal{N}_\bot$ is defined dually by:

$$[\![\bot]\!] \;:=\; [\![1]\!]^{\star}$$

The behavior $[\![A \otimes B]\!] \subseteq \mathcal{N}_{A \otimes B}$ is defined from behaviors $[\![A]\!] \subseteq \mathcal{N}_A$ and $[\![B]\!] \subseteq \mathcal{N}_B$:

Again, $[\![A \,\invamp\, B]\!] \subseteq \mathcal{N}_{A \invamp B}$ is defined dually:

$$[\![A \,\invamp\, B]\!] \;:=\; [\![A^\perp \otimes B^\perp]\!]^{\star}$$

Exponentials. The behavior $[\![!A]\!] \subseteq \mathcal{N}_{!A}$ is defined from $[\![A]\!] \subseteq \mathcal{N}_A$ as follows:

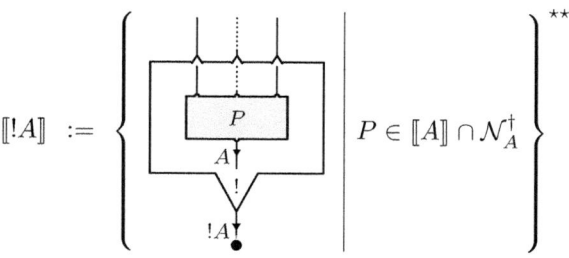

$$[\![!A]\!] \;:=\; \left\{ \;\; \middle| \;\; P \in [\![A]\!] \cap \mathcal{N}_A^\dagger \right\}^{\star\star}$$

And $[\![?A]\!] \subseteq \mathcal{N}_{?A}$ is defined by:

$$[\![?A]\!] \;:=\; [\![!A^\perp]\!]^\star$$

3 Weak Normalisation of Linear Logic

We now chose for $\mathcal{O}$ the set of all weakly normalizing nets. Then, when $P \in \mathcal{N}_A$ and $Q \in \mathcal{N}_{A^\perp}$, $P \star Q$ means that $Cut_A(P, Q)$ reduces to a normal form. Also, $Axiom_A$ being obviously a weakly normalizing net, we can deduce from the definition of duality that $\mathcal{O}_{A^\perp}{}^\star \subseteq \mathcal{O}_A$.

3.1 Reducibility Candidates in **LL**

Definition 5 (Reducibility candidate). *A* reducibility candidate *is a behavior* $\mathcal{P}$ *such that:*

$$\mathcal{O}_{A^\perp}{}^\star \;\subseteq\; \mathcal{P} \;=\; \mathcal{P}^{\star\star} \;\subseteq\; \mathcal{O}_A$$

Lemma 1. *The set of weakly normalizing nets being chosen as observable, for every type A, behavior $[\![A]\!]$ previously defined is a reducibility candidate.*

Proof. The required inclusions hold for base types. Those spread inductively to behaviors built with positive connectors (i.e. 1, $\otimes$, $!$), because the various constructions involved in their definition (including double dual closures) preserve weak normalization. Building behaviors for types with negative connectors (i.e. $\perp$, $\invamp$, $?$) involves duality, but according to the first statement in Property 1, the sets obtained satisfy the same inclusions.

3.2 Reducibility

Definition 6 (Net reducibility). *A net P with interface $\vdash A_1{}^\perp, \ldots, A_n{}^\perp$ is reducible when for every family of nets $(N_i)_{i \in [\![1,n]\!]}$ such that $N_i \in [\![A_i]\!]$, we have:*

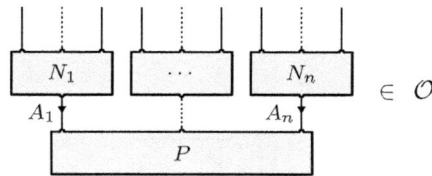

$$\in \mathcal{O}$$

The pictured net will also be denoted shortly as $Cuts(P; N_1, \ldots, N_n)$, sometimes abbreviated further using vectorial notation as $Cuts(P; \boldsymbol{N})$.

Definition 7 (Pointed net reducibility). *A pointed net P with interface $\vdash$ A; $A_1{}^\perp, \ldots, A_n{}^\perp$ is reducible when for every family of nets $(N_i)_{i \in [\![1,n]\!]}$ such that $N_i \in [\![A_i]\!]$, we have:*

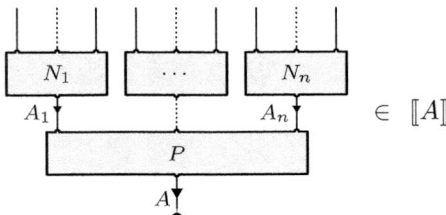

$\in [\![A]\!]$

This last net will be denoted using the same notation $Cuts(P; N_1, \ldots, N_n)$. Notice that when P is a pointed net, no net is plugged on the distinguished port.

Property 3. Reducibility of a net N is equivalent to reducibility of any pointed net obtained from N choosing one of its ports as distinguished.

The weak normalization theorem for *Linear Logic* will follow quickly from the next lemma.

Lemma 2. *Every* LL net *is reducible.*

Proof. By induction on the net N we consider:

(Axiom). We must show that for all $N_1 \in [\![A]\!]$ and $N_2 \in [\![A^\perp]\!]$ the net $N' = Cuts(Axiom_A; N_1, N_2)$ is normalizing. This net is simply $Cut_A(N_1, N_2)$, which normalizes because $[\![A^\perp]\!] = [\![A]\!]^\star$, and therefore $N_1 \star N_2$.

(Cut). We must show that for all reducible $R_1 \in [\![A]\!]$ and $R_2 \in [\![A^\perp]\!]$, the net $N' = Cuts(Cut(R_1, R_2); \mathbf{N})$ normalizes. Quantification over pointed nets N_i is universal, they are supposed to be in their respective candidates: $N_i \in [\![A_i]\!]$ (this kind of quantification will be left implicit in the rest of the proof). In this case, since $Cuts(R_1; \mathbf{N}) \in [\![A]\!]$ and $Cuts(R_2; \mathbf{N}) \in [\![A^\perp]\!]$ the normalization of N' follows.

(Unit). When N is of this form:

The associated pointed net will be denoted as $Unit_\bullet$. Using Property 3, reducibility of N can be written $Unit_\bullet \in [\![1]\!]$, but this holds by definition.

(Co-unit). When N is of the following form, and assuming reducibility of P by induction hypothesis:

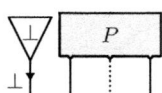

We use $\mathbf{N}^\bullet$ to denote the associated pointed net, the reducibility of N can be written $Cuts(N^\bullet; \mathbf{N}) \in [\![\perp]\!]$. Given that $[\![\perp]\!] = \{ Unit_\bullet \}^\star$, this assertion holds when the net $Cuts(N; Unit_\bullet, \boldsymbol{N})$ normalizes. This net reduces to $Cuts(P; \boldsymbol{N})$ which normalizes by reducibility of P.

(Tensor). When N is of the following form, and assuming reducibility of P and Q as induction hypothesis:

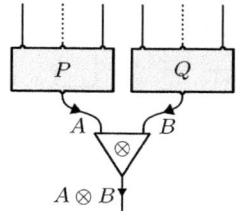

We use $Tensor_\bullet(P, Q)$ to denote the associated pointed net. Reducibility of N can be written $Cuts(Tensor_\bullet(P, Q), \boldsymbol{N}) \in [\![A \otimes B]\!]$, and holds by definition of $[\![A \otimes B]\!]$ because $Cuts(P; \boldsymbol{N}) \in [\![A]\!]$ and $Cuts(Q; \boldsymbol{N}) \in [\![B]\!]$ by reducibility of P and Q.

(Par). When N is of the following form, and assuming reducibility of P as induction hypothesis:

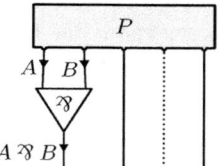

Reducibility of N can be written $Cuts(N^\bullet; \boldsymbol{N}) \in [\![A \,\math14 B]\!]$. Given that $[\![A \,\math14 B]\!] = \{ Tensor_\bullet(R_1, R_2) \mid R_1 \in [\![A^\perp]\!], R_2 \in [\![B^\perp]\!] \}^\star$, this assertion holds when for all $R_1 \in [\![A^\perp]\!]$ and $R_2 \in [\![B^\perp]\!]$, the net $Cuts(N; Tensor_\bullet(R_1, R_2), \boldsymbol{N})$ normalizes. But, the latter reduces to $Cuts(P; R_1, R_2, \boldsymbol{N})$ which normalizes by reducibility of P.

(Promotion). When N is of the following form, and assuming reducibility of P as induction hypothesis:

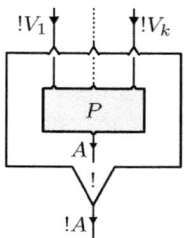

This net is also denoted as $Prom(P)$. Its reducibility can be written: for all $N_0 \in [\![!A]\!]^\star$, for all $N_i \in [\![?V_i^\perp]\!]^\star$, $Cuts(Prom(P); N_0, \boldsymbol{N}) \in \mathcal{O}$. But this holds as soon as for all $N_0 \in [\![!A]\!]^\star$ and for all $N_i \in [\![?V_i^\perp]\!]^\star$ $(i > 1)$ we have:

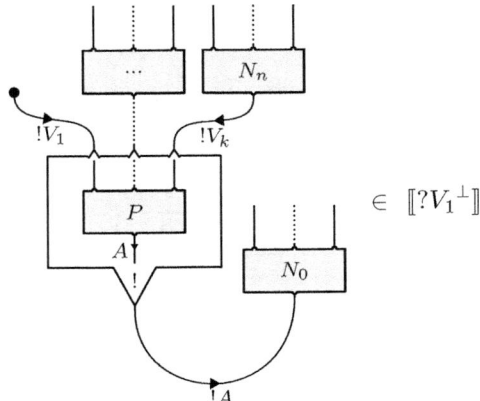

Given that $[\![?V_i^\perp]\!] = \{\, Prom_\bullet(R) \mid R \in [\![V_i]\!] \,\}^\star$, we now need to show that $Cuts($ $Prom(P); N_0, \mathbf{N}) \in \mathcal{O}$, but only for nets $N_1 \in \{\, Prom_\bullet(R) \mid R \in [\![V_1]\!] \,\}$. Let $N_1 = Prom_\bullet(R_1)$ be a net from this set (that is $R_1 \in [\![V_1]\!]$). Let us then consider our net pointed on the second *box input*, and by iteration of this method, our problem boils down to the case where all nets N_i are such that $N_i = Prom_\bullet(R_i)$ for some $R_i \in [\![V_i]\!]$. Finally, we just need to show $Cuts(Prom_\bullet(P); Prom_\bullet(R_i)) \in$ $[\![!A]\!]$. This net reduces by exponential commutation to $Prom_\bullet(P')$ where $P' = Cuts(P; Prom_\bullet(R_i))$. But P being reducible by induction hypothesis, since $R_i \in$ $[\![V_i]\!]$, we obtain $P' \in [\![A]\!]$. Therefore, $Prom_\bullet(P') \in [\![!A]\!]$.

(Dereliction). When N is of the following form, and assuming reducibility of P as induction hypothesis:

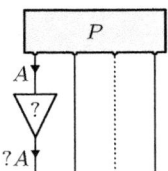

Reducibility of N can be written $Cuts(N^\bullet; \mathbf{N}) \in [\![?A]\!]$. Given that $[\![?A]\!] = \{\, Prom_\bullet(R) \mid R \in [\![A^\perp]\!] \,\}^\star$, this assertion holds as soon as for all $R \in [\![A^\perp]\!]$, the net $Cuts(N; Prom_\bullet(R), \mathbf{N})$ normalizes. But this net reduces to $Cuts(P; R, \mathbf{N})$, which normalizes by reducibility of P.

(Weakening). Idem, the net reduces to $Cuts(P; \mathbf{N})$ plus some *weakenings*.

(Contraction). Idem, the net reduces to $Cuts(P; Prom_\bullet(R), Prom_\bullet(R), \mathbf{N})$ plus some *contractions*.

Theorem 1. *Reduction of LL nets is weakly normalizing.*

Proof. It is a corollary of Lemma 2: every net N is reducible, and we can instantiate the N_i by axioms in the definition of reducibility.

4   Weak Normalisation of Full Differential Linear Logic

We now come to the actual contribution of this paper and extend the proof of the previous theorem to differential constructions.

4.1   Behaviors and Reducibility Candidates in *Full-DiLL*

In presence of differential constructions, behaviors for exponential types must be built differently. First, for every set $\mathcal{S}_A \subseteq \mathcal{N}_A$, we define inductively a set $T(\mathcal{S}_A)$ as the smallest subset of $\mathcal{N}_{!A}$ such that:

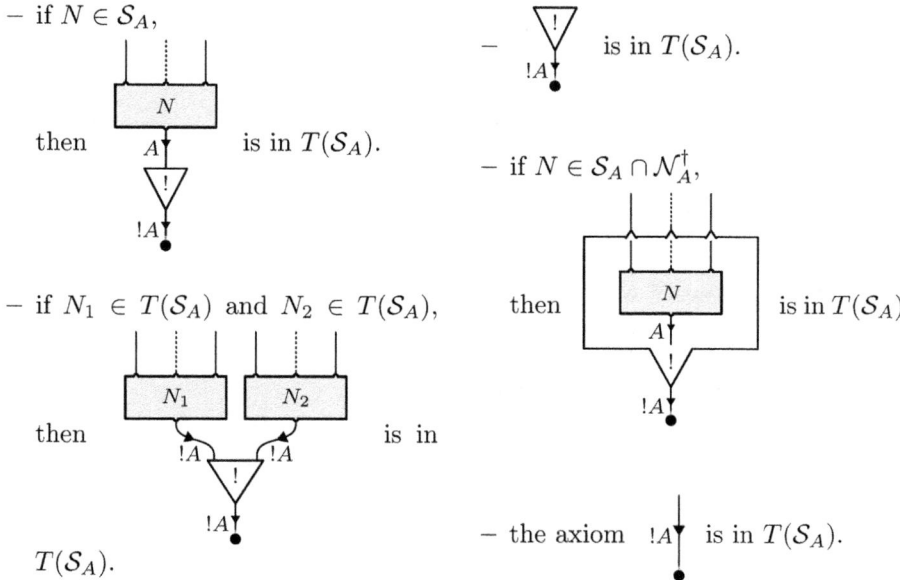

The set $T(\mathcal{S}_A)$ contains nets alike to the one given below, where some *co-derelictions* and *promotions* are found behind a *co-contraction* tree of arbitrary arity. Informally, one could say that going from $\mathcal{S}_A$ to $T(\mathcal{S}_A)$ consists in applying a macro-construction constituted of several exponential operators, but restricted to a single exponential layer.

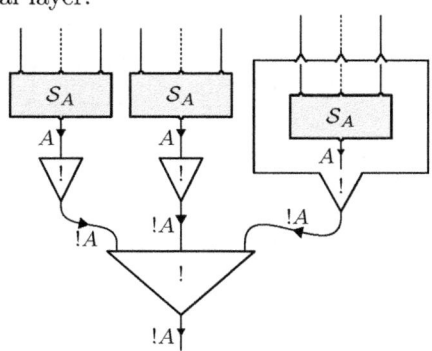

Let $[\![!A]\!] := (T([\![A]\!]))^{\star\star}$ be the new inductive way to define exponential be-haviors (and reducibility candidates) in *Full-DiLL*. It replaces the one given for *LL* in 2.2. Notice that in a differential setting, nets belonging to all those sets we consider are generalized nets (formal sums of simple nets).

4.2   Reduction of Exponential Macro-Constructions

We use rectangular cells labeled ‹ ? › to represent trees of *weakening* and *con-traction* nodes directed upwards:

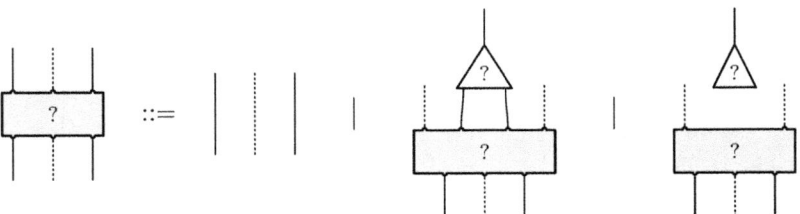

In this definition, some links might not be connected to any nodes. Those links are not required to have exponential types.

Property 4. For every $N \in T(\mathcal{S}_A)$,

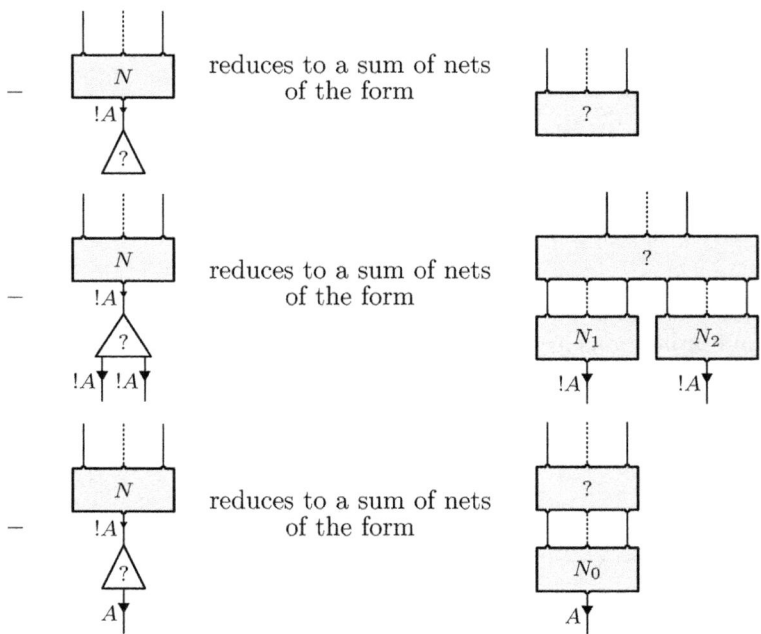

where $N_1, N_2 \in T(\mathcal{S}_A)$ and $N_0 \in \mathcal{S}_A$ (those nets might be different for each member of the sums obtained).

Proof. A safe method is to prove all the three cases simultaneously by induction on N.

4.3   Reducibility

Net reducibility is still expressed by the two properties that were given in Definition 6 and Definition 7. Those definitions would be slightly more complex if second order was explicitly considered (details can be found in [4]). However, this would not interfere with our following developments. As expected, a difficulty to prove normalization of our system comes from the *chain rule* given in 1.2.

Lemma 3. *Take any reducible net R with interface $\vdash A\,;?V_1{}^{\perp},\dots,?V_k{}^{\perp},?U_1,$
$\dots,?U_l$ and let $N_1 \in [\![V_1]\!]$, $\dots$, $N_k \in [\![V_k]\!]$, we have:*

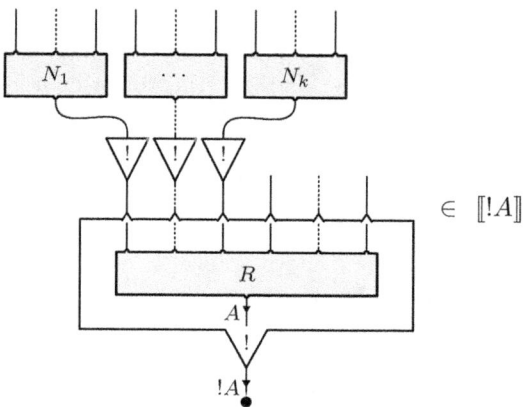

$$\in \ [\![!A]\!]$$

Proof. Recursively on k. No difficulty when $k = 0$. When $k > 0$, we assume that the property holds for any $k' < k$ (generalized recurrence hypothesis). Applying the *chain rule* to deal with the interaction between one *co-dereliction* and the box, and then solving the interactions between generated *contractions* and remaining *co-derelictions*, bring us to a sum of several nets for which the recursion hypothesis applies (the number of *co-derelictions* was diminished strictly below k). Thanks to this hypothesis, one can show that each net of the sum is in $[\![!A]\!]$.

This statement can be generalized to settings where arbitrary exponential constructs face the inputs of a *promotion* box.

Lemma 4. *Let R be a reducible net with interface $\vdash A\,;?V_1{}^{\perp},\dots,?V_k{}^{\perp}$ and let $N_1 \in T([\![V_1]\!])$, $\dots$, $N_k \in T([\![V_k]\!])$, we have:*

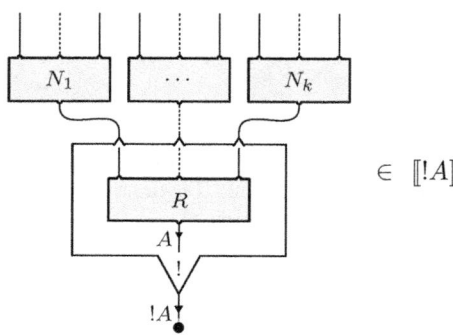

$$\in \ [\![!A]\!]$$

Proof. Again, we will reduce this net. Though, to avoid complications we need to use a specific strategy. Two cases can occur: *(i)* every N_i is a *co-dereliction* or an *axiom*. *(ii)* one of the N_i is not a *co-dereliction* or an *axiom*. The first case is dealt with using Lemma 3. The second case reduces to the first one using exponential commutations: a *co-weakening* or an other *box* will enter the main box in one step, and similarly a *co-contraction* will, but in several steps (the operation needs to be iterated for both nets behind it). Since by global hypothesis all those constructs are in their respective candidates, we can show that the net obtained inside the box after reduction remains reducible, and we actually fall back to the first case.

The inductive definition of $T(\mathcal{S}_A)$ has been chosen so as to obtain this lemma. It is essential in the proof of the following theorem which is the main result of this paper. The fact that we used a specific reduction strategy in the proof of this lemma means that some more work would be needed if we were to prove strong normalisation of the system, as it was done by *Pagani* and *Tortora de Falco* for *LL* in [8]. It is presently not clear whether this would be the only point that would raise difficulties.

Theorem 2. *Every* Full-DiLL net *is reducible.*

Proof. By induction on the net N:

(Axiom, Cut and Multiplicatives). Same as in the proof of Lemma 2.

(Co-weakening). When N is a simple *co-weakening*:

N is part of the syntax $T(\llbracket A \rrbracket)$ and therefore it belongs to $\llbracket !A \rrbracket$ by definition.

(Co-dereliction). When N is a *co-dereliction* construct, assuming $P \in \llbracket A \rrbracket$ by induction hypothesis, we obtain again:

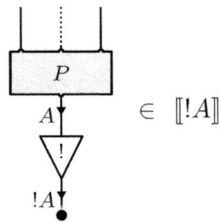

as a direct consequence of the fact that this net belongs to $T(\llbracket A \rrbracket)$.

(Co-contraction). The *co-contraction* raises an issue similar to the one of *promotion* in the proof of Theorem 2. To deal with this case, it is sufficient to show that when $P_1 \in \llbracket !A \rrbracket$, $P_2 \in \llbracket !A \rrbracket$ and $N_0 \in \llbracket !A \rrbracket^\star$ the following net normalizes:

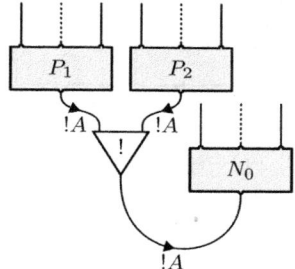

In other words, we have to show that:

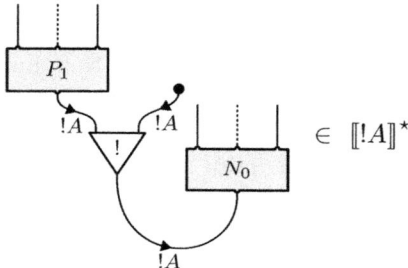

$$\in \ [\![!A]\!]^\star$$

That is to say, by definition of $[\![!A]\!]$, that the following net normalizes for all P_2 $\in T([\![A]\!])$:

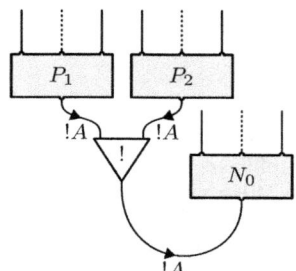

By the same argument, we need to prove this only for $P_1 \in T([\![A]\!])$. Finally, we are reduced to showing that for all $P_1 \in T([\![A]\!])$ and $P_2 \in T([\![A]\!])$:

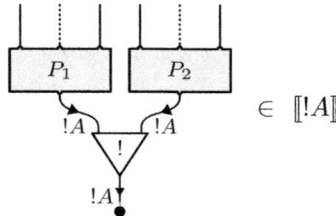

$$\in \ [\![!A]\!]$$

and this is a direct consequence of the definition of $[\![!A]\!]$.

(Promotion). When P is reducible and $N_1 \in [\![!V_1]\!], \ldots, N_k \in [\![!V_k]\!]$, we want:

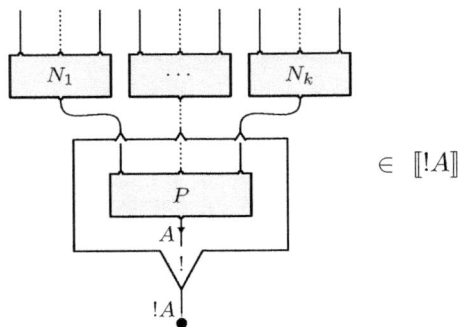

$$\in \quad [\![!A]\!]$$

Always using the same argument, everything boils down to cases where $N_1 \in T([\![V_1]\!]), \ldots, N_k \in T([\![V_k]\!])$, and Lemma 4 can be applied.

(Weakening, Dereliction, Contraction). Nets built with these three constructions interact with exponential macro-constructions in a normalizing way. Property 4 is used to deal with these cases.

(Sums of nets). Let J be a finite set, and for every $j \in J$, let P_j be a reducible net with interface $\vdash A_1^{\perp}, \ldots, A_n^{\perp}$. We want to show that $\sum_{j \in J} P_j$ is also reducible. In fact, we know that for every family $N_i \in [\![A_i]\!]$, the net $Cuts(\sum_{j \in J} P_j; \boldsymbol{N}) = \sum_{j \in J} Cuts(P_j; \boldsymbol{N})$ normalizes because for every $j \in J$, by reducibility of P_j, we know that $Cuts(P_j; \boldsymbol{N})$ normalizes.

Acknowledgments. I want to thank *Thomas Ehrhard* for his constant support, *Christine Tasson* who also helped in proof-reading this paper, and *Paolo Tranquilli* for the fruitful discussions we had some time ago.

References

1. Ehrhard, T., Regnier, L.: The differential lambda-calculus. In: TCS, vol. 309, pp. 1–41. Elsevier Science Publishers Ltd., Amsterdam (2004)
2. Ehrhard, T., Regnier, L.: Differential interaction nets. In: TCS, vol. 364, pp. 166–195. Elsevier Science Publishers Ltd., Amsterdam (2006)
3. Ehrhard, T., Regnier, L.: Uniformity and the Taylor expansion of ordinary lambda-terms. In: TCS, vol. 403, pp. 347–372. Elsevier Science Publishers Ltd., Amsterdam (2008)
4. Girard, J.-Y.: Linear logic. Theoretical Computer Science 50, 1–102 (1987)
5. Lafont, Y.: Interaction nets. In: Principles of Programming Languages, pp. 95–108. ACM, New York (1990)
6. Lafont, Y.: From proof nets to interaction nets. In: Girard, J.-Y., Lafont, Y., Regnier, L. (eds.) Advances in Linear Logic, pp. 225–247. Cambridge University Press, Cambridge (1995)
7. Pagani, M.: The Cut-Elimination Thereom for Differential Nets with Boxes. In: Curien, P.-L. (ed.) TLCA 2009. LNCS, vol. 5608, pp. 219–233. Springer, Heidelberg (2009)
8. Pagani, M., de Falco, L.T.: Strong Normalization Property for Second Order Linear Logic. Theoretical Computer Science 411(2), 410–444 (2010)
9. Tranquilli, P.: Intuitionistic differential nets and lambda-calculus. Theoretical Computer Science 412(20), 1979–1997 (2011)

Controlling Program Extraction in Light Logics

Marc Lasson

ENS Lyon, Université de Lyon, LIP*

marc.lasson@ens-lyon.org

Abstract. We present two refinements, based on program extraction in elementary affine logic and light affine logic, of Krivine & Leivant's system FA$_2$. This system allows higher-order equations to specify the computational content of extracted programs. The user can then prove a generic formula, using these equations as axioms. The system guarantees that the extracted program satisfies the specification and is computable in elementary time (for elementary affine logic) or polynomial time (for light affine logic). Finally, we show that both systems are complete with respect to elementary time and polynomial time functions.

Introduction

Light Linear Logic and Elementary Linear Logic are both variants of linear logic introduced by Jean-Yves Girard in [5] that characterize, through the Curry-Howard correspondence, respectively the class of polynomially computable functions and elementary recursive functions.

There are two usual ways to program in implicit complexity systems based on light logics: either as a type system for a λ-calculus or by extracting programs from proofs in a sequent calculus (see [4] for instance).

The former is used for propositional fragments of Elementary Affine Logic in [9] and of Light Affine Logic in [2]. However, when the programmer provides a λ-term which is not typable, he does not get any feedback helping him to write a typable term implementing the same function. On the other hand, in the latter approach the programmer must keep in mind the underlying computational behaviour of his function during the proof and check later, by external arguments, that the extracted λ-term implements the desired function.

In this paper, we describe a system in which we try to make the second approach a bit more practical. In our system, proofs of the same formula are extracted to extensionally equivalent terms; and programs automatically satisfy the given specification used as axioms during the proof.

FA$_2$ is an intuitionistic second-order logic whose formulas are built upon first-order terms, predicate variables, arrows and two kinds of quantifiers, one on first-order variables and the other on predicate variables. Jean-Louis Krivine & Michel Parigot [6, 7] and Daniel Leivant [8] described a methodology to use this

---

* UMR 5668 CNRS ENS Lyon UCBL INRIA

Work partially supported by project ANR-08-BLANC-0211-01 COMPLICE.

L. Ong (Ed.): TLCA 2011, LNCS 6690, pp. 123–137, 2011.

system for programming with proofs. In this system, the induction principle for integers may be expressed by

$$\forall X, (\forall y, X\,y \Rightarrow X\,(\mathsf{s}\,y)) \Rightarrow X\,0 \Rightarrow X\,x.$$

This formula is written $N[x]$ and it is used to represent integers. The programmer then gives some specifications of a function. For instance for the addition, he may give:

$$\mathrm{plus}(0, y) = y \qquad \text{and} \qquad \mathrm{plus}(s(x), y) = s(\mathrm{plus}(x, y)).$$

Now, if he finds a proof of $\forall x\,y, N[x] \Rightarrow N[y] \Rightarrow N[\mathrm{plus}(x, y)]$ in which he is allowed to rewrite formulas with the specifications, then it is proved that the λ-term extracted from this proof is a program satisfying the specifications. We have adapted the system FA_2 of Leivant and Krivine in two directions:

- We replace the grammar of first-order terms by the whole λ-calculus. We can then extract higher-order functions instead of purely arithmetical functions.
- We ensure complexity bounds by restricting the contraction in its logic.

In the next section, we introduce the grammar for our formulas and describe how we interpret them. In Section 2, we present an intuitionistic proof system $\vdash_{\overline{\lambda}_{LJ}}$ and we show how it can be used for programming. This logic has already been studied in a few papers (mainly to study parametricity theories) [13, 12, 3]. In Section 3, we present an elementary restriction of $\vdash_{\overline{\lambda}_{LJ}}$ that we call $\vdash_{\overline{\lambda}_{EAL}}$ which is to $\vdash_{\overline{\lambda}_{LJ}}$ what (intuitionistic) Elementary Affine Logic (henceforth EAL) is to system $\mathcal{F}$. We show that all functions implementable in $\vdash_{\overline{\lambda}_{EAL}}$ are elementary recursive by collapsing our system back to EAL. Then we give two proofs that $\vdash_{\overline{\lambda}_{EAL}}$ is complete in the sense that all elementary recursive functions are implementable in $\vdash_{\overline{\lambda}_{EAL}}$: one by using the completeness of EAL and the other by invoking, like in [4], Kalmar's characterization of elementary functions. The main interest of the second proof is to illustrate how to program in our systems. Finally we present in Section 4 a polynomial restriction of $\vdash_{\overline{\lambda}_{LJ}}$ that we call $\vdash_{\overline{\lambda}_{LAL}}$ which is to $\vdash_{\overline{\lambda}_{LJ}}$ what (intuitionistic) Light Affine Logic is to system $\mathcal{F}$. And we prove that $\vdash_{\overline{\lambda}_{LAL}}$ is correct and complete with respect to polynomial time using the same techniques used in Section 3.

1   Types, First-Order Terms and Formulas

We assume for the rest of this document that we have three countably infinite disjoint sets of variables. The set of so-called *type variables* whose elements are denoted with letters from the beginning of the Greek alphabet (ie. α, β, ...). The set of *first-order variables* whose elements are denoted with letters from the end of the Latin alphabet (ie. x, y, z). Finally, the set of *second-order variables* whose elements are denoted with uppercase letters from the end of the Latin alphabet (ie. X, Y, Z).

We also assume that we have an injection of second-order variables into type variables and write α_X for the image of a variable X by this injection. This will be useful later when we will send formulas onto system $\mathcal{F}$ types by a forgetful projection $\cdot \mapsto [\![\cdot]\!]$.

Definition 1. *The following grammars define the terms of the system:*

1. *Types are system $\mathcal{F}$ types:*

$$\tau, \sigma, \ldots \; := \; \alpha \mid \forall \alpha, \tau \mid \sigma \to \tau$$

2. *First-order terms are Church-style λ-calculus terms:*

$$s, t, \ldots \; := \; x \mid (s\,t) \mid (t\,\tau) \mid \lambda x : \tau.t \mid \Lambda \alpha.t$$

3. *Finally, second-order formulas are given by the following grammar:*

$$P, Q, \ldots \; := \; X\,t_1\,t_2\ldots t_n \mid P \to Q \mid \forall X : [\tau_1, .., \tau_n],\, P \mid \forall x : \tau,\, P \mid \forall \alpha,\, P$$

These grammars describe terms that will be used in this paper, λ, Λ and the three different $\forall$ behave as binders like in usual calculi. We always consider terms up to α-equivalence and we do not bother with capture problems. We also admit we have six notions of substitution which we assume to be well-behaved with regard to the α-equivalence: $\tau[\sigma/\alpha]$, $t[\tau/\alpha]$, $t[s/x]$, $P[\tau/\alpha]$, $P[t/x]$, and $P[Q/X\,x_1 \ldots x_n]$.

The last substitution is not very usual (the notation comes from [6]): it replaces occurrences of the form $X\,t_1\,\ldots t_n$ by the formula $Q[t_1/x_n]\ldots[t_n/x_n]$ and it is not defined if P contains occurrences of X of the form $X\,t_1 \ldots t_k$ with $k \neq n$. The type system $\vdash_{\text{OK}}$ guarantees that such occurrences cannot appear in a well-typed formula.

We adopt the usual conventions about parentheses: arrows are right associative (it means that we write $A \to B \to C$ instead of $A \to (B \to C)$) and application is left associative (meaning we write $t_1\,t_2\,t_3$ instead of $(t_1\,t_2)\,t_3$). In the following, we allow ourselves not to write the type of first and second order $\forall$ when they can be guessed from the context. We write $\forall X, F$ instead of $\forall X : [],F$ for quantification over relation of arity zero. And we call (second-order) *propositional formula* any formula with no sub-formula of the form $\forall x : \tau, F$ nor $\forall \alpha, F$ and only second-order quantifier of arity zero.

Example 1. Here are some examples of formulas of interest:

1. Leibniz's equality at type τ between two terms t_1 and t_2 $\forall X : [\tau], X\,t_1 \to X\,t_2$ which we write $t_1 =_\tau t_2$ in the remaining of this document.
2. The induction principle for a natural number x: $\forall X : [\text{nat}], (\forall y : \text{nat}, X\,y \to X\,(\text{s }y)) \to (X\,0 \to X\,x)$ which we write $N[x]$ where **nat** will be the type $\forall \alpha, (\alpha \to \alpha) \to \alpha \to \alpha$ of natural numbers in system $\mathcal{F}$ and where s and 0 are first-order variables.
3. And the extensionality principle

$$\forall \alpha \beta, \forall f\,g : \alpha \to \beta, (\forall x : \alpha, f\,x =_\beta g\,x) \to f =_{\alpha \to \beta} g$$

Definition 2. *A context is an ordered list of elements of the form:*

$$\alpha : \text{Type} \;\; or \;\; x : \tau \;\; or \;\; X : [\tau_1, ..., \tau_n].$$

We write $\Sigma \vdash_{\mathrm{OK}} \tau :$ Type to state that all free type variables of τ occur in Σ, $\Sigma \vdash_{\mathrm{OK}} t : \tau$ to state the term t has type τ in Σ (which is exactly the typability in system $\mathcal{F}$), and finally $\Sigma \vdash_{\mathrm{OK}} P :$ Prop to state that P is a well-typed formula in Σ. The rules of the type system defined by $\vdash_{\mathrm{OK}}$ are straight-forward. They are given in the appendices.

Example 2. Formulas in the previous example are well-typed:

1. $\Gamma, x : \tau, y : \tau \vdash_{\mathrm{OK}} x =_\tau y :$ Prop,
2. $\Gamma, \mathsf{s} : \mathsf{nat} \to \mathsf{nat}, 0 : \mathsf{nat}, x : \mathsf{nat} \vdash_{\mathrm{OK}} N[x] :$ Prop,
3. $\vdash_{\mathrm{OK}} \forall \alpha\, \beta, \forall f\, g : \alpha \to \beta, (\forall x : \alpha, f\, x =_\beta g\, x) \to f =_{\alpha \to \beta} g :$ Prop.

Interpretations in standard models

In this section, we build a small realizability model for our proof system which we will use later to prove correctness with respect to the specification of the extracted proofs. We want the model to satisfy the extensionality principle, because we will need to be able to replace in our proofs higher-order terms by other extensionally equal terms. This construction will be very similar to the first model sketched in Section 3 of [10].

We define the set $\mathcal{P}$ of programs to be the set of untyped λ-terms modulo β-reduction. In the following, we interpret terms in $\mathcal{P}$, types by partial equivalence relations (PER) on $\mathcal{P}$ and second-order variables by sets of elements in $\mathcal{P}$ stable under equivalence.

Definition 3. *Let Γ be a well-typed context. A Γ-model consists of three partial functions. The first one is a map from type variables to PERs, the second is a map from first-order variables to $\mathcal{P}$ and the last one is a map from second-order variables to sets of tuples of programs. The collection of Γ-models is defined recursively below.*

- *If Γ is empty, then the only Γ-model is the one built by three empty maps.*
- *If Γ has the form $\Delta, x : \tau$ and if $\mathcal{M} = (\mathcal{M}_0, \mathcal{M}_1, \mathcal{M}_2)$ is a Δ-model, then for any $t \in \llbracket \tau \rrbracket_\mathcal{M}$, $(\mathcal{M}_0, \mathcal{M}_1[x \mapsto t], \mathcal{M}_2)$ is a Γ-model (in the following, we simply write it $\mathcal{M}[x \mapsto t]$).*
- *If Γ has the form $\Delta, \alpha :$ Type and if $\mathcal{M} = (\mathcal{M}_0, \mathcal{M}_1, \mathcal{M}_2)$ is a Δ-model, then for any PER R, $(\mathcal{M}_0[\alpha \mapsto R], \mathcal{M}_1, \mathcal{M}_2)$ is a Γ-model (we write it $\mathcal{M}[\alpha \mapsto R]$).*
- *If Γ has the form $\Delta, X : [\tau_1, ..., \tau_n]$ and if $\mathcal{M} = (\mathcal{M}_0, \mathcal{M}_1, \mathcal{M}_2)$ is a Δ-model, then for any $E \subseteq \llbracket \tau_1 \rrbracket_\mathcal{M} \times ... \times \llbracket \tau_n \rrbracket_\mathcal{M}$ such that E satisfies the stability condition*

$$\text{If } (t_1, ..., t_n) \in E \wedge t_1 \sim_{\tau_1}^\mathcal{M} t'_1 \wedge ... \wedge t_n \sim_{\tau_n}^\mathcal{M} t'_n, \text{ then } (t'_1, ..., t'_n) \in E$$

$(\mathcal{M}_0, \mathcal{M}_1, \mathcal{M}_2[X \mapsto E])$ *is a Γ-model (we write it $\mathcal{M}[X \mapsto E]$).*

Where $\sim_\tau^\mathcal{M}$ is a partial equivalence relation whose domain is written $\llbracket \tau \rrbracket_\mathcal{M}$ defined recursively on the structure of τ,

- $\sim_\alpha^{\mathcal{M}}$ *is equal to* $\mathcal{M}_0(\alpha)$,
- $\sim_{\sigma \to \tau}^{\mathcal{M}}$ *is defined by* $t_1 \sim_{\sigma \to \tau}^{\mathcal{M}} t_2 \Leftrightarrow \forall s_1 s_2, s_1 \sim_\sigma^{\mathcal{M}} s_2 \Rightarrow (t_1\, s_1) \sim_\tau^{\mathcal{M}} (t_2\, s_2)$,
- $\sim_{\forall \alpha, \tau}^{\mathcal{M}} = \bigcap_{R \text{ is PER}} \sim_\tau^{\mathcal{M}[\alpha \mapsto R]}$.

Intuitively $t_1 \sim_\tau^{\mathcal{M}} t_2$ *means the pure* λ*-terms* t_1 *and* t_2 *are of type* τ *and they are extensionally equivalent.*

We can now define the interpretation $[\![t]\!]_{\mathcal{M}}$ of a first-order term t such that $\Gamma \vdash_{\text{OK}} t : \tau$ in a Γ-model $\mathcal{M}$ to be the pure λ-term obtained by replacing all occurrences of free variables by their interpretation in $\mathcal{M}$ and by erasing type information.

Lemma 1. *If we have* $\Gamma \vdash_{\text{OK}} t : \tau$ *and* $\mathcal{M}$ *a* Γ-model, then $[\![t]\!]_{\mathcal{M}} \in [\![\tau]\!]_{\mathcal{M}}$.

We can now define a classical interpretation of formulas in our models.

Definition 4. *Let* P *be a formula such that* $\Gamma \vdash_{\text{OK}} P : \text{Prop}$ *and* $\mathcal{M}$ *be a* Γ-model.

- $\mathcal{M} \models X\, t_1 \ldots t_n$ *iff* $([\![t_1]\!]_{\mathcal{M}}, \ldots, [\![t_n]\!]_{\mathcal{M}}) \in \mathcal{M}(X)$,
- $\mathcal{M} \models P \to Q$ *iff* $\mathcal{M} \models P$ *implies* $\mathcal{M} \models Q$,
- $\mathcal{M} \models \forall X : [\tau_1, \ldots, \tau_n], P$ *iff for all* $E \subseteq [\![\tau_1]\!]_{\mathcal{M}} \times \ldots \times [\![\tau_n]\!]_{\mathcal{M}}$ *satisfying the stability condition,* $\mathcal{M}[X \mapsto E] \models P$,
- $\mathcal{M} \models \forall x : \tau, P$ *iff for all* $t \in [\![\tau]\!]_M$, $\mathcal{M}[x \mapsto t] \models P$,
- $\mathcal{M} \models \forall \alpha, P$ *iff for all PER* R *on* $\mathcal{P}$, $\mathcal{M}[\alpha \mapsto R] \models P$,

If E *is a set of formulas well-typed in* Γ, *for all* Γ-model $\mathcal{M}$, *we write* $\mathcal{M} \models E$ *to mean that* $\mathcal{M} \models Q$ *for all* $Q \in E$. *And if* T *is another set of formulas well-typed in* Γ, *we write* $T \models_\Gamma E$ *if for all* Γ-model $\mathcal{M}$, $\mathcal{M} \models T$ *implies* $\mathcal{M} \models E$ *(and we write* $T \models_\Gamma P$ *in place of* $T \models_\Gamma \{P\}$*).*

One can check that the Leibniz equality is interpreted by the equivalence in the model. Hence we have the following lemmas.

Lemma 2. *If* $\Gamma \vdash_{\text{OK}} t_1 : \tau$, $\Gamma \vdash_{\text{OK}} t_2 : \tau$ *and* $\mathcal{M}$ *is a* Γ-model, then $\mathcal{M} \models t_1 =_\tau t_2$ $\Leftrightarrow [\![t_1]\!]_{\mathcal{M}} \sim_\tau^{\mathcal{M}} [\![t_2]\!]_{\mathcal{M}}$.

Definition 5. *Suppose we have* $\Gamma \vdash_{\text{OK}} P_1 : \text{Prop}$, $\Gamma \vdash_{\text{OK}} t_1 : \tau$ *and* $\Gamma \vdash_{\text{OK}} t_2 : \tau$, *we say that* $P_1 \xrightarrow{t_1 = t_2} P_2$ *if there exists a formula* Q *such that* $\Gamma, x : \tau \vdash_{\text{OK}} Q : \text{Prop}$, $P_1 \equiv Q[t_1/x]$ *and* $P_2 \equiv Q[t_2/x]$.

Lemma 3
   If $\mathcal{M} \models t_1 =_\tau t_2$ *and* $P_1 \xrightarrow{t_1 = t_2} P_2$ *then* $\mathcal{M} \models P_1 \Rightarrow \mathcal{M} \models P_2$.

These models satisfy the extensionality principle:

Lemma 4

$$\mathcal{M} \models \forall \alpha\, \beta, \forall f\, g : \alpha \to \beta, (\forall x : \alpha, f\, x =_\beta g\, x) \to f =_{\alpha \to \beta} g.$$

The following lemmas state that these models are standard in the sense that $[\![\text{nat}]\!]$ is only inhabited by terms equivalent to a Church numeral.

Lemma 5. $\quad - \mathcal{M} \models \forall x : \boldsymbol{nat}, x =_{nat} x \,\boldsymbol{nat}\, s\, 0$

- $t \in \llbracket \boldsymbol{nat} \rrbracket_{\mathcal{M}}$ *implies that there exists an integer* n *such that* $\llbracket t \rrbracket_{\mathcal{M}} \sim_{nat}^{\mathcal{M}} \lceil n \rceil$ *(where* $\lceil n \rceil$ *is the nth Church numeral).*
- *All naturals satisfy the induction principle:* $\mathcal{M} \models \forall x : \boldsymbol{nat}, N[x]$.

Proof. $\quad -$ It is a consequence of the parametricity of the model (see [10] for the parametricity and [13] for the fact that parametricity implies it).

- $\{x, y | x \sim_{nat}^{\mathcal{M}} \lceil n \rceil \& \lceil n \rceil \sim_{nat}^{\mathcal{M}} y$ for some $n \in \mathbb{N}\}$ is a PER than can be used to prove from $t \sim_{nat}^{\mathcal{M}} t$ that $t\, s\, 0 \sim_{nat}^{\mathcal{M}} \lceil n \rceil$ for some n. Therefore $t \sim_{nat}^{\mathcal{M}} \lceil n \rceil$.
- You can prove by induction on n that for all $t \in \llbracket \text{nat} \rrbracket$, $t \sim_{nat}^{\mathcal{M}} \lceil n \rceil$ implies $\mathcal{M} \models N[t]$. $\qquad \square$

Projecting formulas on types

In order to write the rules of our proof system in the next section, we are going to need a way to project second-order formulas on system $\mathcal{F}$ types. The projection of a formula is going to give us the type of its realizers.

Definition 6. *Given a formula* F, *we define the type* $\lfloor F \rfloor$ *recursively built from* F *in the following way.*

$$\lfloor X\, t_1 \dots t_n \rfloor \equiv \alpha_X \qquad \lfloor A \to B \rfloor \equiv \lfloor A \rfloor \to \lfloor B \rfloor \qquad \lfloor \forall \alpha, F \rfloor \equiv \lfloor F \rfloor$$
$$\lfloor \forall x : \alpha, F \rfloor \equiv \lfloor F \rfloor \qquad \lfloor \forall X : [\tau_1, ..., \tau_n], F \rfloor \equiv \forall \alpha_X, \lfloor F \rfloor$$

For instance, $\lfloor t_1 =_\tau t_2 \rfloor \equiv \forall \alpha, \alpha \to \alpha$ and $\lfloor N[x] \rfloor \equiv \lfloor \forall X : [\text{nat}], (\forall y : \text{nat}, X\, y \to X\, (s\, y)) \to X\, 0 \to X\, x \rfloor \equiv \forall \alpha, (\alpha \to \alpha) \to \alpha \to \alpha \equiv \text{nat}$.

2   The Proof System

Sequents are of the form $\Gamma; \Delta \vdash t : P$ where Γ is a context, Δ is an unordered set of assignments of the form $x : Q$ where t is a first-order term, x a first-order variable and P and Q are formulas. Our system has two parameters: Σ and $\mathbf{H}$.

Σ is a well-typed context with the types of functions we want to implement. In this section and in Section 3, we use the set

$$\Sigma = \{ \, 0 : \text{nat}, s : \text{nat} \to \text{nat}, \text{pred} : \text{nat} \to \text{nat}, \text{mult} : \text{nat} \to \text{nat} \to \text{nat},$$
$$\text{minus} : \text{nat} \to \text{nat} \to \text{nat}, \text{plus} : \text{nat} \to \text{nat} \to \text{nat},$$
$$\text{sum} : (\text{nat} \to \text{nat}) \to \text{nat} \to \text{nat}, \text{prod} : (\text{nat} \to \text{nat}) \to \text{nat} \to \text{nat} \, \}.$$

And $\mathbf{H}$ is a set of equational formulas of the form $\forall x_1 : \tau_1, ..., \forall x_n : \tau_n, t_1 =_\tau t_2$ well-typed in Σ. In this section and in Section 3 we take $\mathbf{H}$ to be equal to:

$$\{0 =_{\text{nat}} \Lambda \alpha. \lambda f : \alpha \to \alpha. x : \alpha. x, \quad s =_{\text{nat} \to \text{nat}} \lambda n : \text{nat}. \Lambda \alpha. \lambda f : \alpha \to \alpha. x : \alpha. n\, \alpha\, f(f\, x),$$
$$\forall x\, y : \text{nat}, \text{plus}\; x\, (s\; y) =_{\text{nat}} s\, (\text{plus}\; x\; y), \quad \forall x : \text{nat}, \text{plus}\; x\; 0 =_{\text{nat}} x,$$
$$\forall x\, y : \text{nat}, \text{mult}\; x\, (s\; y) =_{\text{nat}} \text{plus}\; x\, (\text{mult}\; x\; y), \quad \forall x : \text{nat}, \text{mult}\; x\; 0 =_{\text{nat}} 0,$$
$$\forall x : \text{nat}, \text{pred}\, (s\; x) =_{\text{nat}} x, \quad \text{pred}\; 0 =_{\text{nat}} 0,$$
$$\forall x\, y : \text{nat}, \text{minus}\; x\, (s\; y) =_{\text{nat}} \text{pred}\, (\text{minus}\; x\; y), \quad \forall x : \text{nat}, \text{minus}\; x\; 0 =_{\text{nat}} x,$$
$$\forall x : \text{nat}, \forall f : \text{nat} \to \text{nat}, \text{sum}\; f\, (s\; x) =_{\text{nat}} \text{plus}\, (\text{sum}\; f\; x)\, (f\; x),$$
$$\forall f : \text{nat} \to \text{nat}, \text{sum}\; f\; 0 =_{\text{nat}} 0,$$
$$\forall x : \text{nat}, \forall f : \text{nat} \to \text{nat}, \text{prod}\; f\, (s\; x) =_{\text{nat}} \text{mult}\, (\text{prod}\; f\; x)\, (f\; x),$$
$$\forall f : \text{nat} \to \text{nat}, \text{prod}\; f\; 0 =_{\text{nat}} s\; 0\}.$$

The proof system described by the following rules is called *the core affine λ-system* (where λ is here for recalling that formulas contain higher-order individuals) and will be used by all the systems introduced later.

$$\frac{\Sigma, \Gamma \vdash_{\text{OK}} P : \text{Prop}}{\Gamma; x : P \vdash x : P} \ \text{AXIOM} \qquad \frac{\Gamma; \Delta \vdash t : Q \qquad \Sigma, \Gamma \vdash_{\text{OK}} P : \text{Prop}}{\Gamma; \Delta, x : P \vdash t : Q} \ \substack{\text{WEAKENING} \\ x \notin \Delta}$$

$$\frac{\Gamma; \Delta_1 \vdash t_1 : P \qquad \Gamma; \Delta_2, y : Q \vdash t_2 : R}{\Gamma; \Delta_1, x : P \to Q, \Delta_2 \vdash t_2[(x\,t_1)/y] : R} \ \text{APPLICATION}$$

$$\frac{\Gamma; \Delta, x : P \vdash t : Q}{\Gamma; \Delta \vdash \lambda x : \lfloor P \rfloor.t : P \to Q} \ \text{ABSTRACTION}$$

$$\frac{\Gamma; \Delta_1 \vdash t_1 : P \qquad \Gamma; x : P, \Delta_2 \vdash t_2 : R}{\Gamma; \Delta_1, \Delta_2 \vdash t_2[t_1/x] : R} \ \text{CUT}$$

$$\frac{\Gamma, \alpha : \text{Type}; \Delta \vdash t : P}{\Gamma; \Delta \vdash t : \forall \alpha, P} \ \forall^{\alpha}_{\text{intro}} \qquad \frac{\Gamma; \Delta, x : P[\tau/\alpha] \vdash t : Q \qquad \Sigma, \Gamma \vdash_{\text{OK}} \tau : \text{Type}}{\Gamma; \Delta, x : (\forall \alpha, P) \vdash t : Q} \ \forall^{\alpha}_{\text{elim}}$$

$$\frac{\Gamma; \Delta, x : (P[a/y]) \vdash t : Q \qquad \Sigma, \Gamma \vdash_{\text{OK}} a : \tau}{\Gamma; \Delta x : (\forall y : \tau, P) \vdash t : Q} \ \forall^{1}_{\text{elim}}$$

$$\frac{\Gamma, x : \tau; \Delta \vdash t : P}{\Gamma; \Delta \vdash t : \forall x : \tau, P} \ \forall^{1}_{\text{intro}} \qquad \frac{\Gamma, X : [\tau_1, ..., \tau_n]; \Delta \vdash t : P}{\Gamma; \Delta \vdash (\Lambda \alpha_X.t) : \forall X : [\tau_1, ..., \tau_n], P} \ \forall^{2}_{\text{intro}}$$

$$\frac{\Gamma; \Delta, y : P[Q/X\,x_1...x_n] \vdash t : R \qquad \Sigma, \Gamma, x_1 : \tau_1, ..., x_n : \tau_n \vdash_{\text{OK}} Q : \text{Prop}}{\Gamma; \Delta, x : (\forall X : [\tau_1, ..., \tau_n], P) \vdash t[x \lfloor Q \rfloor/y] : R} \ \forall^{2}_{\text{elim}}$$

$$\mathbf{H} \models_{\Sigma, \Gamma} t_1 =_\tau t_2 \text{ and } P_1 \xrightarrow{t_1 =_\tau t_2} P_2 \quad \frac{\Gamma; \Delta \vdash t : P_1}{\Gamma; \Delta \vdash t : P_2} \ \text{EQUALITY}$$

The core affine λ-system (parametrized by Σ and H)

By adding to the core system the following rule, you obtain the *intuitionistic system* $\vdash_{\lambda_\text{LJ}}$ which is a sequent-calculus presentation of a second-order predicate logic with higher-order individuals. It is essentially the same logic as the one described in [13] or [12].

$$\frac{\Gamma; \Delta, x : P, x : P \vdash_{\lambda_\text{LJ}} t : Q}{\Gamma; \Delta, x : P \vdash_{\lambda_\text{LJ}} t : Q} \ \text{UNRESTRICTED CONTRACTION}$$

The intuitionistic λ-system

Our proof systems are well-behaved with respect to our notion of model.

Lemma 6. *If $\Gamma; x_1 : P_1, ..., x_n : P_n \vdash_{\lambda_\text{LJ}} t : P$, then $\mathbf{H} \cup \{P_1, ..., P_n\} \models_\Gamma P$.*

Proof. The proof consists of an induction on the structure of the proof $\Gamma; x_1 : P_1, ..., x_n : P_n \vdash_{\lambda_\text{LJ}} t : P$ and an intensive use of substitution lemmas. $\square$

The following lemma gives us the type of proof-terms.

Lemma 7. *If $\Gamma; \Delta \vdash t : P$, then $\lfloor \Gamma \rfloor, \lfloor \Delta \rfloor \vdash_{\text{OK}} t : \lfloor P \rfloor$ where $\lfloor \Gamma \rfloor$ is obtained from Γ by replacing occurrences of "$X : [\tau_1, ..., \tau_n]$" by "$\alpha_X : \text{Type}$" and letting others unchanged and $\lfloor \Delta \rfloor = \{x : \lfloor P \rfloor; (x : P) \in \Delta\}$.*

And the converse holds for the propositional fragment of the intuitionistic system (it tells us that the propositional fragment is system $\mathcal{F}$):

Lemma 8. *If Δ only contains propositional formula and P is propositional, $X_1 : [], ..., X_n : []; \Delta \vdash_{\overline{\lambda}\text{LJ}} t : P$ if and only if $\alpha_{X_1} : \text{Type}, ..., \alpha_{X_n} : \text{Type}, \lfloor \Delta \rfloor \vdash t : \lfloor P \rfloor$.*

A simple realizability theory

Definition 7. *Given a formula F and a term t, we can define the formula $t \Vdash F$ by induction on F in the following way.*

$$t \Vdash X\,t_1 \ldots t_n \equiv X\,t_1 \ldots t_n\,t \qquad t \Vdash P \to Q \equiv \forall x : \lfloor P \rfloor, x \Vdash P \to (t\,x) \Vdash Q$$
$$t \Vdash \forall X : [\tau_1, ..., \tau_n], P \equiv \forall \alpha_X, \forall X : [\tau_1, ..., \tau_n, \alpha_X], t\,\alpha_X \Vdash P$$
$$t \Vdash \forall x : \tau, P \equiv \forall x : \tau, t \Vdash P \qquad t \Vdash \forall \alpha, P \equiv \forall \alpha, t \Vdash P$$

Lemma 9. *(Adequacy theorem)*

If $\Gamma; x_1 : P_1, ..., x_n : P_n \vdash_{\overline{\lambda}\text{LJ}} t : P$, then
$$\Gamma, x_1 : \lfloor P_1 \rfloor, ..., x_n : \lfloor P_n \rfloor; x_1 : (x_1 \Vdash P_1), ..., x_n : (x_n \Vdash P_n) \vdash_{\overline{\lambda}\text{LJ}} t : (t \Vdash P).$$

Proof. The result comes easily with an induction on the structure of the proof of $\Gamma; x_1 : P_1, ..., x_n : P_n \vdash t : P$. □

Programming with proofs

Definition 8. *Let $D[x]$ be a formula such that $\Gamma, x : \tau \vdash_{\text{OK}} D[x] : \text{Prop}$ for some τ. We say that $D[x]$ is* data-type *of parameter x of type $\lfloor D \rfloor$ relatively to a Γ-model $\mathcal{M}$ if we have both $\mathcal{M} \models \forall r\,x : \lfloor D \rfloor, (r \Vdash D[x]) \to r =_\tau x$ and $\mathcal{M} \models \forall x : \lfloor D \rfloor, x \Vdash D[x]$ (or equivalently the converse $\forall r\,x : \lfloor D \rfloor, r =_\tau x \to (r \Vdash D[x])$).*

Lemma 10. *$N[x]$ is a data-type in all Σ-models.*

Proof. The proof is similar to the one for FA_2 in [6] or [13]. □

Lemma 11. *If $A[x]$ and $B[y]$ are two data-types in a Γ-model $\mathcal{M}$, so is $F[f] \equiv \forall x : \lfloor A \rfloor, A[x] \to B[f\,x]$.*

Proof. It is a consequence of the definition and extensionality. □

The following theorem states that if we can find a model $\mathcal{M}$ satisfying **H** (informally it means that we know our specifications to be implementable), then the program extracted from any proof of a data-type is equal, in the model, to the parameter of the data-type.

Theorem 1. *If $D[x]$ is a data-type and $\Gamma; \vdash t : D[x]$ then for all Σ, Γ-model $\mathcal{M}$ such that $\mathcal{M} \models \mathbf{H}$, we have $\mathcal{M} \models x =_{\lfloor D \rfloor} t$.*

Proof. $\Gamma; \vdash t : D[x]$ implies $\Gamma; \vdash t : t \Vdash D[x]$ and then $\mathcal{M} \models t \Vdash D[x]$ and since $D[x]$ is a data-type we conclude $\mathcal{M} \Vdash t =_{\lfloor D \rfloor} x$. $\qquad\qquad\square$

We call a *proof of totality* any proof of a datatype. If π is the program extracted from any proof of totality of plus ($\vdash \pi : \forall x, N[x] \rightarrow \forall y, N[y] \rightarrow N[\text{plus } x\,y]$), then the previous theorem states that π will be equivalent to any instantiation of plus that satisfies the axioms in $\mathbf{H}$.

3   Elementary Time Characterisation

In this section, we are going to show how we can constrain the unrestricted contraction of $\vdash_{\lambda_{\mathrm{LJ}}}$ in order to obtain a light logic which ensures a complexity bound for elementary time. First, we extend the set of formulas with a new modality ! to obtain the following grammar for formulas.

$$P, Q, \ldots := X\, t_1\, t_2 \ldots t_n \mid P \rightarrow Q \mid \forall X : [\tau_1, .., \tau_n], P \mid \forall x : \tau, P \mid \forall \alpha, P \mid !P$$

In order to ease the reading of formulas we write $!^k P$ instead of $!...!P$ with k modalities. We extend the projection on types by ignoring the modality $\lfloor !P \rfloor = \lfloor P \rfloor$. And we obtain the *elementary λ-system* $\vdash_{\lambda_{\mathrm{EAL}}}$ by adding the following two rules to the core system (where $!\Delta$ is the context obtained from Δ by prefixing all formulas with a modality). Note that in this system, the arrow $\rightarrow$ is affine (it is usually written $\multimap$ in the literature).

$$\frac{\Gamma; \Delta, x : !P, x : !P \vdash_{\lambda_{\mathrm{EAL}}} t : Q}{\Gamma; \Delta, x : !P \vdash_{\lambda_{\mathrm{EAL}}} t : Q} \ \text{Contraction} \qquad \frac{\Gamma; \Delta \vdash t : P}{\Gamma; !\Delta \vdash_{\lambda_{\mathrm{EAL}}} t : !P} \ \text{Promotion}$$

The elementary λ-system

The elementary λ-system $\vdash_{\lambda_{\mathrm{EAL}}}$ naturally embeds into the intuitionistic λ-system by forgetting all modalities. For all formula F, we write F^- to denote the formula where we have erased all modalities and $\Delta^- = \{x : P^- | (x : P) \in \Delta\}$.

Lemma 12. *If $\Gamma; \Delta \vdash_{\lambda_{\mathrm{EAL}}} t : P$ then $\Gamma; \Delta^- \vdash_{\lambda_{\mathrm{LJ}}} t : P^-$.*

Correctness

We can bring $\vdash_{\lambda_{\mathrm{EAL}}}$ back to Elementary Affine Logic in order to prove that extracted programs are elementary bounded. First, one should notice that the propositional fragment of $\vdash_{\lambda_{\mathrm{EAL}}}$ is exactly the usual elementary affine logic (for instance the one described in [4]). If Δ is a propositional context and P a propositional formula, we simply write $\Delta \vdash_{\mathrm{EAL}} t : P$ instead of $\Gamma; \Delta \vdash_{\lambda_{\mathrm{EAL}}} t : P$ where $\Gamma = \langle X : []; X \text{ occurring in } \Delta \rangle$.

Definition 9. *Given a formula F, we define the propositional formula F° recursively built from F in the following way.*

$$(X\, t_1 \ldots t_n)^\circ = X \quad (A \to B)^\circ = A^\circ \to B^\circ \qquad (\forall \alpha, F)^\circ = F^\circ$$
$$(\forall x : \alpha, F)^\circ = F^\circ \qquad (!F)^\circ = !F^\circ \qquad (\forall X : [\tau_1, ..., \tau_n], F)^\circ = \forall X, F^\circ$$

And for contexts: $(x_1 : P_1, ..., x_n : P_n)^\circ = x_1 : P_1^\circ, ..., x_n : P_n^\circ$.

Since program extraction concerns only the second-order structure of the proof we have the following lemma.

Lemma 13. *If $\Gamma; \Delta \vdash_{\lambda EAL} t : P$, then $\Delta^\circ \vdash_{EAL} t : P^\circ$.*

Definition 10. *We say that a program $t \in \mathcal{P}$ represents a (set-theoretical) total function f if for all integers $m_1, ..., m_n$, the term $(t \lceil m_1 \rceil ... \lceil m_n \rceil)$ may be normalized to the Church numeral $\lceil f(m_1, ..., m_n) \rceil$.*

We say that $t \in \mathbb{E}$ if it represents a total function f belonging to the set of elementary computable functions.

We denote by $\mathcal{N}[x]$ the following decoration of $N[x]$ with modalities

$$\mathcal{N}[x] \equiv \forall X : [\mathrm{nat}], !(\forall y : \mathrm{nat}, X\, y \to X\,(\mathrm{s}\ y)) \to !(X\, 0 \to X\, x)$$

and simply by $\mathcal{N}$ the formula $\forall X, !(X \to X) \to !(X \to X) = (\mathcal{N}[x])^\circ$.

The following lemma is the classical correction result for EAL (see [4] for a proof for instance).

Lemma 14. *If $\vdash_{EAL} t : !^{k_1}\mathcal{N} \to ... \to !^{k_n}\mathcal{N} \to !^k\mathcal{N}$ then $[\![t]\!] \in \mathbb{E}$ (where $[\![t]\!]$ is the untyped λ-term obtained by erasing type information from t).*

Finally by combining lemmas 12, 13, and 14 we prove the desired correction theorem.

Theorem 2. *(Correction theorem) If we have*

$$\vdash_{\lambda EAL} t_1 : \forall x_1...x_n : \boldsymbol{nat}, !^{k_1}\mathcal{N}[x_1] \to ... \to !^{k_n}\mathcal{N}[x_k] \to !^k\mathcal{N}[t_2\, x_1 ... x_n]$$

then $[\![t_1]\!]^{\mathcal{M}} \in \mathbb{E}$ and moreover we have that $\mathcal{M} \models t_1 =_{nat \to ... \to nat} t_2$ for all $\mathcal{M} \models \mathbf{H}$.

Completeness

In this section we give two proofs of the fact that all elementary recursive functions may be extracted from a proof of totality.

In order to ease the reading of the paper, we omit term annotations (the "$x :$ " in Δ and "$t :$" on the right-hand side of the symbol $\vdash$) since, given a proof tree, these decorations are unique up to renaming of variables. We also let the typing context Γ and proofs of the typing sequents $\vdash_{OK}$ implicit. We use double lines in proof trees to represent multiple easy steps and we also use vertical dots to mean that the proof can be easily completed.

First proof of completeness: using the completeness of EAL. The following theorem gives us a link between typable terms in EAL and provably total functions in our system. And if we assume the completeness of EAL (which is proved in [4]), it gives us directly that all elementary recursive functions may be extracted from a proof of totality.

Theorem 3. *Let t such that $\vdash_{\text{EAL}} t : \mathcal{N} \to ... \to \mathcal{N} \to !^k\mathcal{N}$, then*

$$\vdash_{\lambda\text{EAL}} \forall x_1...x_n : \textbf{nat}, \mathcal{N}[x_1] \to ... \to \mathcal{N}[x_n] \to !^{k+1}\mathcal{N}[t\,x_1\,...\,x_n].$$

Proof. First one should notice that the adequacy theorem is true in $\vdash_{\lambda\text{EAL}}$ (when we define $t \Vdash !F$ by $!(t \Vdash F)$), therefore $\vdash_{\lambda\text{EAL}} t : \mathcal{N} \to ... \to \mathcal{N} \to !^k\mathcal{N}$ implies $\vdash_{\lambda\text{EAL}} (t \Vdash \mathcal{N} \to ... \to \mathcal{N} \to !^k\mathcal{N})$ (*). We are going to need the two simple lemmas below:

1. We have $\vdash_{\lambda\text{EAL}} \forall r, (r \Vdash \mathcal{N}) \to \mathcal{N}[r]$.
   The idea of the proof is that $(r \Vdash \mathcal{N})$ is equal to

   $$\forall \alpha, \forall X : [\alpha], \forall f : \alpha \to \alpha, !(\forall y : \alpha, X\,y \to X\,(f\,y)) \to !(\forall z : \alpha, X\,z \to X\,(r\,\alpha\,f\,z))$$

   and by taking $\alpha = \textbf{nat}$, $f = \textbf{s}$ and $z = 0$, we obtain $\mathcal{N}[r\,\textbf{nat}\,\textbf{s}\,0]$. And since $\mathcal{M} \models \forall x : \textbf{nat}, x\,\textbf{nat}\,\textbf{s}\,0 = x$ we can deduce $\mathcal{N}[r]$.
2. And we have $\vdash_{\lambda\text{EAL}} \forall r, \mathcal{N}[r] \to !(r \Vdash \mathcal{N})$.

$$
\cfrac{
  \cfrac{
    \cfrac{\vdots\ \pi_1}{\vdash \forall y, y \Vdash \mathcal{N} \to (\textbf{s}\,y) \Vdash \mathcal{N}}
    \quad
    \cfrac{\vdots\ \pi_2 \qquad \cfrac{\vdash 0 \Vdash \mathcal{N} \qquad r \Vdash \mathcal{N} \vdash r \Vdash \mathcal{N}}{0 \Vdash \mathcal{N} \to r \Vdash \mathcal{N} \vdash r \Vdash \mathcal{N}}}{}
  }{}
}{}
$$

where π_1 and π_2 use the rule EQUALITY with

$$\textbf{H} \models_\Gamma (0\,\alpha\,f\,z) =_\alpha z \text{ and } \textbf{H} \models_\Gamma (\textbf{s}\,y\,\alpha\,f\,z) =_\alpha (y\,\alpha\,f\,(f\,z))$$

and $\Gamma = \langle y : \textbf{nat}, \alpha : \text{Type}, f : \alpha \to \alpha, z : \alpha \rangle$.

Now to prove the sequent $\vdash \forall x_1...x_n, \mathcal{N}[x_1] \to ... \to \mathcal{N}[x_n] \to !^{k+1}\mathcal{N}[t\,x_1\,...\,x_n]$, it is enough, thanks to 2, to find a proof of $\vdash \forall x_1...x_n, !(x_1 \Vdash \mathcal{N}) \to ... \to !(x_n \Vdash \mathcal{N}) \to !^{k+1}\mathcal{N}[t\,x_1\,...\,x_n]$ or using PROMOTION $\vdash \forall x_1...x_n, \mathcal{N}[x_1] \to ... \to \mathcal{N}[x_n] \to !^k\mathcal{N}[t\,x_1\,...\,x_n]$ and then by invoking 1, we have to prove $\vdash \forall x_1...x_n, (x_1 \Vdash \mathcal{N}) \to ... \to (x_n \Vdash \mathcal{N}) \to (t\,x_1\,...\,x_n) \Vdash !^k\mathcal{N}$ which is equivalent to (*). $\qquad\square$

Second proof of completeness: encoding Kalmar's functions. The characterization due to Kalmar [11] states that elementary recursive functions is the smallest class of functions containing some base functions (constants, projections, addition, multiplication and subtraction) and stable by a composition scheme, by bounded sum and bounded product. In the remaining of the section, we show how we can implement this functions and these schemes in $\vdash_{\lambda\text{EAL}}$.

– It is very easy to find a proof of $\vdash \mathcal{N}[0]$ and a proof $\vdash \forall x, \mathcal{N}[x] \to \mathcal{N}[\mathsf{s}\, x]$. We can obtain a proof $\vdash \mathcal{N}[\mathsf{s}\, 0]$ by composing them.

– The following proof gives us the addition (in order to make it fit the space we divide it in two parts). We use "$x + y$" as a notation for the term (plus $x\, y$).

$$\pi \quad \cfrac{\cfrac{\vdots}{\mathcal{N}[y], !F \vdash\, !(X\, 0 \to X y)} \quad X\, y \to X(x+y), X\, 0 \to X y \vdash X\, 0 \to X(x+y)}{\cfrac{\mathcal{N}[x], \mathcal{N}[y], !F, !F \vdash\, !(X\, 0 \to X(x+y))}{\vdash \forall x\, y : \mathsf{nat}, \mathcal{N}[x] \to \mathcal{N}[y] \to \mathcal{N}[x+y]}}$$

$$\cfrac{\cfrac{\cfrac{\cfrac{\vdots}{!F \vdash\, !(\forall z, X(z+y) \to X(\mathsf{s}\,(z+y)))}}{!F \vdash\, !(\forall z, X(z+y) \to X((\mathsf{s}\, z)+y))} \quad \cfrac{!(X\, y \to X(x+y)) \vdash\, !(X\, y \to X(x+y))}{!(X\, (0+y) \to X(x+y)) \vdash\, !(X\, y \to X(x+y))}}{!(\forall z, X(z+y) \to X((\mathsf{s}\, z)+y)) \to !(X\, (0+y) \to X(x+y)), !F \vdash\, !(X\, y \to X(x+y))}}{\mathcal{N}[x], !F \vdash\, !(X\, y \to X(x+y))}$$
$$\pi$$

Where $F = \forall z, X\, z \to X(\mathsf{s}\, z)$. Note that we have used in the branch below the EQUALITY rule with $\mathbf{H} \models \forall x\, y, (\mathsf{s}\, x)+y = \mathsf{s}\,(x+y)$ and $\mathbf{H} \models \forall y, 0+y = y$. We extract the usual λ-term for addition $\lambda n\, m : \mathsf{nat}.\Lambda\alpha.\lambda f : \alpha \to \alpha.\lambda x : \alpha.n\, f\,(m\, f\, x)$.

– By iterating the addition, it is very easy to find a proof of $\forall x\, y : \mathsf{nat}, \mathcal{N}[x] \to \mathcal{N}[y] \to\, !\mathcal{N}[\mathrm{mult}\, x\, y]$. In order to build the scheme of bounded product, we need to find a proof of $\forall x\, y : \mathsf{nat}, \mathcal{N}[x] \to \mathcal{N}[y] \to \mathcal{N}[\mathrm{mult}\, x\, y]$. Such a proof has been found and checked using a proof assistant based on our system, but it is too big to fit in here. The λ-term extracted from this proof is $\lambda n\, m : \mathsf{nat}.\Lambda\alpha.\lambda f : \alpha \to \alpha.n\, \alpha\,(m\,(\alpha \to \alpha)\,(\lambda g : \alpha \to \alpha.\lambda x : \alpha.f\,(g\, x)))\,(\lambda x : \alpha.x)$.

– We can implement the predecessor by proving $\vdash \forall x, \mathcal{N}[x] \to \mathcal{N}[\mathrm{pred}\, x]$. The proof is not so easy: you have to instantiate the second-order quantifier with $x \mapsto (X\,(\mathrm{pred}\, x) \to X\, x) \otimes X(\mathrm{pred}\, x)$ where $P \otimes Q \equiv \forall X, (P \to Q \to X) \to X$ is the usual second-order encoding of the tensor. It corresponds to a standard technique for implementing the predecessor in λ-calculus.

– Then it is easy to implement the subtraction by proving $\vdash \forall x\, y, \mathcal{N}[x] \to \mathcal{N}[y] \to\, !\mathcal{N}[\mathrm{minus}\, x\, y]$ with the induction principle $\mathcal{N}[y]$.

– By iterating a proof of the totality of the successor on a proof of $\mathcal{N}[0]$, it is easy to prove that $\vdash \forall x, \mathcal{N}[x] \to\, !\mathcal{N}[x]$. Such proof is called a *coercion* and it allow us to replace occurrences of $\mathcal{N}[x]$ at a negative position by $!\mathcal{N}[x]$. Using this we can now bring every proof of totality $\vdash \forall x_1, ..., x_n, !^{k_1}\mathcal{N}[x_1] \to ... \to\, !^{k_n}\mathcal{N}[x_n] \to\, !^k\mathcal{N}[f\, x_1 ... x_n]$ to a "normal form" $\vdash \forall x_1, ..., x_n, \mathcal{N}[x_1] \to ... \to \mathcal{N}[x_n] \to\, !^k\mathcal{N}[f\, x_1 ... x_n]$.

– Using CONTRACTIONS, one can prove that the following rules with $p+1$ premisses is admissible (where $s = \sum_{i=1}^{q} k_i$ and $g\, \overline{x}$ stands for $g\, x_1 ... x_q$). It does encode the composition.

$$\cfrac{\overset{\text{proof for } g_i}{\mathcal{N}[x_1], ..., \mathcal{N}[x_q] \vdash\, !^{k_i}\mathcal{N}[g_i\, \overline{x}]} \quad \overset{\text{proof for } f}{\mathcal{N}[g_1\, \overline{x}], ..., \mathcal{N}[g_p\, \overline{x}] \vdash\, !^k\mathcal{N}[f\,(g_1\, \overline{x})...(g_p\, \overline{x})]}}{\vdash \forall x_1 ... x_n, \mathcal{N}[x_1] \to ... \to \mathcal{N}[x_q] \to\, !^{s+k+1}\mathcal{N}[f\,(g_1\, x_1 ... x_q)...(g_p\, x_1 ... x_q)]}$$

− The bounded sum is implemented by a proof of $!!(\forall y, \mathcal{N}[y] \to {!}^k \mathcal{N}[f\,y]) \to \forall n, \mathcal{N}[n] \to {!}^{k+2}\mathcal{N}[\text{sum}\,f\,n]$. The key idea in this proof is to use the induction principle of $\mathcal{N}[n]$ with the predicate $x \mapsto \mathcal{N}[x] \otimes {!}^k \mathcal{N}[\text{sum}\,f\,x]$. Let H be the formula $\forall y, \mathcal{N}[y] \to {!}^k \mathcal{N}[f\,y]$ and K_1 be the formula

$$\forall y, !(\mathcal{N}[y] \otimes {!}^k \mathcal{N}[\text{sum}\,fy]) \to !(\mathcal{N}[\mathbf{s}\,y] \otimes {!}^k \mathcal{N}[\text{sum}\,f\,(\mathbf{s}\,y)])$$

and K_2 the formula $!(\mathcal{N}[0] \otimes {!}^k \mathcal{N}[\text{sum}\,f\,0]) \to !(\mathcal{N}[n] \otimes {!}^k \mathcal{N}[\text{sum}\,f\,n])$.

$$\frac{\dfrac{\dfrac{\vdots\ \dot{\pi}}{\dfrac{!!H \vdash !K_1}{\dfrac{K_1 \to K_2, !!H \vdash !K_2}{\mathcal{N}[n], !!H \vdash !K_2}}} \qquad \dfrac{\vdots}{K_2 \vdash {!}^{k+1}\mathcal{N}[\text{sum}\,f\,n]}}{!!H, \mathcal{N}[n] \vdash {!}^{k+2}\mathcal{N}[\text{sum}\,f\,n]}}{\vdash !!(\forall y, \mathcal{N}[y] \to {!}^k \mathcal{N}[f\,y]) \to \forall n, \mathcal{N}[n] \to {!}^{k+2}\mathcal{N}[\text{sum}\,f\,n]}$$

Where π uses a proof of the totality of the addition $\vdash \forall xy, \mathcal{N}[x] \to \mathcal{N}[y] \to \mathcal{N}[\text{plus}\,x\,y]$ and a proof of $\vdash \mathcal{N}[0]$.
− Finally, we obtain the bounded product by replacing proofs for zeros by proof for ones and the proof for addition by a proof for multiplication.

4 Polynomial-Time Characterisation

In this final section, we show that it is possible to have the same approach for polynomial time. We are therefore going to add more constraints in order to obtain a polynomial bound on the time complexity. In this section, we work with binary words instead of Church numerals, so we take $\mathbf{\Sigma} = \{\mathtt{I} : \mathtt{wrd} \to \mathtt{wrd}; \mathtt{O} : \mathtt{wrd} \to \mathtt{wrd}; \varepsilon : \mathtt{wrd}\}$ and

$$\begin{aligned}\mathbf{H} = \{\ &\mathtt{I} =_{\mathtt{wrd}\to\mathtt{wrd}} \lambda w : \mathtt{wrd}.\varLambda\alpha\lambda(f : \alpha \to \alpha)(g : \alpha \to \alpha)(x : \alpha).f\,(w\,\alpha\,f\,g\,x),\\ &\mathtt{O} =_{\mathtt{wrd}\to\mathtt{wrd}} \lambda w : \mathtt{wrd}.\varLambda\alpha\lambda(f : \alpha \to \alpha)(g : \alpha \to \alpha)(x : \alpha).g\,(w\,\alpha\,f\,g\,x),\\ &\varepsilon =_{\mathtt{wrd}} \varLambda\alpha\lambda(f : \alpha \to \alpha)(g : \alpha \to \alpha)(x : \alpha).x\}\end{aligned}$$

where $\mathtt{wrd} = \forall\alpha, (\alpha \to \alpha) \to (\alpha \to \alpha) \to \alpha \to \alpha$ is a type for representing binary words for instance $\lceil 0100 \rceil = \mathtt{O}\,(\mathtt{I}\,(\mathtt{O}\,(\mathtt{O}\,\varepsilon)))$. As for integers, there is an induction principle $W[w]$

$$\forall X : [\mathtt{wrd}], (\forall y : \mathtt{wrd}, X\,y \to X\,(\mathtt{O}\,y)) \to (\forall y : \mathtt{wrd}, X\,y \to X\,(\mathtt{I}\,y)) \to X\,\varepsilon \to X\,w$$

associated to $\mathtt{wrd}$. And one can prove that $W[w]$ is a data-type (the proof is similar to that of Lemma 10 showing that $N[x]$ is a data-type).

We now extend the set of formulas with another modality $\S$ to obtain the following grammar for formulas.

$$P, Q, ... := X\,t_1\,t_2\,...\,t_n \mid P \to Q \mid \forall X : [\tau_1, .., \tau_n], P \mid \forall x : \tau, P \mid \forall\alpha, P \mid !P \mid \S P$$

We extend the projection $\lfloor F \rfloor$ to the new modality by ignoring it $\lfloor \S F \rfloor = \lfloor F \rfloor$. We build the *light affine λ-system* $\vdash_{\text{LAL}}$ by adding the following three rules to the core affine system (compared to $\vdash_{\text{EAL}}$ we restricted the promotion rule to at most one hypothesis on the left side and we added a promotion for dealing with $\S$).

$$\frac{\Gamma; x : P \vdash_{\overline{\lambda}\text{LAL}} t : Q}{\Gamma; x : !P \vdash_{\overline{\lambda}\text{LAL}} t : !Q} \qquad \frac{\Gamma; \emptyset \vdash_{\overline{\lambda}\text{LAL}} t : Q}{\Gamma; \emptyset \vdash_{\overline{\lambda}\text{LAL}} t : !Q} \qquad \frac{\Gamma; \Delta, x : !P, x : !P \vdash_{\overline{\lambda}\text{EAL}} t : Q}{\Gamma; \Delta, x : !P \vdash_{\overline{\lambda}\text{EAL}} t : Q}$$

$$\frac{\Gamma; x_1 : P_1, ..., x_n : P_n \vdash_{\overline{\lambda}\text{LAL}} t : P}{\Gamma; x_1 : !P_1, ..., x_k : !P_k, x_{k+1} : \S P_{k+1}, ..., x_n : \S P_n \vdash_{\overline{\lambda}\text{LAL}} t : \S P}$$

The light affine λ-system

Correction

We can bring $\vdash_{\overline{\lambda}\text{LAL}}$ back on Light Affine Logic in order to prove that extracted programs are polynomially time bounded. First, one should notice that the propositional fragment of $\vdash_{\overline{\lambda}\text{LAL}}$ is exactly the usual intuitionistic light affine logic (for instance the one describe in [1]). If Δ is a propositional context and P a propositional formula, we simply write $\Delta \vdash_{\text{LAL}} t : P$ instead of $\Gamma; \Delta \vdash_{\overline{\lambda}\text{LAL}} t : P$ where $\Gamma = \langle X : []; X \text{ occurring in } \Delta \rangle$. We extend the notation F° to the new modality with $(\S F)^\circ = \S F^\circ$ and we have:

Lemma 15. *If $\Gamma; \Delta \vdash_{\overline{\lambda}\text{LAL}} t : P$, then $\Delta^\circ \vdash_{\text{LAL}} t : P^\circ$.*

Definition 11. *We say that a program $t \in \mathcal{P}$ represents a (set-theoretical) total function f on binary words if for all word w, the term $(t \lceil w \rceil)$ may be normalized into $\lceil f(w) \rceil$. We say that $t \in \mathbb{P}$ if it represents a total function f and there exists a Turing machine that computes f in polynomial time.*

It is important to realize that $t \in \mathbb{P}$ does not imply that t reduces in polynomial time but only that the function implemented by t is computable in polynomial time. In order to obtain an effective way to compute the function in polynomial time one could use instead of t the proof-net associated to the corresponding LAL proof.

We denote by $\mathcal{W}[x]$ the following decoration of $W[x]$ with modalities

$$\forall X:[\text{wrd}], !(\forall y:\text{wrd}, X \, y \to X \, (\text{O} \, y)) \to !(\forall y:\text{wrd}, X \, y \to X \, (\text{I} \, y)) \to \S(X \, \varepsilon \to X \, w)$$

and simply by $\mathcal{W}$ the formula $\forall X, !(X \to X) \to !(X \to X) \to \S(X \to X) = (\mathcal{W}[x])^\circ$.

This lemma is the classical correction result for ILAL (see [1]).

Lemma 16. *If $\vdash_{\text{LAL}} t : !^{k_1}\mathcal{W} \to ... \to !^{k_n}\mathcal{W} \to \S^k\mathcal{W}$ then $[\![t]\!] \in \mathbb{P}$.*

And we obtain:

Theorem 4. *(Correctness theorem) If we have*

$$\vdash_{\overline{\lambda}\text{LAL}} t : \forall x_1...x_n : \text{wrd}, !^{k_1}\mathcal{W}[x_1] \to ... \to !^{k_n}\mathcal{W}[x_k] \to \S^k\mathcal{W}[f \, x_1 ... x_n]$$

then $[\![t]\!]^{\mathcal{M}} \in \mathbb{P}$ and moreover we have that $\mathcal{M} \models t =_{\text{wrd}\to...\to\text{wrd}} f$ for all $\mathcal{M} \models \mathbf{H}$.

Completeness: using the completeness of ILAL

It is known[1] that all functions computable in polynomial time can be represented by a λ-term typable in $\vdash_{\text{LAL}}$ with a type of the form $\mathcal{W} \to \S^k\mathcal{W}$. And thanks to the following theorem, we can use this to prove the completeness of $\vdash_{\overline{\lambda}\text{LAL}}$.

Theorem 5. $\vdash_{\text{LAL}} t : \mathcal{W} \to \S^k \mathcal{W}$ *implies* $\vdash_{\overline{\lambda}\text{LAL}} \forall x : \boldsymbol{wrd}, \mathcal{W}[x] \to \S^{k+1} \mathcal{W}[t\,x]$.

Proof. As for $\vdash_{\overline{\lambda}\text{EAL}}$, the adequacy theorem is true in $\vdash_{\overline{\lambda}\text{EAL}}$ (we define $t \Vdash \S F$ by $\S(t \Vdash F)$), therefore $\vdash_{\overline{\lambda}\text{EAL}} t : \mathcal{W} \to \S^k \mathcal{W}$ implies $\vdash_{\overline{\lambda}\text{EAL}} (t \Vdash \mathcal{W} \to \dots \to \mathcal{W} \to \S^k \mathcal{W})$ (*). We are going to need the two simple lemmas below:

1. We have $\vdash_{\overline{\lambda}\text{EAL}} \forall r, (r \Vdash \mathcal{W}) \to \mathcal{W}[r]$ (the proof is similar as in Theorem 3).
2. And we have $\vdash_{\overline{\lambda}\text{EAL}} \forall r, \mathcal{W}[r] \to \S(r \Vdash \mathcal{W})$.

$$
\begin{array}{ccc}
\vdots & \vdots & \vdots \\
 & & \pi_3 \\
\pi_1 & \pi_2 & \dfrac{\vdash \varepsilon \Vdash \mathcal{W} \qquad r \Vdash \mathcal{W} \vdash r \Vdash \mathcal{W}}{\varepsilon \Vdash \mathcal{W} \to r \Vdash \mathcal{W} \vdash r \Vdash \mathcal{W}} \\
\dfrac{\vdash H_1}{\vdash !H_1} & \dfrac{\vdash H_2}{\vdash !H_2} & \dfrac{}{\S(\varepsilon \Vdash \mathcal{W} \to r \Vdash \mathcal{W}) \vdash \S(r \Vdash \mathcal{W})}
\end{array}
$$

$$
\dfrac{\mathcal{W}\, r \vdash \S(r \Vdash \mathcal{W})}{\vdash \forall r, \mathcal{W}\, r \to \S(r \Vdash \mathcal{W})}
$$

with $H_1 = \forall y, y \Vdash \mathcal{W} \to (\mathrm{O}\,y) \Vdash \mathcal{W}$ and $H_2 = \forall y, y \Vdash \mathcal{W} \to (\mathrm{I}\,y) \Vdash \mathcal{W}$ where π_1, π_2 and π_3 use the rule EQUALITY with $\mathbf{H} \models_\Gamma (\varepsilon\,\alpha\,f\,g\,z) =_\alpha z$, $\mathbf{H} \models_\Gamma (\mathrm{O}\,y\,\alpha\,f\,g\,z) =_\alpha f\,(y\,\alpha\,f\,g\,z))$, $\mathbf{H} \models_\Gamma (\mathrm{I}\,y\,\alpha\,f\,g\,z) =_\alpha g\,(y\,\alpha\,f\,g\,z))$, and $\Gamma = \alpha : \text{Type}, f : \alpha \to \alpha, g : \alpha \to \alpha, z : \alpha$

The end of the proof is similar to the end of the proof of theorem 3.   □

References

[1] Asperti, A., Roversi, L.: Intuitionistic Light Affine Logic. ACM Trans. Comput. Log. 3(1), 137–175 (2002)
[2] Baillot, P.: Type inference for light affine logic via constraints on words. Theoretical Computer Science 328, 289–323 (2004)
[3] Bernardy, J.-P., Lasson, M.: Realizability and parametricity in pure type systems. In: Hofmann, M. (ed.) FOSSACS 2011. LNCS, vol. 6604, pp. 108–122. Springer, Heidelberg (2011)
[4] Danos, V., Joinet, J.B.: Linear Logic & Elementary Time. Information and Computation 183 (2001)
[5] Girard, J.Y.: Light Linear Logic. Information and Computation 143(2), 175–204 (1998)
[6] Krivine, J.L.: Lambda-calculus, types and models. Ellis Horwood, England (1993)
[7] Krivine, J.L., Parigot, M.: Programming with Proofs. Elektronische Informationsverarbeitung und Kybernetik 26(3), 149–167 (1990)
[8] Leivant, D.: Reasoning about Functional Programs and Complexity Classes Associated with Type Disciplines. In: FOCS, pp. 460–469. IEEE, Los Alamitos (1983)
[9] Martini, S., Coppola, P.: Typing lambda terms in elementary logic with linear constraints. In: Abramsky, S. (ed.) TLCA 2001. LNCS, vol. 2044, pp. 76–90. Springer, Heidelberg (2001)
[10] Plotkin, G.D., Abadi, M., Cardelli, L.: Subtyping and parametricity. In: LICS, pp. 310–319. IEEE Computer Society, Los Alamitos (1994)
[11] Rose, H.: Sub-recursion: functions and hierarchy. Oxford University Press, Oxford (1984)
[12] Takeuti, I.: An Axiomatic System of Parametricity. In: de Groote, P., Hindley, J.R. (eds.) TLCA 1997. LNCS, vol. 1210, pp. 354–372. Springer, Heidelberg (1997)
[13] Wadler, P.: The Girard-Reynolds isomorphism (second edition). Theor. Comput. Sci. 375(1-3), 201–226 (2007)

An Elementary Affine λ-Calculus
with Multithreading and Side Effects*

Antoine Madet and Roberto M. Amadio

Laboratoire PPS, Université Paris Diderot
{madet,amadio}@pps.jussieu.fr

Abstract. Linear logic provides a framework to control the complexity of higher-order functional programs. We present an extension of this framework to programs with multithreading and side effects focusing on the case of elementary time. Our main contributions are as follows. First, we introduce a modal call-by-value λ-calculus with multithreading and side effects. Second, we provide a combinatorial proof of termination in elementary time for the language. Third, we introduce an elementary affine type system that guarantees the standard subject reduction and progress properties. Finally, we illustrate the programming of iterative functions with side effects in the presented formalism.

Keywords: Elementary Linear logic. Resource Bounds. Lambda Calculus. Regions. Side Effects.

1   Introduction

There is a well explored framework based on Linear Logic to control the complexity of higher-order functional programs. In particular, *light logics* [11,10,3] have led to a polynomial light affine λ-calculus [14] and to various type systems for the standard λ-calculus guaranteeing that a well-typed term has a bounded complexity [9,8,5]. Recently, this framework has been extended to a higher-order process calculus [12] and a functional language with recursive definitions [4]. In another direction, the notion of *stratified region* [7,1] has been used to prove the termination of higher-order multithreaded programs with side effects.

Our general goal is to extend the framework of light logics to a higher-order functional language with multithreading and side effects by focusing on the case of elementary time [10]. The key point is that termination does not rely anymore on stratification but on the notion of depth which is standard in light logics. Indeed, light logics suggest that complexity can be tamed through a fine analysis of the way the depth of the occurrences of a λ-term can vary during reduction.

Our core functional calculus is a λ-calculus extended with a constructor '!' (the modal operator of linear logic) marking duplicable terms and a related let !

---

* Work partially supported by project ANR-08-BLANC-0211-01 "COMPLICE" and the Future and Emerging Technologies (FET) programme within the Seventh Framework Programme for Research of the European Commission, under FET-Open grant number: 243881 (project CerCo).

L. Ong (Ed.): TLCA 2011, LNCS 6690, pp. 138–152, 2011.

Table 1. Overview of the λ-calculi considered

Functional	$\lambda^! \supset \lambda^!_\delta \supset \lambda^!_{EA}$
$\cap$	
Concurrent	$\lambda^{!R} \supset \lambda^{!R}_\delta \supset \lambda^{!R}_{EA}$

destructor. The depth of an occurrence in a λ-term is the number of $!'s$ that must be crossed to reach the occurrence. In Section 2, following previous work on an affine-intuitionistic system [2], we extend this functional core with parallel composition and operations producing side effects on an 'abstract' notion of state. In Section 3, we analyse the impact of side-effects operations on the depth of the occurrences. Based on this analysis, we propose a formal system called *depth system* that controls the depth of the occurrences and which is a variant of a system proposed in [14]. In Section 4, we show that programs well-formed in the depth system are guaranteed to terminate in elementary time. The proof is based on an original combinatorial analysis of the depth system. In particular, as a corollary of this analysis one can derive an elementary bound for the functional fragment under an arbitrary reduction strategy ([10] assumes a specific reduction strategy while [14] relies on a standardization theorem). In Section 5, we refine the depth system with a second order (polymorphic) elementary affine type system and show that the resulting system enjoys subject reduction and progress (besides termination in elementary time). Finally, in Section 6, we discuss the expressivity of the resulting type system. On the one hand we check that the usual encoding of elementary functions goes through. On the other hand, and more interestingly, we provide examples of iterative (multithreaded) programs with side effects. The λ-calculi introduced are summarized in Table 1. For each concurrent language there is a corresponding functional fragment and each language (functional or concurrent) refines the one on its left hand side. The elementary complexity bounds are obtained for the $\lambda^!_\delta$ and $\lambda^{!R}_\delta$ calculi while the progress property and the expressivity results refer to their typed refinements $\lambda^!_{EA}$ and $\lambda^{!R}_{EA}$, respectively. Proofs are available in the technical report [13].

2   A Modal λ-Calculus with Multithreading and Regions

In this section we introduce a call-by-value modal λ-calculus endowed with parallel composition and operations to read and write *regions*. We call it $\lambda^{!R}$. A region is an *abstraction* of a set of dynamically generated values such as imperative references or communication channels. We regard $\lambda^{!R}$ as an abstract, highly non-deterministic language which entails complexity bounds for more concrete languages featuring references or channels (we will give an example of such a language in Section 6). To this end, it is enough to map the dynamically generated values to their respective regions and observe that the reductions in the concrete languages are simulated in $\lambda^{!R}$ (see, *e.g.*, [2]). The purely functional fragment, called $\lambda^!$, is very close to the *light affine λ-calculus* of Terui [14] where

the paragraph modality '§' used for polynomial time is dropped and where the
'!' modality is relaxed as in elementary linear logic [10].

2.1   Syntax

The syntax of the language is described in Table 2. We have the usual set of
variable $x, y, \ldots$ and a set of regions $r, r', \ldots$. The set of values V contains the
unit constant $*$, variables, regions, λ-abstraction and modal values $!V$ which
are marked with the *bang* operator '!'. The set of terms M contains values,
application, modal terms $!M$, a let! operator, $\mathsf{set}(r, V)$ to write the value V at
region r, $\mathsf{get}(r)$ to fetch a value from region r and $(M \mid N)$ to evaluate M and
N in parallel. A store S is the composition of several stores $(r \leftarrow V)$ in parallel.
A program P is a combination of terms and stores. Evaluation contexts follow a
call-by-value discipline. Static contexts C are composed of parallel compositions.
Note that stores can only appear in a static context, thus $M(M' \mid (r \leftarrow V))$ is
not a legal term.

Table 2. Syntax of programs: $\lambda^{!R}$

$x, y, \ldots$	(Variables)
$r, r', \ldots$	(Regions)
$V ::= * \mid r \mid x \mid \lambda x.M \mid !V$	(Values)
$M ::= V \mid MM \mid !M \mid \text{let } !x = M \text{ in } M$	
$\qquad \mathsf{set}(r, V) \mid \mathsf{get}(r) \mid (M \mid M)$	(Terms)
$S ::= (r \leftarrow V) \mid (S \mid S)$	(Stores)
$P ::= M \mid S \mid (P \mid P)$	(Programs)
$E ::= [\,] \mid EM \mid VE \mid !E \mid \text{let } !x = E \text{ in } M$	(Evaluation Contexts)
$C ::= [\,] \mid (C \mid P) \mid (P \mid C)$	(Static Contexts)

We define $!^0 M = M$, $!^{n+1}M = !(!^n M)$, $!^n(P \mid P) = (!^n P \mid !^n P)$, and $!^n(r \leftarrow V) = (r \leftarrow V)$. In the terms $\lambda x.M$ and $\text{let } !x = N \text{ in } M$ the occurrences of x in
M are bound. The set of free variables of M is denoted by $\mathsf{FV}(M)$. The number
of free occurrences of x in M is denoted by $\mathsf{FO}(x, M)$. $M[V/x]$ denotes the term
M in which each free occurrence of x has been substituted by the value V (we
insist for substituting values because in general the language is not closed under
arbitrary substitutions). As usual, we abbreviate $(\lambda z.N)M$ with $M; N$, where z
is not free in N.

Each program has an *abstract syntax tree* as exemplified in Figure 1(a). A
path starting from the root to a node of the tree denotes an *occurrence* of the
program that is denoted by a word $w \in \{0, 1\}^*$ (see Figure 1(b)).

2.2   Operational Semantics

The operational semantics of the language is described in Table 3. Programs
are considered up to a structural equivalence $\equiv$ which is the least equivalence
relation preserved by static contexts, and which contains the equations for α-
renaming and for the commutativity and associativity of parallel composition.

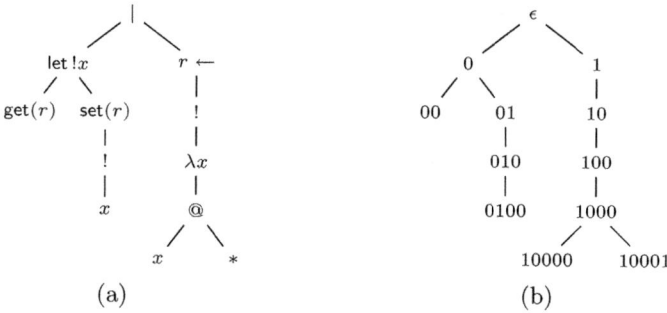

Fig. 1. Syntax tree and addresses of $P = $ let $!x = \text{get}(r)$ in $\text{set}(r, !x) \mid (r \leftarrow !(\lambda x.x*))$

Table 3. Semantics of $\lambda^{!R}$ programs

$$
\begin{array}{rcl}
P \mid P' & \equiv & P' \mid P \qquad \text{(Commutativity)} \\
(P \mid P') \mid P'' & \equiv & P \mid (P' \mid P'') \quad \text{(Associativity)}
\end{array}
$$

$$
\begin{array}{rcll}
E[(\lambda x.M)V] & & \rightarrow & E[M[V/x]] \\
E[\text{let } !x = !V \text{ in } M] & & \rightarrow & E[M[V/x]] \\
E[\text{set}(r, V)] & & \rightarrow & E[*] \qquad \mid (r \leftarrow V) \\
E[\text{get}(r)] & \mid (r \leftarrow V) \rightarrow & E[V] \\
E[\text{let } !x = \text{get}(r) \text{ in } M] \mid (r \leftarrow !V) & \rightarrow & E[M[V/x]] \mid (r \leftarrow !V)
\end{array}
$$

The reduction rules apply modulo structural equivalence and in a static context C. In the sequel, $\xrightarrow{*}$ denotes the reflexive and transitive closure of $\rightarrow$.

The let! operator is 'filtering' modal terms and 'destructs' the bang of the value $!V$ after substitution. When writing to a region, values are accumulated rather than overwritten (remember that $\lambda^{!R}$ is an abstract language that can simulate more concrete ones where values relating to the same region are associated with distinct addresses). On the other hand, reading a region amounts to select non-deterministically one of the values associated with the region. We distinguish two rules to read a region. The first *consumes* the value from the store, like when reading a communication channel. The second *copies* the value from the store, like when reading a reference. Note that in this case the value read must be duplicable (of the shape $!V$).

Example 1. Program P of Figure 1 reduces as follows:

$$
P \quad \rightarrow \quad \text{set}(r, !(\lambda x.x*)) \mid (r \leftarrow !(\lambda x.x*)) \quad \rightarrow \quad * \mid (r \leftarrow !(\lambda x.x*)) \mid (r \leftarrow !(\lambda x.x*))
$$

3   Depth System

In this section, we analyse the interaction between the depth of the occurrences and side effects. This leads to the definition of a depth system and the notion of *well-formed* program. As a first step, we introduce a naive definition of *depth*.

Definition 1 (naive depth). *The depth $d(w)$ of an occurrence w is the number of ! labels that the path leading to the end node crosses. The depth $d(P)$ of a program P is the maximum depth of its occurrences.*

With reference to Figure 1, $d(0100) = d(100) = d(1000) = d(10000) = d(10001) = 1$, whereas other occurrences are at depth 0. In particular, occurrences 010 and 10 are at depth 0; what matters in computing the depth of an occurrence is the number of !'s that precede strictly the end node. Thus $d(P) = 1$.

By considering that deeper occurrences have less weight than shallow ones, the usual proof of termination in elementary time [10] relies on the observation that when reducing a redex at depth i the following holds:

(1) the depth of the term does not increase,
(2) the number of occurrences at depth $j < i$ does not increase,
(3) the number of occurrences at depth i strictly decreases,
(4) the number of occurrences at depth $j > i$ may be increased by a multi-
    plicative factor k bounded by the number of occurrences at depth $i + 1$.

If we consider the functional core of our language (*i.e.* by removing all operators dealing with regions, stores and multithreading), it is not difficult to check that the properties above can be guaranteed by the following requirements: (1) in $\lambda x.M$, x may occur at most once in M and at depth 0, (2) in let $!x = M$ in N, x may occur arbitrarily many times in N and at depth 1.

However, we observe that side effects may increase the depth or generate occurrences at lower depth than the current redex, which violates Property (1) and (2) respectively. Then to find a suitable notion of depth, it is instructive to consider the following program examples where $M_r = $ let $!z = \mathsf{get}(r)$ in $!(z*)$.

(A) $E[\mathsf{set}(r, !V)]$ (B) $\lambda x.\mathsf{set}(r, x); !\mathsf{get}(r)$
(C) $!(M_r) \mid (r \leftarrow !(\lambda y.M_{r'})) \mid (r' \leftarrow !(\lambda y.*))$ (D) $!(M_r) \mid (r \leftarrow !(\lambda y.M_r))$

(A) Suppose the occurrence $\mathsf{set}(r, !V)$ is at depth $\delta > 0$ in E. Then when evaluat-
    ing such a term we always end up in a program of the shape $E[*] \mid (r \leftarrow !V)$
    where the occurrence $!V$, previously at depth δ, now appears at depth 0.
    This contradicts Property (2).
(B) If we apply this program to V we obtain $!V$, hence Property (1) is violated
    because from a program of depth 0, we reduce to a program of depth 1. We
    remark that this is because the read and write operations do not execute at
    the same depth.
(C) According to our definition, this program has depth 2, however when we
    reduce it we obtain a term $!^3*$ which has depth 3, hence Property (1) is
    violated. This is because the occurrence $\lambda y.M_{r'}$ originally at depth 1 in the
    store, ends up at depth 2 in the place of z applied to $*$.
(D) If we accept circular stores, we can even write diverging programs whose
    depth is increased by 1 every two reduction steps.

Given these remarks, the rest of this section is devoted to a revised notion of depth and to a related set of inference rules called depth system. Every program which is valid in the depth system will terminate in elementary time. First, we introduce the following contexts:

$$\Gamma = x_1 : \delta_1, \ldots, x_n : \delta_n \qquad R = r_1 : \delta_1, \ldots, r_n : \delta_n$$

where δ_i is a natural number. We write $dom(\Gamma)$ and $dom(R)$ for the sets $\{x_1, \ldots, x_n\}$ and $\{r_1, \ldots, r_n\}$ respectively. We write $R(r_i)$ for the depth δ_i associated with r_i in the context R. Then, we revisit the notion of depth as follows.

Definition 2 (revised depth). *Let P be a program, R a region context where $dom(R)$ contains all the regions of P and $d_n(w)$ the naive depth of an occurrence w of P. If w does not appear under an occurrence $r \leftarrow$ (a store), then the revised depth $d_r(w)$ of w is $d_n(w)$. Otherwise, $d_r(w)$ is $R(r) + d_n(w)$. The revised depth $d_r(P)$ of the program is the maximum revised depth of its occurrences.*

Note that the revised depth is relative to a fixed region context. In the sequel we write $d(\_)$ for $d_r(\_)$. On functional terms, this notion of depth is equivalent to the one given in Definition 1. However, if we consider the program of Figure 1, we now have $d(10) = R(r)$ and $d(100) = d(1000) = d(10000) = d(10001) = R(r) + 1$.

A judgement in the depth system has the shape $R; \Gamma \vdash^\delta P$ and it should be interpreted as follows: the free variables of $!^\delta P$ may only occur at the depth specified by the context Γ, where depths are computed according to R. The inference rules of the depth system are presented in Table 4. We comment on the rules. The variable rule says that the current depth of a free variable is specified by the context. A region and the constant $*$ may appear at any depth. The λ-abstraction rule requires that the occurrence of x in M is at the same depth as the formal parameter; moreover it occurs at most once so that no duplication is possible at the current depth (Property (3)). The application rule says that

Table 4. Depth system for programs: $\lambda_\delta^{!R}$

$$\frac{}{R; \Gamma, x : \delta \vdash^\delta x} \qquad \frac{}{R; \Gamma \vdash^\delta r} \qquad \frac{}{R; \Gamma \vdash^\delta *}$$

$$\frac{\mathsf{FO}(x, M) \leq 1 \quad R; \Gamma, x : \delta \vdash^\delta M}{R; \Gamma \vdash^\delta \lambda x.M} \qquad \frac{R; \Gamma \vdash^\delta M_i \quad i = 1, 2}{R; \Gamma \vdash^\delta M_1 M_2}$$

$$\frac{R; \Gamma \vdash^{\delta+1} M}{R; \Gamma \vdash^\delta\, !M} \qquad \frac{R; \Gamma \vdash^\delta M_1 \quad R; \Gamma, x : (\delta + 1) \vdash^\delta M_2}{R; \Gamma \vdash^\delta \mathsf{let}\ !x = M_1\ \mathsf{in}\ M_2}$$

$$\frac{}{R, r : \delta; \Gamma \vdash^\delta \mathsf{get}(r)} \qquad \frac{R, r : \delta; \Gamma \vdash^\delta V}{R, r : \delta; \Gamma \vdash^\delta \mathsf{set}(r, V)}$$

$$\frac{R, r : \delta; \Gamma \vdash^\delta V}{R, r : \delta; \Gamma \vdash^0 (r \leftarrow V)} \qquad \frac{R; \Gamma \vdash^\delta P_i \quad i = 1, 2}{R; \Gamma \vdash^\delta (P_1 \mid P_2)}$$

we may only apply a term to another one if they are at the same depth. The let! rule requires that the bound occurrences of x are one level deeper than the current depth; note that there is no restriction on the number of occurrences of x since duplication would happen one level deeper than the current depth. The bang rule is better explained in a bottom-up way: crossing a modal occurrence increases the current depth by one. The key cases are those of read and write: the depth of these two operations is specified by the region context. The current depth of a store is always 0, however, the depth of the value in the store is specified by R (note that it corresponds to the revised definition of depth). We remark that R is constant in a judgement derivation.

Definition 3 (well-formedness). *A program P is* well-formed *if for some R, Γ, δ a judgement $R; \Gamma \vdash^\delta P$ can be derived.*

Example 2. The program of Figure 1 is well-formed with the following derivation where $R(r) = 0$:

$$
\cfrac{R; \Gamma \vdash^0 \mathsf{get}(r) \qquad \cfrac{\cfrac{R; \Gamma, x : 1 \vdash^1 x}{R; \Gamma, x : 1 \vdash^0 !x}}{R; \Gamma, x : 1 \vdash^0 \mathsf{set}(r, !x)}}{\cfrac{R; \Gamma \vdash^0 \mathsf{let}\ !x = \mathsf{get}(r)\ \mathsf{in}\ \mathsf{set}(r, !x) \qquad \cfrac{\vdots}{R; \Gamma \vdash^0 (r \leftarrow !(\lambda x.x*))}}{R; \Gamma \vdash^0 \mathsf{let}\ !x = \mathsf{get}(r)\ \mathsf{in}\ \mathsf{set}(r, !x)\ |\ (r \leftarrow !(\lambda x.x*))}}
$$

On the other hand, the following term is not well-formed: $P = \lambda x.\mathsf{let}\ !y = x\ \mathsf{in}\ !(y!(yz))$.

Indeed, the second occurrence of y in $!(y!(yz))$ is one level too deep, hence reduction may increase the depth by one. For example, $P!!V$ of depth 2 reduces to $!(!V!(!V)z)$ of depth 3.

We reconsider the troublesome programs with side effects. Program (A) is well-formed with judgement (i):

$$
\begin{array}{lll}
R; \Gamma \vdash^0 E[\mathsf{set}(r, !V)] & \text{with } R = r : \delta & (i) \\
R; \Gamma \vdash^0 !M_r\ |\ (r \leftarrow !(\lambda y.M_{r'}))\ |\ (r' \leftarrow !(\lambda y.*)) & \text{with } R = r : 1, r' : 2 & (ii)
\end{array}
$$

Indeed, the occurrence $!V$ is now preserved at depth δ in the store. Program (B) is not well-formed since the read operation requires $R(r) = 1$ and the write operations require $R(r) = 0$. Program (C) is well-formed with judgement (ii); indeed its depth does not increase anymore because $!M_r$ has depth 2 but since $R(r) = 1$ and $R(r') = 2$, $(r \leftarrow !(\lambda y.M_{r'}))$ has depth 3 and $(r' \leftarrow !(\lambda y.*))$ has depth 2. Hence program (C) has already depth 3. Finally, it is worth noticing that the diverging program (D) is not well-formed since $\mathsf{get}(r)$ appears at depth 1 in $!M_r$ and at depth 2 in the store.

Theorem 1 (properties on the depth system). *The following properties hold:*

1. *If $R; \Gamma \vdash^\delta M$ and x occurs free in M then $x : \delta'$ belongs to Γ and all occurrences of x in $!^\delta M$ are at depth δ'.*
2. *If $R; \Gamma \vdash^\delta P$ then $R; \Gamma, \Gamma' \vdash^\delta P$.*

3. If $R; \Gamma, x : \delta' \vdash^\delta M$ and $R; \Gamma \vdash^{\delta'} V$ then $R; \Gamma \vdash^\delta M[V/x]$ and
   $d(!^\delta M[V/x]) \leq max(d(!^\delta M), d(!^{\delta'} V))$.
4. If $R; \Gamma \vdash^0 P$ and $P \to P'$ then $R; \Gamma \vdash^0 P'$ and $d(P) \geq d(P')$.

4   Elementary Bound

In this section, we prove that well-formed programs terminate in elementary time. To this end, we define a measure on programs based on the number of occurrences at each depth.

Definition 4 (measure). *Given a program P and $0 \leq i \leq d(M)$, let $\omega_i(P)$ be the number of occurrences in P of depth i increased by 2 (so $\omega_i(P) \geq 2$). We define $\mu_n^i(P)$ for $n \geq i \geq 0$ as follows:*

$$\mu_n^i(P) = (\omega_n(P), \ldots, \omega_{i+1}(P), \omega_i(P))$$

We write $\mu_n(P)$ for $\mu_n^0(P)$. We order the vectors of $n + 1$ natural number with the (well-founded) lexicographic order $>$ from right to left.

To simplify the proofs of the following properties, we assume the occurrences labelled with $|$ and $r \leftarrow$ do not count in the measure and that $set(r)$ counts for two occurrences, such that the measure strictly decreases on the rule $E[set(r, V)] \to E[*] \mid (r \leftarrow V)$.

Following this assumption, we derive a termination property by observing that the measure strictly decreases during reduction.

Proposition 1 (termination). *If P is well-formed, $P \to P'$ and $n \geq d(P)$ then $\mu_n(P) > \mu_n(P')$.*

Proof. By a case analysis on the reduction rules. The crucial cases are those that increase the number of occurrences, namely both let! reductions: the one that is functional and the one that copies from the store. Thanks to the design of our depth system, we observe that both rules generate duplication of occurrences in the same way, hence we may only consider the functional case as an illustration where $P = E[\text{let } !x =!V \text{ in } M] \to P' = E[M[V/x]]$.

Let the occurrence of the redex let $!x =!V$ in M be at depth i. The restrictions on the formation of terms require that x may only occur in M at depth 1 and hence in P at depth $i + 1$. We have that $\omega_i(P') = \omega_i(P) - 2$ because the let! node disappears. Clearly, $\omega_j(P) = \omega_j(P')$ if $j < i$. The number of occurrences of x in M is bounded by $k = \omega_{i+1}(P) \geq 2$. Thus if $j > i$ then $\omega_j(P') \leq k \cdot \omega_j(P)$. Let's write, for $0 \leq i \leq n$, $\mu_n^i(P) \cdot k = (\omega_n(P) \cdot k, \omega_{n-1}(P) \cdot k, \ldots, \omega_i(P) \cdot k)$. Then we have:

$$\mu_n(P') \leq (\mu_n^{i+1}(P) \cdot k, \omega_i(P) - 2, \mu_{i-1}(P)) \tag{1}$$

and finally $\mu_n(P) > \mu_n(P')$.

We now want to show that termination is actually in elementary time. We recall that a function f on integers is elementary if there exists a k such that for any n, $f(n)$ can be computed in time $\mathcal{O}(t(n, k))$ where:

$$t(n, 0) = 2^n, \qquad t(n, k + 1) = 2^{t(n,k)} .$$

Definition 5 (tower functions). *We define a family of tower functions* $t_\alpha(x_1, \ldots, x_n)$ *by induction on n where we assume $\alpha \geq 1$ and $x_i \geq 2$:*

$$t_\alpha() = 0$$
$$t_\alpha(x_1, x_2, \ldots, x_n) = (\alpha \cdot x_1)^{2^{t_\alpha(x_2, \ldots, x_n)}} \quad n \geq 1$$

Then we need to prove the following crucial lemma.

Lemma 1 (shift). *Assuming $\alpha \geq 1$ and $\beta \geq 2$, the following property holds for the tower functions with $x, \mathbf{x}$ ranging over numbers greater or equal to 2:*

$$t_\alpha(\beta \cdot x, x', \mathbf{x}) \leq t_\alpha(x, \beta \cdot x', \mathbf{x})$$

Now, by a closer look at the shape of the lexicographic ordering during reduction, we are able to compose the decreasing measure with a tower function.

Theorem 2 (elementary bound). *Let P be a well-formed program with $\alpha = d(P)$ and let t_α denote the tower function with $\alpha + 1$ arguments. Then if $P \to P'$ then $t_\alpha(\mu_\alpha(P)) > t_\alpha(\mu_\alpha(P'))$.*

Proof. We exemplify the proof for $\alpha = 2$ and the crucial case where

$$P = \mathsf{let}\ !x = !V\ \mathsf{in}\ M \to P' = M[V/x]$$

Let $\mu_2(P) = (x, y, z)$ such that $x = \omega_2(P)$, $y = \omega_1(P)$ and $z = \omega_0(P)$. We want to show that: $t_2(\mu_2(P')) < t_2(\mu_2(P))$. We have:

$$t_2(\mu_2(P')) \leq t_2(x \cdot y, y \cdot y, z - 2) \text{ by inequality (1)}$$
$$\leq t_2(x, y^3, z - 2) \quad \text{ by Lemma 1}$$

Hence we are left to show that: $t_2(y^3, z-2) < t_2(y, z)$, *i.e.*, $(2y^3)^{2^{2(z-2)}} < (2y)^{2^{2z}}$. We have: $(2y^3)^{2^{2(z-2)}} \leq (2y)^{3 \cdot 2^{2(z-2)}}$. Thus we need to show: $3 \cdot 2^{2(z-2)} < 2^{2z}$ which is true. Hence $t_2(\mu_2(P')) < t_2(\mu_2(P))$.

This shows that the number of reduction steps of a program P is bounded by an elementary function where the height of the tower only depends on $d(P)$. We also note that if $P \xrightarrow{*} P'$ then $t_\alpha(\mu_\alpha(P))$ bounds the size of P'. Thus we can conclude with the following corollary.

Corollary 1. *The normalisation of programs of bounded depth can be performed in time elementary in the size of the terms.*

We remark that if P is a purely functional term then the elementary bound holds under an arbitrary reduction strategy.

5 An Elementary Affine Type System

The depth system entails termination in elementary time but does *not* guarantee that programs 'do not go wrong'. In particular, the introduction and elimination of bangs during evaluation may generate programs that deadlock, *e.g.*,

$$\mathsf{let}\ !y = (\lambda x.x)\ \mathsf{in}\ !(yy) \tag{2}$$

Table 5. Types and contexts

$t, t', \ldots$	(Type variables)
$\alpha ::= \mathbf{B} \mid A$	(Types)
$A ::= t \mid \mathbf{1} \mid A \multimap \alpha \mid !A \mid \forall t.A \mid \mathsf{Reg}_r A$	(Value-types)
$\Gamma ::= x_1 : (\delta_1, A_1), \ldots, x_n : (\delta_n, A_n)$	(Variable contexts)
$R ::= r_1 : (\delta_1, A_1), \ldots, r_n : (\delta_n, A_n)$	(Region contexts)

$$\frac{}{R \downarrow t} \qquad \frac{}{R \downarrow 1} \qquad \frac{}{R \downarrow \mathbf{B}} \qquad \frac{R \downarrow A \quad R \downarrow \alpha}{R \downarrow (A \multimap \alpha)}$$

$$\frac{R \downarrow A}{R \downarrow !A} \qquad \frac{r : (\delta, A) \in R}{R \downarrow \mathsf{Reg}_r A} \qquad \frac{R \downarrow A \quad t \notin R}{R \downarrow \forall t.A}$$

$$\frac{\forall r : (\delta, A) \in R \quad R \downarrow A}{R \vdash} \qquad \frac{R \vdash \quad R \downarrow \alpha}{R \vdash \alpha} \qquad \frac{\forall x : (\delta, A) \in \Gamma \quad R \vdash A}{R \vdash \Gamma}$$

is well-formed but the evaluation is stuck. In this section we introduce an elementary affine type system ($\lambda_{EA}^{!R}$) that guarantees that programs cannot deadlock (except when trying to read an empty store).

The upper part of Table 5 introduces the syntax of types and contexts. Types are denoted with $\alpha, \alpha', \ldots$. Note that we distinguish a special behaviour type $\mathbf{B}$ which is given to the entities of the language which are not supposed to return a value (such as a store or several terms in parallel) while types of entities that may return a value are denoted with A. Among the types A, we distinguish type variables $t, t', \ldots$, a terminal type $\mathbf{1}$, an affine functional type $A \multimap \alpha$, the type $!A$ of terms of type A that can be duplicated, the type $\forall t.A$ of polymorphic terms and the type $\mathsf{Reg}_r A$ of the region r containing values of type A. Hereby types may depend on regions.

In contexts, natural numbers δ_i play the same role as in the depth system. Writing $x : (\delta, A)$ means that the variable x ranges on values of type A and may occur at depth δ. Writing $r : (\delta, A)$ means that addresses related to region r contain values of type A and that read and writes on r may only happen at depth δ. The typing system will additionally guarantee that whenever we use a type $\mathsf{Reg}_r A$ the region context contains an hypothesis $r : (\delta, A)$.

Because types depend on regions, we have to be careful in stating in Table 5 when a region-context and a type are compatible ($R \downarrow \alpha$), when a region context is well-formed ($R \vdash$), when a type is well-formed in a region context ($R \vdash \alpha$) and when a context is well-formed in a region context ($R \vdash \Gamma$). A more informal way to express the condition is to say that a judgement $r_1 : (\delta_1, A_1), \ldots, r_n : (\delta_n, A_n) \vdash \alpha$ is well formed provided that: (1) all the region names occurring in the types $A_1, \ldots, A_n, \alpha$ belong to the set $\{r_1, \ldots, r_n\}$, (2) all types of the shape $\mathsf{Reg}_{r_i} B$ with $i \in \{1, \ldots, n\}$ and occurring in the types $A_1, \ldots, A_n, \alpha$ are such that $B = A_i$. We notice the following substitution property on types.

Proposition 2. *If $R \vdash \forall t.A$ and $R \vdash B$ then $R \vdash A[B/t]$.*

Table 6. An elementary affine type system: $\lambda_{EA}^{!R}$

$$\frac{R \vdash \Gamma \quad x : (\delta, A) \in \Gamma}{R; \Gamma \vdash^\delta x : A} \qquad \frac{R \vdash \Gamma}{R; \Gamma \vdash^\delta * : 1} \qquad \frac{R \vdash \Gamma \quad r : (\delta', A) \in R}{R; \Gamma \vdash^\delta r : \mathsf{Reg}_r A}$$

$$\frac{\mathsf{FO}(x, M) \leq 1 \\ R; \Gamma, x : (\delta, A) \vdash^\delta M : \alpha}{R; \Gamma \vdash^\delta \lambda x. M : A \multimap \alpha} \qquad \frac{R; \Gamma \vdash^\delta M : A \multimap \alpha \quad R; \Gamma \vdash^\delta N : A}{R; \Gamma \vdash^\delta MN : \alpha}$$

$$\frac{R; \Gamma \vdash^{\delta+1} M : A}{R; \Gamma \vdash^\delta !M : !A} \qquad \frac{R; \Gamma \vdash^\delta M : !A \quad R; \Gamma, x : (\delta + 1, A) \vdash^\delta N : B}{R; \Gamma \vdash^\delta \mathsf{let}\ !x = M \ \mathsf{in}\ N : B}$$

$$\frac{R; \Gamma \vdash^\delta M : A \quad t \notin (R; \Gamma)}{R; \Gamma \vdash^\delta M : \forall t. A} \qquad \frac{R; \Gamma \vdash^\delta M : \forall t. A \quad R \vdash B}{R; \Gamma \vdash^\delta M : A[B/t]}$$

$$\frac{r : (\delta, A) \in R \quad R \vdash \Gamma}{R; \Gamma \vdash^\delta \mathsf{get}(r) : A} \qquad \frac{r : (\delta, A) \in R \\ R; \Gamma \vdash^\delta V : A}{R; \Gamma \vdash^\delta \mathsf{set}(r, V) : 1} \qquad \frac{r : (\delta, A) \in R \\ R; \Gamma \vdash^\delta V : A}{R; \Gamma \vdash^0 (r \leftarrow V) : \mathbf{B}}$$

$$\frac{R; \Gamma \vdash^\delta P : \alpha \quad R; \Gamma \vdash^\delta S : \mathbf{B}}{R; \Gamma \vdash^\delta (P \mid S) : \alpha} \qquad \frac{P_i \text{ not a store } i = 1, 2 \quad R; \Gamma \vdash^\delta P_i : \alpha_i}{R; \Gamma \vdash^\delta (P_1 \mid P_2) : \mathbf{B}}$$

Example 3. One may verify that $r : (\delta, 1 \multimap 1) \vdash \mathsf{Reg}_r(1 \multimap 1)$ can be derived while the following judgements cannot: $r : (\delta, 1) \vdash \mathsf{Reg}_r(1 \multimap 1)$, $r : (\delta, \mathsf{Reg}_r 1) \vdash 1$.

A typing judgement takes the form: $R; \Gamma \vdash^\delta P : \alpha$. It attributes a type α to the program P at depth δ, in the region context R and the context Γ. Table 6 introduces an elementary affine type system *with regions*. One can see that the δ's are treated as in the depth system. Note that a region r may occur at any depth. In the let! rule, M should be of type $!A$ since x of type A appears one level deeper. A program in parallel with a store should have the type of the program since we might be interested in the value the program reduces to; however, two programs in parallel cannot reduce to a single value, hence we give them a behaviour type. The polymorphic rules are straightforward where $t \notin (R; \Gamma)$ means t does not occur free in a type of R or Γ.

Example 4. The well-formed program (C) can be given the following typing judgement: $R; \_ \vdash^0 !(M_r) \mid (r \leftarrow !(\lambda y. M_{r'})) \mid (r' \leftarrow !(\lambda y.*)) : !!1$ where: $R = r : (1, !(1 \multimap 1)), r' : (2, !(1 \multimap 1))$. Also, we remark that the deadlocking program (2) admits no typing derivation.

Theorem 3 (subject reduction and progress). *The following properties hold.*

1. (Well-formedness) *Well-typed programs are well-formed.*
2. (Weakening) *If* $R; \Gamma \vdash P : \alpha$ *and* $R \vdash^{\delta} \Gamma, \Gamma'$ *then* $R; \Gamma, \Gamma' \vdash^{\delta} P : \alpha$.
3. (Substitution) *If* $R; \Gamma, x : (\delta', A) \vdash^{\delta} M : \alpha$ *and* $R; \Gamma' \vdash^{\delta'} V : A$ *and* $R \vdash \Gamma, \Gamma'$ *then* $R; \Gamma, \Gamma' \vdash^{\delta} M[V/x] : \alpha$.
4. (Subject Reduction) *If* $R; \Gamma \vdash^{\delta} P : \alpha$ *and* $P \to P'$ *then* $R; \Gamma \vdash^{\delta} P' : \alpha$.
5. (Progress) *Suppose* P *is a closed typable program which cannot reduce. Then* P *is structurally equivalent to a program*

$$M_1 \mid \cdots \mid M_m \mid S_1 \mid \cdots \mid S_n \quad m, n \geq 0$$

where M_i *is either a value or can be decomposed as a term* $E[\mathsf{get}(r)]$ *such that no value is associated with the region* r *in the stores* $S_1, \ldots, S_n$.

6   Expressivity

In this section, we consider two results that illustrate the expressivity of the elementary affine type system. First we show that all elementary functions can be represented and second we develop an example of iterative program with side effects.

Completeness. The representation result just relies on the functional core of the language $\lambda_{EA}^{!}$. Building on the standard concept of Church numeral, Table 7 provides a representation for natural numbers and the multiplication function. We denote with $\mathbb{N}$ the set of natural numbers. The precise notion of representation is spelled out in the following definitions where by strong β-reduction we mean that reduction under λ's is allowed.

Definition 6 (number representation). *Let* $\emptyset \vdash^{\delta} M : \mathsf{N}$. *We say* M *represents* $n \in \mathbb{N}$, *written* $M \Vdash n$, *if, by using a strong β-reduction relation,* $M \xrightarrow{*} \overline{n}$.

Definition 7 (function representation). *Let* $\emptyset \vdash^{\delta} F : (\mathsf{N}_1 \multimap \ldots \multimap \mathsf{N}_k) \multimap {!}^p\mathsf{N}$ *where* $p \geq 0$ *and* $f : \mathbb{N}^k \to \mathbb{N}$. *We say* F *represents* f, *written* $F \Vdash f$, *if for all* M_i *and* $n_i \in \mathbb{N}$ *where* $1 \leq i \leq k$ *such that* $\emptyset \vdash^{\delta} M_i : N$ *and* $M_i \Vdash n_i$, $FM_1 \ldots M_k \Vdash f(n_1, \ldots, n_k)$.

Table 7. Representation of natural numbers and the multiplication function

$$\mathsf{N} = \forall t.{!}(t \multimap t) \multimap {!}(t \multimap t) \qquad \text{(type of numerals)}$$

$$\overline{n} : \mathsf{N} \qquad \text{(numerals)}$$
$$\overline{n} = \lambda f.\mathsf{let} \; {!}f = f \; \mathsf{in} \; {!}(\lambda x.f(\cdots(fx)\cdots))$$

$$\mathsf{mult} : \mathsf{N} \multimap (\mathsf{N} \multimap \mathsf{N}) \qquad \text{(multiplication)}$$
$$\mathsf{mult} = \lambda n.\lambda m.\lambda f.\mathsf{let} \; {!}f = f \; \mathsf{in} \; n(m{!}f)$$

Table 8. Representation of lists

$$\mathsf{List}\ A = \forall t.!(A \multimap t \multimap t) \multimap !(t \multimap t) \qquad \text{(type of lists)}$$

$$[u_1, \ldots, u_n]\ :\ \mathsf{List}\ A \qquad\qquad\qquad \text{(list represent.)}$$
$$[u_1, \ldots, u_n] = \lambda f.\mathsf{let}\ !f = f\ \mathsf{in}\ !(\lambda x.fu_1(fu_2 \ldots (fu_n x))$$

$$\mathsf{list\_it}\ :\ \forall u.\forall t.!(u \multimap t \multimap t) \multimap \mathsf{List}\ u \multimap !t \multimap !t \quad \text{(iterator on lists)}$$
$$\mathsf{list\_it} = \lambda f.\lambda l.\lambda z.\mathsf{let}\ !z = z\ \mathsf{in}\ \mathsf{let}\ !y = lf\ \mathsf{in}\ !(yz)$$

Elementary functions are characterized as the smallest class of functions containing zero, successor, projection, subtraction and which is closed by composition and bounded summation/product. These functions can be represented in the sense of Definition 7 by adapting the proofs from Danos and Joinet [10].

Theorem 4 (completeness). *Every elementary function is representable in* $\lambda_{EA}^!$.

Iteration with Side Effects. We rely on a slightly modified language where reads, writes and stores relate to concrete addresses rather than to abstract regions. In particular, we introduce terms of the form $\nu x\ M$ to generate a fresh address name x whose scope is M. One can then write the following program:

$$\nu x\ ((\lambda y.\mathsf{set}(y, V))x) \xrightarrow{*} \nu x * \mid (x \leftarrow V)$$

where x and y relate to a region r, *i.e.* they are of type $\mathsf{Reg}_r A$. Our type system can be easily adapted by associating region types with the address names. Next we show that it is possible to program the iteration of operations producing a side effect on an inductive data structure. Specifically, in the following we show how to iterate, possibly in parallel, an update operation on a list of addresses of the store. The examples have been tested on a running implementation of the language.

Following Church encodings, we define the representation of lists and the associated iterator in Table 8. Here is the function multiplying the numeral pointed by an address at region r:

$$\mathsf{update}\ :\ !\mathsf{Reg}_r\mathsf{N} \multimap !\mathbf{1} \multimap !\mathbf{1}$$
$$\mathsf{update} = \lambda x.\mathsf{let}\ !x = x\ \mathsf{in}\ \lambda z.!((\lambda y.\mathsf{set}(x, y))(\mathsf{mult}\ \overline{2}\ \mathsf{get}(x)))$$

Consider the following list of addresses and stores:

$$[!x, !y, !z] \mid (x \leftarrow \overline{m}) \mid (y \leftarrow \overline{n}) \mid (z \leftarrow \overline{p})$$

Note that the bang constructors are needed to match the type $!\mathsf{Reg}_r\mathsf{N}$ of the argument of update. Then we define the iteration as:

$$\mathsf{run} : !!\mathbf{1} \qquad \mathsf{run} = \mathsf{list\_it}\ !\mathsf{update}\ [!x, !y, !z]\ !!*$$

Notice that it is well-typed with $R = r : (2, \mathsf{N})$ since both the read and the write appear at depth 2. Finally, the program reduces by updating the store as expected:

$$\mathsf{run} \mid (x \leftarrow \overline{m}) \mid (y \leftarrow \overline{n}) \mid (z \leftarrow \overline{p}) \xrightarrow{*} !!\mathbf{1} \mid (x \leftarrow \overline{2m}) \mid (y \leftarrow \overline{2n}) \mid (z \leftarrow \overline{2p})$$

Building on this example, suppose we want to write a program with three concurrent threads where each thread multiplies by 2 the memory cells pointed by a list. Here is a function waiting to apply a functional f to a value x in three concurrent threads:

$$\text{gen\_threads} : \forall t.\forall t'.!(t \multimap t') \multimap !t \multimap \mathbf{B}$$
$$\text{gen\_threads} = \lambda f.\text{let } !f = f \text{ in } \lambda x.\text{let } !x = x \text{ in } !(fx) \mid !(fx) \mid !(fx)$$

We define the functional F as run but parametric in the list:

$$F : \text{List } !\text{Reg}_r.\mathsf{N} \multimap !!1 \qquad F = \lambda l.\text{list\_it } !\text{update } l \; !!*$$

And the final term is simply:

$$\text{run\_threads} : \mathbf{B} \qquad \text{run\_threads} = \text{gen\_threads } !F \; ![!x, !y, !z]$$

where $R = r : (3, !\mathsf{N})$. Our program then reduces as follows:

$$\text{run\_threads} \quad \mid (x \leftarrow \overline{m}) \quad \mid (y \leftarrow \overline{n}) \quad \mid (z \leftarrow \overline{p})$$
$$\xrightarrow{*} \; !!!1 \mid !!!1 \mid !!!1 \mid (x \leftarrow \overline{8m}) \mid (y \leftarrow \overline{8n}) \mid (z \leftarrow \overline{8p})$$

Note that different thread interleavings are possible but in this particular case the reduction is confluent.

7 Conclusion

We have introduced a type system for a higher-order functional language with multithreading and side effects that guarantees termination in elementary time thus providing a significant extension of previous work that had focused on purely functional programs.

In the proposed approach, the depth system plays a key role and allows for a relatively simple presentation. In particular we notice that we can dispense with the notion of *stratified region* that arises in recent work on the termination of higher-order programs with side effects [1,7]. Hence, it becomes possible to type circular stores in $\lambda_{EA}^{!R}$ like *e.g.* $(r \leftarrow \lambda x.\text{set}(r, x); \text{get}(r))$ whereas the stratification condition precludes them. Note that this store is going to be consumed when r is read. However, concerning duplicable stores (*i.e.* of the shape $(r \leftarrow !V)$), the value V is implicitly stratified by the difference of depth between V and r. We note that we can also dispense with the distinction between affine and intuitionistic hypotheses [6,2].

As a future work, we would like to adapt our approach to polynomial time. In another direction, one could ask if it is possible to program in a simplified language without bangs and then try to infer types or depths.

Acknowledgements. We would like to thank Patrick Baillot for numerous helpful discussions and a careful reading on a draft version of this paper, plus the anonymous referees for their feedbacks.

References

1. Amadio, R.M.: On stratified regions. In: Hu, Z. (ed.) APLAS 2009. LNCS, vol. 5904, pp. 210–225. Springer, Heidelberg (2009)
2. Amadio, R.M., Baillot, P., Madet, A.: An affine-intuitionistic system of types and effects: confluence and termination. Technical report, Laboratoire PPS (2009), http://hal.archives-ouvertes.fr/hal-00438101/
3. Asperti, A., Roversi, L.: Intuitionistic light affine logic. ACM Trans. Comput. Log. 3(1), 137–175 (2002)
4. Baillot, P., Gaboardi, M., Mogbil, V.: A polytime functional language from light linear logic. In: Gordon, A.D. (ed.) ESOP 2010. LNCS, vol. 6012, pp. 104–124. Springer, Heidelberg (2010)
5. Baillot, P., Terui, K.: A feasible algorithm for typing in elementary affine logic. In: Urzyczyn, P. (ed.) TLCA 2005. LNCS, vol. 3461, pp. 55–70. Springer, Heidelberg (2005)
6. Barber, A.: Dual intuitionistic linear logic. Technical Report ECS-LFCS-96-347, The Laboratory for Foundations of Computer Science, University of Edinburgh (1996)
7. Boudol, G.: Typing termination in a higher-order concurrent imperative language. Inf. Comput. 208(6), 716–736 (2010)
8. Coppola, P., Dal Lago, U., Ronchi Della Rocca, S.: Light logics and the call-by-value lambda calculus. Logical Methods in Computer Science 4(4) (2008)
9. Coppola, P., Martini, S.: Optimizing optimal reduction: A type inference algorithm for elementary affine logic. ACM Trans. Comput. Log. 7, 219–260 (2006)
10. Danos, V., Joinet, J.-B.: Linear logic and elementary time. Inf. Comput. 183(1), 123–137 (2003)
11. Girard, J.-Y.: Light linear logic. Inf. Comput. 143(2), 175–204 (1998)
12. Lago, U.D., Martini, S., Sangiorgi, D.: Light logics and higher-order processes. In: EXPRESS 2010. EPTCS, vol. 41, pp. 46–60 (2010)
13. Madet, A., Amadio, R.M.: Elementary affine λ-calculus with multithreading and side effects. Technical report, Laboratoire PPS (2011), http://hal.archives-ouvertes.fr/hal-00569095/
14. Terui, K.: Light affine lambda calculus and polynomial time strong normalization. Archive for Mathematical Logic 46(3-4), 253–280 (2007)

Böhm's Theorem for Resource Lambda Calculus through Taylor Expansion*

Giulio Manzonetto[1] and Michele Pagani[2]

[1] Intelligent Systems, Radboud University
g.manzonetto@cs.ru.nl
[2] Laboratoire LIPN, CNRS UMR7030 Université Paris 13
michele.pagani@lipn.univ-paris13.fr

Abstract. We study the resource calculus, an extension of the λ-calculus allowing to model resource consumption. We achieve an internal separation result, in analogy with Böhm's theorem of λ-calculus. We define an equivalence relation on the terms, which we prove to be the maximal non-trivial congruence on normalizable terms respecting β-reduction. It is significant that this equivalence extends the usual η-equivalence and is related to Ehrhard's Taylor expansion – a translation mapping terms into series of finite resources.

Keywords: differential linear logic, resource lambda-calculus, separation property, Böhm-out technique.

Introduction

Böhm's theorem in the λ-calculus. Böhm's theorem [1] is a fundamental result in the untyped λ-calculus [2] stating that, given two closed distinct $\beta\eta$-normal λ-terms M and N, there exists a sequence of λ-terms $\vec{L}$, such that $M\vec{L}$ β-reduces to the first projection $\lambda xy.x$ and $N\vec{L}$ β-reduces to the second projection $\lambda xy.y$. The original issue motivating this result was the quest for solutions of systems of equations between λ-terms: given closed terms $M_1, N_1, \ldots, M_n, N_n$, is there a λ-term S such that $SM_1 \equiv_\beta N_1 \wedge \cdots \wedge SM_n \equiv_\beta N_n$ holds? The answer is trivial for $n = 1$ (just take $S = \lambda z.N_1$ for a fresh variable z) and Böhm's theorem gives a positive answer for $n = 2$ and M_1, M_2 distinct $\beta\eta$-normal forms (apply the theorem to M_1, M_2 and set $S = \lambda f.f\vec{L}N_1N_2$). The result has been then generalized in [3] to treat every finite family $M_1, \ldots, M_n$ of pairwise distinct $\beta\eta$-normal forms. This generalization is non-trivial since each M_i may differ from the other ones at distinct addresses of its syntactic tree.

As an important consequence of Böhm's theorem we have that the $\beta\eta$-equivalence is the maximal non-trivial congruence on normalizable terms extending the β-equivalence. The case of non-normalizable terms has been addressed by Hyland in [4]. Indeed, the $\beta\eta$-equivalence is not maximal on such terms, and one must consider the congruence $\mathcal{H}^*$ equating two λ-terms whenever they have

---

* This work is partly supported by NWO Project 612.000.936 CALMOC and the chaire CNRS "Logique linéaire et calcul".

L. Ong (Ed.): TLCA 2011, LNCS 6690, pp. 153–168, 2011.

the same Böhm tree up to possibly infinite η-expansions [2, §16.2]. Then one proves that for all closed λ-terms $M \not\equiv_{\mathcal{H}^\star} N$ there is a sequence of λ-terms $\vec{L}$ such that $M\vec{L}$ β-reduces to the identity $\lambda x.x$ while $N\vec{L}$ is unsolvable (i.e., it does not interact with the environment [2, §8.3]) or *vice versa*. This property is called *semi-separation* because of the asymmetry between the two values: the identity, on the one hand, and any unsolvable term, on the other hand. In fact, non-normalizable terms represent partial functions, and one cannot hope to separate a term less defined than another one without sending the first to an unsolvable term (corresponding to the empty function). Despite the fact that Hyland's semi-separability is weaker than the full separability achieved by Böhm's theorem, it is sufficient to entail that $\mathcal{H}^\star$ is the maximal non-trivial congruence on λ-terms extending β-equivalence and equating all unsolvable terms [2, Thm. 16.2.6].

The resource λ-calculus. We study Böhm's theorem in the resource λ-calculus (Λ^r for short), which is an extension of the λ-calculus along two directions. First, Λ^r is resource sensitive. Following Girard's linear logic [5], the λ-calculus application can be written as $M(N^!)$ emphasizing the fact that the argument N is actually infinitely available for the function M, i.e. it can be erased or copied as many times as needed during the evaluation. Λ^r extends this setting by allowing also applications of the form $M(N^n)$ where N^n denotes a finite resource that must be used exactly n-times during the evaluation. If the number n does not match the needs of M then the application evaluates to the empty sum 0, expressing the absence of a result. In fact, 0 is a β-normal form giving a notion of unsolvable different from the λ-calculus one represented by looping terms[1]. The second feature of Λ^r is the non-determinism. Indeed, the argument of an application, instead of being a single term, is a bag of resources, each being either finite or infinitely available. In the evaluation several possible choices arise, corresponding to the different possibilities of distributing the resources among the occurrences of the formal parameter. The outcome is a finite formal sum of terms collecting all possible results.

Boudol has been the first to extend the λ-calculus with a resource sensitive application [8]. His resource calculus was designed to study Milner's encoding of the lazy λ-calculus into the π-calculus [9,10]. Some years later, Ehrhard and Regnier introduced the differential λ-calculus [11], drawing on insights gained from the quantitative semantics of linear logic, denoting proofs/terms as smooth (i.e. infinitely differentiable) functions. As remarked by the authors, the differential λ-calculus is quite similar to Boudol's calculus, the resource sensitive application $M(N^n)$ corresponds to applying the n-th derivative of M at 0 to N. This intuition was formalized by Tranquilli, who defined the present syntax of Λ^r and showed a Curry-Howard correspondence between this calculus and Ehrhard and Regnier's differential nets [12]. The main differences between Boudol's calculus and Λ^r are that the former is equipped with explicit substitution and lazy operational semantics, while the latter is a true extension of the regular λ-calculus.

---

[1] Denotational models of Λ^r distinguishing between 0 and the usual unsolvable terms are built in [6]. For more details on the notion of solvability in Λ^r see [7].

Since we cannot separate M from $M + M$ we will conveniently suppose that the sum on Λ^r is idempotent as in [13]; this amounts to say that we only check whether a term appears in a result, not how many times it appears.

A resource conscious Böhm's theorem. A notable outcome of Ehrhard and Regnier's work has been to develop the λ-calculus application as an infinite series of finite applications, $M(N^!) = \sum_{n=0}^{\infty} \frac{1}{n!} M(N^n)$, in analogy with the Taylor expansion of the entire functions. In [14], the authors relate the Böhm tree of a λ-term with its Taylor expansion, giving the intuition that the latter is a resource conscious improvement of the former. Following this intuition, we achieve the main result of this paper, namely a separation property in Λ^r that can be seen as a resource sensitive Böhm's theorem (Theorem 2). Such a result states that for all closed β-normal M, N having η-different Taylor expansion, there is a sequence $\vec{L}$, such that $M\vec{L}$ β-reduces to $\lambda x.x$ and $N\vec{L}$ β-reduces to 0, or *vice versa*.

This theorem reveals a first sharp difference between Λ^r and the λ-calculus, as our result is more similar to Hyland's semi-separation than Böhm's theorem, even if we consider the β-normal forms. This is due to the empty sum 0, the unsolvable β-normal form, outcome of the resource consciousness of Λ^r.

Taylor expansion is a semantical notion, in the sense that it is an infinite series of finite terms. It is then notable that we give a syntactic characterization of the Taylor equality introducing the τ-equivalence in Definition 3 (Proposition 1). As expected, our semi-separability is strong enough to entail that the $\eta\tau$-equivalence induces the maximal non-trivial congruence on β-normalizable terms extending the β-equivalence (Corollary 1).

A crucial ingredient in the classic proof of Böhm's theorem is the fact that it is possible to erase subterms in order to pull out of the terms their structural difference. This is not an easy task in Λ^r, since the finite resources must be consumed and cannot be erased. In this respect, our technique has some similarities with the one developed to achieve the separation for the λI-calculus (i.e., the λ-calculus without weakening, [2, §10.5]). Moreover, since the argument of an application is a bag of resources, comparing a difference between two terms may turn into comparing the differences between two multisets of terms, and this problem presents analogies with that of separating a finite set of terms [3].

Basic definitions and notations. We let $\mathbf{N}$ denote the set of natural numbers. Given a set $\mathcal{X}$, $\mathcal{M}_f(\mathcal{X})$ is the set of all finite multisets over $\mathcal{X}$. Given a reduction $\xrightarrow{r}$ we let $\xleftarrow{r}$, $\xrightarrow{r*}$ and $\equiv_r$ denote its transpose, its transitive-reflexive closure and its symmetric-transitive-reflexive closure, respectively.

An operator $F(-)$ (resp. $F(-,-)$) is *extended by linearity* (resp. *bilinearity*) by setting $F\left(\Sigma_i A_i\right) = \Sigma_i F(A_i)$ (resp. $F\left(\Sigma_i A_i, \Sigma_j B_j\right) = \Sigma_{i,j} F(A_i, B_j)$).

1   Resource Calculus

Syntax. The *resource calculus* has three syntactic categories: *terms* that are in functional position, *bags* that are in argument position and represent unordered lists of resources, and *finite formal sums* that represent the possible results of

$$
\begin{array}{llll}
\Lambda^r: & M, N, L & ::= x \mid \lambda x.M \mid MP & \text{terms} \\
\Lambda^b: & P, Q, R & ::= [M_1, \ldots, M_n, \mathbb{M}^!] & \text{bags} \\
\Lambda^e: & A, B & ::= M \mid P & \text{expressions} \\
& \mathbb{M}, \mathbb{N} \in \mathbf{2}\langle \Lambda^r \rangle \quad \mathbb{P}, \mathbb{Q} \in \mathbf{2}\langle \Lambda^b \rangle \quad \mathbb{A}, \mathbb{B} \in \mathbf{2}\langle \Lambda^e \rangle := \mathbf{2}\langle \Lambda^r \rangle \cup \mathbf{2}\langle \Lambda^b \rangle & \text{sums}
\end{array}
$$

(a) Grammar of terms, resources, bags, expressions, sums.

$$
\lambda x.(\textstyle\sum_i M_i) := \sum_i \lambda x.M_i \qquad\qquad \mathbb{M}(\textstyle\sum_i P_i) := \sum_i \mathbb{M}P_i
$$

$$
(\textstyle\sum_i M_i)\mathbb{P} := \sum_i M_i\mathbb{P} \qquad\qquad [(\textstyle\sum_i M_i)] \cdot \mathbb{P} := \sum_i [M_i] \cdot \mathbb{P}
$$

(b) Notation on $\mathbf{2}\langle \Lambda^e \rangle$.

$$
y\langle N/x \rangle := \begin{cases} N & \text{if } y = x, \\ 0 & \text{otherwise,} \end{cases} \qquad \begin{array}{l} (\lambda y.M)\langle N/x \rangle := \lambda y.(M\langle N/x \rangle), \\ (MP)\langle N/x \rangle := M\langle N/x \rangle P + M(P\langle N/x \rangle), \end{array}
$$

$$
[\mathbb{M}^!]\langle N/x \rangle := [\mathbb{M}\langle N/x \rangle, \mathbb{M}^!], \qquad ([M] \cdot P)\langle N/x \rangle := [M\langle N/x \rangle] \cdot P + [M] \cdot P\langle N/x \rangle,
$$

(c) Linear substitution, in the abstraction case we suppose $y \notin \mathrm{FV}(N) \cup \{x\}$.

Fig. 1. Syntax, notations and linear substitution of resource calculus

a computation. Figure 1(a) provides the grammar for generating the set Λ^r of terms and the set Λ^b of bags, together with their typical metavariables.

A *bag* $[\vec{M}, \mathbb{M}^!]$ is a compound object, consisting of a multiset of *linear resources* $[\vec{M}]$ and a set of terms $\mathbb{M}$ presented in additive notation (see the discussion on sets and sums below) representing the *reusable resources*. Roughly speaking, the linear resources in $\vec{M}$ must be used exactly once during a reduction, while the reusable ones in $\mathbb{M}$ can be used ad libitum (hence, following the linear logic notation, $\mathbb{M}$ is decorated with a ! superscript).

We shall deal with bags as if they were multisets presented in multiplicative notation, defining union by $[\vec{M}, \mathbb{M}^!] \cdot [\vec{N}, \mathbb{N}^!] := [\vec{M}, \vec{N}, (\mathbb{M} + \mathbb{N})^!]$. This operation is commutative, associative and has the *empty bag* $1 := [0^!]$ as neutral element. To avoid confusion with application we will never omit the dot ".". To lighten the notations we write $[L_1, \ldots, L_k]$ for the bag $[L_1, \ldots, L_k, 0^!]$, and $[M^k]$ for the bag $[M, \ldots, M]$ containing k copies of M. Such a notation allows to decompose a bag in several ways, and this will be used throughout the paper.

As for example:

$$
[x, y, (x+y)^!] = [x] \cdot [y, (x+y)^!] = [x^!] \cdot [x, y, y^!] = [x, y] \cdot [(x+y)^!] = [x, x^!] \cdot [y, y^!].
$$

Expressions (whose set is denoted by Λ^e) are either terms or bags and will be used to state results holding for both categories.

Let $\mathbf{2}$ be the semiring $\{0, 1\}$ with $1 + 1 = 1$ and multiplication defined in the obvious way. For any set $\mathcal{X}$, we write $\mathbf{2}\langle \mathcal{X} \rangle$ for the free $\mathbf{2}$-module generated by $\mathcal{X}$, so that $\mathbf{2}\langle \mathcal{X} \rangle$ is isomorphic to the finite powerset of $\mathcal{X}$, with addition corresponding to union, and scalar multiplication defined in the obvious way.

However we prefer to keep the algebraic notations for elements of $2\langle\mathcal{X}\rangle$, hence set union will be denoted by $+$ and the empty set by 0. This amounts to say that $2\langle\Lambda^r\rangle$ (resp. $2\langle\Lambda^b\rangle$) denotes the set of finite formal sums of terms (resp. bags), with an idempotent sum. We also set $2\langle\Lambda^e\rangle = 2\langle\Lambda^r\rangle \cup 2\langle\Lambda^b\rangle$. This is an abuse of notation, as $2\langle\Lambda^e\rangle$ here does not denote the 2-module generated over $\Lambda^e = \Lambda^r \cup \Lambda^b$ but rather the union of the two 2-modules; this amounts to say that sums may be taken only in the same sort.

The *size of* $\mathbb{A} \in 2\langle\Lambda^e\rangle$ is defined inductively by: $\text{size}(\Sigma_i A_i) = \Sigma_i \text{size}(A_i)$, $\text{size}(x) = 1$, $\text{size}(\lambda x.M) = \text{size}(M) + 1$, $\text{size}(MP) = \text{size}(M) + \text{size}(P) + 1$, $\text{size}([M_1,\ldots,M_k,\mathbb{M}^!]) = \Sigma_{i=1}^k \text{size}(M_i) + \text{size}(\mathbb{M}) + 1$.

Notice that the grammar for terms and bags does not include any sums, but under the scope of a $(\cdot)^!$. However, as syntactic sugar – and *not* as actual syntax – we extend all the constructors to sums as shown in Figure 1(b). In fact all constructors except the $(\cdot)^!$ are (multi)linear, as expected. The intuition is that a reusable sum $(M + N)^!$ represents a resource that can be used several times and each time one can choose non-deterministically M or N.

Observe that in the particular case of empty sums, we get $\lambda x.0 := 0$, $M0 := 0$, $0P := 0$, $[0] := 0$ and $0 \cdot P := 0$, but $[0^!] = 1$. Thus 0 annihilates any term or bag, except when it lies under a $(\cdot)^!$. As an example of this *extended (meta-)syntax*, we may write $(x_1+x_2)[y_1+y_2,(z_1+z_2)^!]$ instead of $x_1[y_1,(z_1+z_2)^!]+x_1[y_2,(z_1+z_2)^!] + x_2[y_1,(z_1+z_2)^!] + x_2[y_2,(z_1+z_2)^!]$. This kind of meta-syntactic notation is discussed thoroughly in [14].

The α-equivalence and the set $\text{FV}(\mathbb{A})$ of free variables are defined as in ordinary λ-calculus. From now on expressions are considered up to α-equivalence. Concerning specific terms we set:

$$\mathbf{I} := \lambda x.x, \qquad \mathbf{X}_n := \lambda x_1 \ldots \lambda x_n \lambda x.x[x_1^!]\ldots[x_n^!] \text{ for } n \in \mathbf{N},$$

where $\mathbf{X}_n$ is called the *n-th Böhm permutator*.

Due to the presence of two kinds of resources, we need two different notions of substitutions: the usual λ-calculus substitution and a linear one, which is particular to differential and resource calculi (see [14,15]).

Definition 1 (Substitutions). *We define the following substitution operations.*

1. *$A\{N/x\}$ is the usual capture-free substitution of N for x in A. It is extended to sums as in $\mathbb{A}\{\mathbb{N}/x\}$ by linearity in $\mathbb{A}$, and using the notations of Figure 1(b) for $\mathbb{N}$.*
2. *$A\langle N/x\rangle$ is the linear substitution defined inductively in Figure 1(c). It is extended to $\mathbb{A}\langle\mathbb{N}/x\rangle$ by bilinearity in both $\mathbb{A}$ and $\mathbb{N}$.*

Intuitively, linear substitution replaces the resource to *exactly one* linear free occurrence of the variable. In presence of multiple occurrences, all possible choices are made and the result is the sum of them. E.g., $(x[x])\langle\mathbf{I}/x\rangle = \mathbf{I}[x] + x[\mathbf{I}]$.

Example 1. Notice the difference between $[x,x^!]\{M + N/x\} = [(M + N),(M + N)^!] = [M,(M + N)^!] + [N,(M + N)^!]$ and $[x,x^!]\langle M + N/x\rangle = [x,x^!]\langle M/x\rangle + [x,x^!]\langle N/x\rangle = [M,x^!] + [x,M,x^!] + [N,x^!] + [x,N,x^!]$.

Linear substitution bears resemblance to differentiation, as shown clearly in Ehrhard and Regnier's differential λ-calculus [11]. For instance, it enjoys the following Schwarz's Theorem, whose proof is rather classic and is omitted.

Lemma 1 (Schwarz's Theorem [14,11]). *Given* $\mathbb{A} \in 2\langle \Lambda^e \rangle$, $\mathbb{M}, \mathbb{N} \in 2\langle \Lambda^r \rangle$ *and* $y \notin \mathrm{FV}(\mathbb{M}) \cup \mathrm{FV}(\mathbb{N})$ *we have* $\mathbb{A}\langle \mathbb{M}/y \rangle \langle \mathbb{N}/x \rangle = \mathbb{A}\langle \mathbb{N}/x \rangle \langle \mathbb{M}/y \rangle + \mathbb{A}\langle \mathbb{M}\langle \mathbb{N}/x \rangle /y \rangle$. *In particular, if* $x \notin \mathrm{FV}(\mathbb{M})$ *the two substitutions commute.*

$$\frac{M \mathrel{R} M}{\lambda x.M \mathrel{R} \lambda x.M}\ \text{lam} \qquad \frac{M \mathrel{R} M}{MP \mathrel{R} MP}\ \text{appl} \qquad \frac{P \mathrel{R} \mathbb{P}}{MP \mathrel{R} M\mathbb{P}}\ \text{appr}$$

$$\frac{M \mathrel{R} M}{[M] \cdot P \mathrel{R} [M] \cdot P}\ \text{lin} \qquad \frac{M \mathrel{R} M}{[M^!] \cdot P \mathrel{R} [M^!] \cdot P}\ \text{bng} \qquad \frac{A \mathrel{R} A}{A + \mathbb{B} \mathrel{R} A + \mathbb{B}}\ \text{sum}$$

(a) Rules defining the context closure of a relation $\mathsf{R} \subseteq \Lambda^e \times 2\langle \Lambda^e \rangle$.

$$\frac{M \mathrel{R} \mathbb{N}}{\lambda x.M \mathrel{R} \lambda x.\mathbb{N}}\ \text{lam} \qquad \frac{M \mathrel{R} \mathbb{N} \quad \mathbb{P} \mathrel{R} \mathbb{Q}}{M\mathbb{P} \mathrel{R} \mathbb{N}\mathbb{Q}}\ \text{app}$$

$$\frac{M \mathrel{R} \mathbb{N} \quad \mathbb{P} \mathrel{R} \mathbb{Q}}{[M] \cdot \mathbb{P} \mathrel{R} [\mathbb{N}] \cdot \mathbb{Q}}\ \text{lin} \qquad \frac{M \mathrel{R} \mathbb{N}}{[M^!] \mathrel{R} [\mathbb{N}^!]}\ \text{bng} \qquad \frac{A \mathrel{R} \mathbb{B} \quad A' \mathrel{R} \mathbb{B}'}{A + A' \mathrel{R} \mathbb{B} + \mathbb{B}'}\ \text{sum}$$

(b) Rules defining a compatible relation $\mathsf{R} \subseteq 2\langle \Lambda^e \rangle \times 2\langle \Lambda^e \rangle$.

Fig. 2. Definition of context closure and compatible relation

Operational semantics. Given a relation $\mathsf{R} \subseteq \Lambda^e \times 2\langle \Lambda^e \rangle$ its *context closure* is the smallest relation in $2\langle \Lambda^e \rangle \times 2\langle \Lambda^e \rangle$ containing R and respecting the rules of Figure 2(a). The main notion of reduction of resource calculus is β-*reduction*, which is defined as the context closure of the following rule:

$$(\beta) \qquad (\lambda x.M)[L_1, \ldots, L_k, \mathbb{N}^!] \xrightarrow{\beta} M\langle L_1/x \rangle \cdots \langle L_k/x \rangle \{\mathbb{N}/x\}.$$

Notice that the β-rule is independent of the ordering of the linear substitutions, as shown by the Schwarz's Theorem above. We say that $A \in \Lambda^e$ is in β-*normal form* (β-nf, for short) if there is no $\mathbb{A}$ such that $A \xrightarrow{\beta} \mathbb{A}$. A sum $\mathbb{A}$ *is in* β-nf if all its summands are. Notice that 0 is a β-nf. It is easy to check that a term M is in β-nf iff $M = \lambda x_1 \ldots x_n.y P_1 \ldots P_k$ for $n, k \geq 0$ and, for every $1 \leq i \leq k$, all resources in P_i are in β-nf. The variable y in M is called *head-variable*.

The regular λ-calculus [2] can be embedded into the resource one by translating every application MN into $M[N^!]$. In this fragment the β-reduction defined above coincides with the usual one. Hence the resource calculus has usual looping terms like $\mathbf{\Omega} := (\lambda x.x[x^!])[(\lambda x.x[x^!])^!]$, but also terms like $\mathbf{I}1$ or $\mathbf{I}[y, y]$ reducing to 0 because there is a mismatch between the number of linear resources needed by the functional part of the application and the number it actually receives.

$$(\textstyle\sum_i A_i)^\circ := \bigcup_i A_i^\circ \qquad x^\circ := \{x\} \qquad (\lambda x.M)^\circ := \lambda x.M^\circ \qquad (MP)^\circ := M^\circ P^\circ$$

$$([\mathbf{M}^!])^\circ := \mathcal{M}_f(\mathbf{M}^\circ) \qquad ([M] \cdot P)^\circ := [M^\circ] \cdot P^\circ$$

Fig. 3. Taylor expansion $\mathbb{A}^\circ$ of $\mathbb{A}$

Theorem 1 (Confluence [15]). *The β-reduction is Church-Rosser on Λ^r.*

The resource calculus is *intensional*, indeed just like in the λ-calculus there are different programs having the same extensional behaviour. In order to achieve an internal separation, we need to consider the η-*reduction* that is defined as the contextual closure of the following rule:

$$(\eta) \qquad\qquad \lambda x.M[x^!] \overset{\eta}{\to} M, \text{ if } x \notin \mathrm{FV}(M).$$

The Taylor expansion. The *finite resource calculus* is the fragment of resource calculus having only linear resources (every bag has the set of reusable resources empty). The terms (resp. bags, expressions) of this sub-calculus are called *finite* and their set is denoted by Λ_f^r (resp. Λ_f^b, Λ_f^e). Notice that the bags of Λ_f^b are actually finite multisets. It is easy to check that the above sets are closed under β-reduction, while η-reduction cannot play any role here.

In Definition 2, we describe the Taylor expansion as a map $(\cdot)^\circ$ from $\mathbf{2}\langle \Lambda^r \rangle$ (resp. $\mathbf{2}\langle \Lambda^b \rangle$) to possibly infinite sets of finite terms (resp. finite bags). The Taylor expansion defined in [11,14], in the context of λ-calculus, is a translation developing every application as an infinite series of finite applications with rational coefficients. In our context, since the coefficients are in $\mathbf{2}$, the Taylor expansion of an expression is a (possibly infinite) set of finite expressions. Indeed, for all sets $\mathcal{X}$, the set of the infinite formal sums $\mathbf{2}\langle \mathcal{X} \rangle_\infty$ with coefficients in $\mathbf{2}$ is isomorphic to the powerset of $\mathcal{X}$. Our Taylor expansion corresponds to the support[2] of the Taylor expansion taking rational coefficients given in [11,14].

To lighten the notations, we adopt for sets of expressions the same abbreviations as introduced for finite sums in Figure 1(b). For example, given $\mathcal{M} \subseteq \Lambda_f^r$ and $\mathcal{P}, \mathcal{Q} \subseteq \Lambda_f^b$ we have $\lambda x.\mathcal{M} = \{\lambda x.M \mid M \in \mathcal{M}\}$ and $\mathcal{P} \cdot \mathcal{Q} = \{P \cdot Q \mid P \in \mathcal{P},\ Q \in \mathcal{Q}\}$.

Definition 2. *Let $\mathbb{A} \in \mathbf{2}\langle \Lambda^e \rangle$. The Taylor expansion of $\mathbb{A}$ is the set $\mathbb{A}^\circ \subseteq \Lambda_f^e$ which is defined (by structural induction on $\mathbb{A}$) in Figure 3.*

Example 2. As previously announced, the Taylor expansion of an expression A can be infinite, e.g., $(\lambda x.x[x^!])^\circ = \{\lambda x.x[x^n] \mid n \in \mathbf{N}\}$. Different terms may share the same Taylor expansion: $(x[(z[y^!])^!])^\circ = (x[(z1 + z[y, y^!])^!])^\circ$. The presence of linear resources permits situations where $M^\circ \subsetneq N^\circ$, for example $M := x[x, x^!]$, $N := x[x^!]$. The presence of non-determinism allows to build terms like

---

[2] I.e., the set of those finite terms appearing in the series with a non-zero coefficient.

$M_1 := x[(y+z)^!], M_2 := x[(y+h)^!]$ such that $M_1^\circ \cap M_2^\circ = \{x[y^n] \mid n \in \mathbf{N}\}$ is infinite. However the intersection can also be finite as in $N_1 := x[y, z^!], N_2 := x[z, y^!]$ where $N_1^\circ \cap N_2^\circ = \{x[y, z]\}$.

2 A Syntactic Characterization of Taylor Equality

In Λ^r, there are distinct $\beta\eta$-nf's which are inseparable, in contrast with what we have in the regular λ-calculus. In fact, all normal forms having equal Taylor expansions are inseparable, since the first author proved in [16] that there is a non-trivial denotational model of resource calculus equating all terms having the same Taylor expansion. For example, $x[(z[y^!])^!]$ and $x[(z1 + z[y, y^!])^!]$ are distinct inseparable $\beta\eta$-nf's since they have the same Taylor expansion.

Because of its infinitary nature, the property of having the same Taylor expansion is more semantical than syntactical. In this section, we provide an alternative syntactic characterization (Definition 3, Proposition 1).

A relation $\mathsf{R} \subseteq \mathbf{2}\langle \Lambda^e \rangle \times \mathbf{2}\langle \Lambda^e \rangle$ is *compatible* if it satisfies the rules in Figure 2(b). The *congruence generated by* a relation R, denoted $\equiv_\mathsf{R}$, is the smallest compatible equivalence relation containing R. Given two relations $\mathsf{R}, \mathsf{S} \subseteq \mathbf{2}\langle \Lambda^e \rangle \times \mathbf{2}\langle \Lambda^e \rangle$, we write $\equiv_\mathsf{RS}$ for the congruence generated by their union $\mathsf{R} \cup \mathsf{S}$.

Definition 3. *The* Taylor equivalence $\equiv_\tau$ *is the congruence generated by:*

$$(\tau) \qquad\qquad [M^!] \equiv_\tau 1 + [M, M^!]$$

Moreover, we set $\mathbb{A} \sqsubseteq_\tau \mathbb{B}$ *iff* $\mathbb{A} + \mathbb{B} \equiv_\tau \mathbb{B}$.

It is not difficult to check that $\sqsubseteq_\tau$ is a compatible preorder. We now prove that it captures exactly the inclusion between Taylor expansions (Proposition 1).

Example 3. We have $z[x, x^!] + z1 \equiv_\tau z[x^!] \sqsubseteq_\tau z[x, x^!] + z[y^!]$, while $x[x^!] \not\equiv_\tau x[y^!] \not\sqsubseteq_\tau x[y, y^!]$.

Note that all elements of A° share the same minimum structure, called here *skeleton*, obtained by taking 0 occurrences of every reusable resource.

Definition 4. *Given* $A \in \Lambda^e$, *its* skeleton $\mathfrak{s}(A) \in \Lambda_\mathsf{f}^e$ *is obtained by erasing all the reusable resources occurring in* A. *That is, inductively:*

$$\mathfrak{s}(x) := x, \quad \mathfrak{s}(\lambda x.M) := \lambda x.\mathfrak{s}(M), \quad \mathfrak{s}(MP) := \mathfrak{s}(M)\mathfrak{s}(P),$$

$$\mathfrak{s}([M_1, \ldots, M_n, \mathbb{M}^!]) := [\mathfrak{s}(M_1), \ldots, \mathfrak{s}(M_n)].$$

Obviously $\mathfrak{s}(A) \in A^\circ$. In general it is false that $\mathfrak{s}(A) \in B^\circ$ entails $A \sqsubseteq_\tau B$. Take for instance $A := x[x^!]$ and $B := x[y^!]$; indeed $\mathfrak{s}(A) = x1 \in \{x[y^n] \mid n \in \mathbf{N}\} = B^\circ$. The above implication becomes however true when A is "expanded enough".

Definition 5. *Given* $k \in \mathbf{N}$, *we say that* $A \in \Lambda^e$ *is* k-expanded *if it is generated by the following grammar:*

$$M, N ::= x \mid \lambda x.M \mid MP \qquad\qquad P, Q ::= 1 \mid [M] \cdot P \mid [M^k, M^!] \cdot P$$

Example 4. The terms x, $\lambda y.x[y]$ and $x1$ are k-expanded for all $k \in \mathbf{N}$. The bag $[y^4, x^3, (x+y)^!]$ is 3-expanded since it can be decomposed as $[x^3, x^!] \cdot [y^4, y^!]$, while it is not 4-expanded.

Lemma 2. *Let $A \in \Lambda^e$ be k-expanded for some $k \in \mathbf{N}$. Then for every $B \in \Lambda^e$ such that $\mathrm{size}(B) \leq k$, we have that $\mathfrak{s}(A) \in B^\circ$ entails $A \sqsubseteq_\tau B$.*

Proof. By induction on A. The only significant case is when A is a bag, which splits in three subcases, depending on how such a bag can be decomposed.

CASE I ($A = [M^k, M^!] \cdot P$, P k-EXPANDED). By definition $\mathfrak{s}(A) = [\mathfrak{s}(M)^k] \cdot \mathfrak{s}(P)$. From $\mathfrak{s}(A) \in B^\circ$, we deduce that $B = Q_1 \cdot Q_2 \cdot Q_3$ where $Q_1 = [L_1, \ldots, L_\ell]$, $Q_2 = [(N_1 + \cdots + N_n)^!]$ for some $\ell, n \geq 0$ and Q_1, Q_2, Q_3 are such that $[\mathfrak{s}(M)^\ell] \in Q_1^\circ$, $[\mathfrak{s}(M)^{k-\ell}] \in Q_2^\circ$ and $\mathfrak{s}(P) \in (Q_2 \cdot Q_3)^\circ$. From $\mathrm{size}(B) \leq k$ we get $\ell < k$ and this entails $n > 0$. We then have that $\mathfrak{s}(M) \in L_i^\circ$, for every $i \leq \ell$, and there is a j such that $\mathfrak{s}(M) \in N_j^\circ$. By induction hypothesis, we have $M \sqsubseteq_\tau L_i$ for every $i \leq \ell$, $M \sqsubseteq_\tau N_j$ and $P \sqsubseteq_\tau Q_2 \cdot Q_3$. Hence, by compatibility of $\sqsubseteq_\tau$, we derive $[M^\ell] \sqsubseteq_\tau Q_1$ and $[M^!] \sqsubseteq_\tau [N_j^!]$, that entails $[M^k, M^!] \sqsubseteq_\tau [M^\ell, M^!] \sqsubseteq_\tau Q_1 \cdot [N_j^!]$. From this, using $[N_j^!] \cdot Q_2 \equiv_\tau Q_2$ and $P \sqsubseteq_\tau Q_2 \cdot Q_3$, we conclude $A \sqsubseteq_\tau Q_1 \cdot [N_j^!] \cdot Q_2 \cdot Q_3 \equiv_\tau B$.

CASE II ($A = [M] \cdot P$, P WITHOUT REUSABLE RESOURCES). By definition $\mathfrak{s}(A) = [\mathfrak{s}(M)] \cdot \mathfrak{s}(P)$. Suppose $\mathfrak{s}(A) \in B^\circ$, then two subcases are possible.

Case $B = [N] \cdot Q$ such that $\mathfrak{s}(M) \in N^\circ$ and $\mathfrak{s}(P) \in Q^\circ$. By induction hypothesis $M \sqsubseteq_\tau N$ and $P \sqsubseteq_\tau Q$. By compatibility, $A = [M] \cdot P \sqsubseteq_\tau [N] \cdot Q = B$.

Case $B = [N^!] \cdot Q$ such that $\mathfrak{s}(M) \in N^\circ$ and $\mathfrak{s}(P) \in ([N^!] \cdot Q)^\circ$. Then by induction hypothesis, $M \sqsubseteq_\tau N$ and $P \sqsubseteq_\tau [N^!] \cdot Q$, and, always by compatibility of $\sqsubseteq_\tau$, we conclude that $A = [M] \cdot P \sqsubseteq_\tau [N, N^!] \cdot Q \sqsubseteq_\tau B$.

CASE III ($A = 1$). From $\mathfrak{s}(A) = 1 \in B^\circ$, we deduce B is a bag containing only reusable resources, hence trivially $A = 1 \sqsubseteq_\tau B$. $\square$

Lemma 3. *For all $A \in \Lambda^e$ and $k \in \mathbf{N}$ there is a k-expanded $\mathbb{A}$ such that $A \equiv_\tau \mathbb{A}$.*

Proof. By immediate structural induction on A. The crucial case is when $A = [M^!] \cdot P$. Then by induction hypothesis $M \equiv_\tau \mathbb{M}$ and $P \equiv_\tau \mathbb{P}$ for some k-expanded $\mathbb{M}, \mathbb{P}$. Let us explicit the first sum into $\mathbb{M} = M_1 + \cdots + M_m$. Then, we have:

$$[M^!] \cdot P \equiv_\tau [\mathbb{M}^!] \cdot \mathbb{P} \equiv_\tau [M_1^k, \ldots, M_m^k, \mathbb{M}^!] \cdot \mathbb{P} + \sum_{n_1=0}^{k-1} \cdots \sum_{n_m=0}^{k-1} [M_1^{n_1}, \ldots, M_m^{n_m}] \cdot \mathbb{P}.$$

Note that all summands in the last sum are k-expanded, since so are $\mathbb{M}, \mathbb{P}$. $\square$

Proposition 1. *For all $\mathbb{A}, \mathbb{B} \in 2\langle \Lambda^e \rangle$ we have that $\mathbb{A} \sqsubseteq_\tau \mathbb{B}$ iff $\mathbb{A}^\circ \subseteq \mathbb{B}^\circ$.*

Proof. ($\Rightarrow$) By a trivial induction on a derivation of $\mathbb{A} + \mathbb{B} \equiv_\tau \mathbb{B}$, remarking that all rules defining $\equiv_\tau$ preserve the property of having equal Taylor expansion.

($\Leftarrow$) By induction on the number of terms in the sum $\mathbb{A}$. If $\mathbb{A} = 0$, then clearly $\mathbb{A} \sqsubseteq_\tau \mathbb{B}$ by the reflexivity of $\equiv_\tau$. If $\mathbb{A} = A + \mathbb{A}'$, then by induction hypothesis we have $\mathbb{A}' \sqsubseteq_\tau \mathbb{B}$. As for A, let $k \geq \max(\mathrm{size}(A), \mathrm{size}(\mathbb{B}))$, by Lemma 3, we have a k-expanded sum $\mathbb{A}'' = A_1 + \cdots + A_a \equiv_\tau A$. From $\mathbb{A}'' \equiv_\tau A$ and the already proved left-to-right direction of the proposition we get $(A'')^\circ = A^\circ \subseteq \mathbb{B}^\circ$. This means that $A_i^\circ \subseteq \mathbb{B}^\circ$ for all $i \leq a$. In particular $\mathfrak{s}(A_i) \in B_{j_i}^\circ$, for a particular summand B_{j_i} of $\mathbb{B}$. Since we are supposing that A_i is k-expanded and $\mathrm{size}(B_{j_i}) \leq k$, we can apply Lemma 2, so arguing $A_i \sqsubseteq_\tau B_{j_i} \sqsubseteq_\tau \mathbb{B}$. Since this holds for every $i \leq a$, we get $\mathbb{A}'' \sqsubseteq_\tau \mathbb{B}$. Then we can conclude $\mathbb{A} \equiv_\tau \mathbb{A}' + \mathbb{A}'' \sqsubseteq_\tau \mathbb{B} + \mathbb{B} \equiv_\tau \mathbb{B}$. $\qquad\square$

We conclude that $\equiv_\tau$ deserves the name of Taylor equivalence.

3   Separating a Finite Term from Infinitely Many Terms

We now know that two $\beta\eta$-nf's $\mathbb{M}, \mathbb{N}$ such that $\mathbb{M} \equiv_\tau \mathbb{N}$ are inseparable; hence, in order to achieve an internal separation, we need to consider the $\eta\tau$-difference. One may hope that $\mathbb{M} \not\equiv_{\eta\tau} \mathbb{N}$ is equivalent to first compute the η-nf of two β-normal $\mathbb{M}, \mathbb{N}$ and then check whether they are τ-different. Unfortunately, this is false as shown in the following counterexample (which is a counterexample to the fact that η-reduction is confluent modulo $\equiv_\tau$ [17, §14.3]).

Example 5. For all variables a, b we define the sums $\mathbb{M}_{a,b} := v[\lambda z.a1, \lambda z.b[z^!]] + v[\lambda z.a[z^!], \lambda z.b[z^!]]$ and $\mathbb{N}_{a,b} := v[\lambda z.a1, b] + v[a, b]$. Note that $\mathbb{M}_{a,b} \xrightarrow{\eta*} \mathbb{N}_{a,b}$, hence:

$$\mathbb{N}_{x,y} \xleftarrow{\eta*} \mathbb{M}_{x,y} \equiv_\tau v[\lambda z.x[z^!], \lambda z.y[z^!]] \equiv_\tau \mathbb{M}_{y,x} \xrightarrow{\eta*} \mathbb{N}_{y,x}.$$

Now, if $x \neq y$, then $\mathbb{N}_{x,y}$ and $\mathbb{N}_{y,x}$ are two inseparable but distinct $\beta\eta$-nf's such that $\mathbb{N}_{x,y} \not\equiv_\tau \mathbb{N}_{y,x}$.

The above example shows that, to study the $\eta\tau$-difference, we cannot simply analyze the structure of the η-nf's. Our solution will be to introduce a relation $\preceq_s$ such that $\mathbb{M} \not\equiv_{\eta\tau} \mathbb{N}$ entails either $\mathbb{M} \not\preceq_s \mathbb{N}$ or *vice versa* (Definition 7). Basically, this approach corresponds to first compute the Taylor expansion of $\mathbb{M}, \mathbb{N}$ and then compute their η-nf pointwise, using the following partial η-reduction on Λ_f^r.

Definition 6. *The* partial η-reduction $\xrightarrow{\varphi}$ *is the contextual closure of the rule* $\lambda x.M[x^n] \xrightarrow{\varphi} M$ *if* $x \notin \mathrm{FV}(M)$.

We define the relation $\preceq_s$ corresponding to the *Smith extension* of $\xrightarrow{\varphi*}$ and $\preceq_\tau$ corresponding to the relation "$\eta \sqsubseteq \eta$" of [2, Def. 10.2.32] keeping in mind the analogy between Taylor expansions and Böhm trees discussed in [14].

Definition 7. *Given* $\mathbb{M}, \mathbb{N} \in 2\langle \Lambda^r \rangle$ *we define:*

- $\mathbb{M} \preceq_s \mathbb{N}$ *iff* $\forall M \in \mathbb{M}^\circ, \exists N \in \mathbb{N}^\circ$ *such that* $M \xrightarrow{\varphi*} N$.
- $\mathbb{M} \preceq_\tau \mathbb{N}$ *iff* $\exists \mathbb{M}' \xrightarrow{\eta*} \mathbb{M}, \exists \mathbb{N}' \xrightarrow{\eta*} \mathbb{N}$ *such that* $\mathbb{M}' \sqsubseteq_\tau \mathbb{N}'$.

It is easy to check that $\preceq_s$ is a preorder, and we conjecture that also $\preceq_\tau$ is (we will not prove it because unnecessary for the present work).

Remark 1. It is clear that $\mathbb{M} \xrightarrow{\eta} \mathbb{N}$ implies $\mathbb{M} \preceq_s \mathbb{N}$. More generally, by transitivity of $\preceq_s$ we have that $\mathbb{M} \xrightarrow{\eta*} \mathbb{N}$ entails $\mathbb{M} \preceq_s \mathbb{N}$.

Lemma 4. *Let* $\mathbb{M}, \mathbb{N} \in \mathbf{2}\langle \Lambda^r \rangle$. *Then* $\mathbb{M} \preceq_\tau \mathbb{N}$ *and* $\mathbb{N} \preceq_\tau \mathbb{M}$ *entails* $\mathbb{M} \equiv_{\eta\tau} \mathbb{N}$

Proof. By hypothesis we have $\mathbb{M}', \mathbb{M}'' \xrightarrow{\eta*} \mathbb{M}$, $\mathbb{N}', \mathbb{N}'' \xrightarrow{\eta*} \mathbb{N}$, such that $\mathbb{M}' \sqsubseteq_\tau \mathbb{N}'$ and $\mathbb{N}'' \sqsubseteq_\tau \mathbb{M}''$. Then, $\mathbb{N}' \equiv_\tau \mathbb{M}' + \mathbb{N}' \xrightarrow{\eta*} \mathbb{M} + \mathbb{N}$, hence $\mathbb{N} \equiv_{\eta\tau} \mathbb{M} + \mathbb{N}$. Symmetrically, $\mathbb{M}'' \equiv_\tau \mathbb{M}'' + \mathbb{N}'' \xrightarrow{\eta*} \mathbb{M} + \mathbb{N}$, hence $\mathbb{M} \equiv_{\eta\tau} \mathbb{M} + \mathbb{N}$, and we conclude $\mathbb{M} \equiv_{\eta\tau} \mathbb{N}$. □

Notice that in the above proof it is crucial the idempotency of the sum.

Lemma 5. *Let* $\mathbb{M}, \mathbb{N} \in \mathbf{2}\langle \Lambda^r \rangle$. *Then* $\mathbb{M} \preceq_s \mathbb{N}$ *entails* $\mathbb{M} \preceq_\tau \mathbb{N}$.

Proof. (Outline) Define $\text{size}^{\text{sl}}(A)$ as the size of A, except for the bag case, where we set $\text{size}^{\text{sl}}([M_1, \ldots, M_m, (\Sigma_{i=m+1}^{m+k} M_i)^!]) := 1 + \max_i(\text{size}^{\text{sl}}(M_i))$. Notice that $\sup_{A \in \mathbb{A}^\circ}(\text{size}^{\text{sl}}(A)) \leq \text{size}(A)$ (in contrast, $\sup_{A \in \mathbb{A}^\circ}(\text{size}(A))$ can be unbounded). Extend the relation $\preceq_s$ on sets $\mathcal{A}, \mathcal{B}$ of finite terms in the obvious way: $\mathcal{A} \preceq_s \mathcal{B}$ iff $\forall M \in \mathcal{A}, \exists N \in \mathcal{B}$ such that $M \xrightarrow{\varphi*} N$. Then, the lemma follows from the following claim (take $\mathcal{A} = \mathbb{M}^\circ$ and $\mathbb{B} = \mathbb{N}$).

Claim 1. For all $\mathcal{A} \subseteq \Lambda_{\text{f}}^e$ and all $\mathbb{B} \in \mathbf{2}\langle \Lambda^e \rangle$, if $\sup_{A \in \mathcal{A}}(\text{size}^{\text{sl}}(A))$ is finite and $\mathcal{A} \preceq_s \mathbb{B}^\circ$, then there exists $\mathbb{B}' \xrightarrow{\eta*} \mathbb{B}$ such that $\mathcal{A} \subseteq \mathbb{B}'$.

The claim is proved by induction on the triplet made of $\sup_{A \in \mathcal{A}}(\text{size}^{\text{sl}}(A))$, $\sup_{A \in \mathcal{A}}(\text{size}^{\text{sl}}(A)) - \inf_{B \in \mathbb{B}^\circ}(\text{size}^{\text{sl}}(B))$ and $\text{size}(\mathbb{B})$ lexicographically ordered. □

To sum up, $\mathbb{M} \not\equiv_{\eta\tau} \mathbb{N}$ implies that, say, $\mathbb{M} \not\preceq_s \mathbb{N}$, which means $\exists M \in \mathbb{M}^\circ$ such that $\forall N \in \mathbb{N}^\circ$ we have $M \xrightarrow{\varphi*} N$. Hence, what Lemma 7 below does, is basically to separate such finite term M from all $\mathbb{N}$'s satisfying the condition $\forall N \in \mathbb{N}^\circ$ $M \xrightarrow{\varphi*} N$ (that are infinitely many). For technical reasons, we will need to suppose that M has a number of λ-abstractions greater than (or equal to) the number of λ-abstractions of $\mathbb{N}$, at any depth where its syntactic tree is defined.

Definition 8. *Let* $M, N \in \Lambda^r$ *be* β-*nf's of the shape* $M = \lambda x_1 \ldots \lambda x_a.y P_1 \ldots P_p$, $N = \lambda x_1 \ldots \lambda x_b.z Q_1 \ldots Q_q$. *We say that* M *is* λ-*wider than* N *if* $a \geq b$ *and each (linear or reusable) resource* L *in* P_i *is* λ-*wider than every (linear or reusable) resource* $L' \in Q_i$, *for all* $i \leq q$. *Given* $\mathbb{M}, \mathbb{N}$ *in* β-*nf we say that* $\mathbb{M}$ *is* λ-*wider than* $\mathbb{N}$ *iff each summand of* $\mathbb{M}$ *is* λ-*wider than all summands of* $\mathbb{N}$.

Notice that empty bags in M make it λ-wider than N, independently from the corresponding bags in N.

Example 6. The term $y1$ is λ-wider than zQ for any bag Q. The term $x[\lambda y.\mathbf{I}, \mathbf{I}]1$ is λ-wider than $x[\mathbf{I}][\mathbf{I}]$ but not than himself.

Remark 2. If $\mathbb{M}$ is λ-wider than $\mathbb{N}$ then every $M \in \mathbb{M}^\circ$ is λ-wider than $\mathbb{N}$.

Lemma 6. *For all* $\mathbb{M}, \mathbb{N}$ *in* β-*nf there is a* β-*normal* $\mathbb{M}'$ *such that* $\mathbb{M}' \xrightarrow{\eta*} \mathbb{M}$ *and* $\mathbb{M}'$ *is* λ-*wider than both* $\mathbb{M}$ *and* $\mathbb{N}$.

Proof. (Outline) For $m > \text{size}(\mathbb{M})$, define $\mathcal{E}_m^h(M)$ by induction as follows:

$$\mathcal{E}_m^0(\mathbb{A}) = \mathbb{A} \qquad\qquad \mathcal{E}_m^{h+1}(A + \mathbb{A}) = \mathcal{E}_m^{h+1}(A) + \mathcal{E}_m^{h+1}(\mathbb{A})$$
$$\mathcal{E}_m^{h+1}(\lambda x_1 \ldots \lambda x_n.y\vec{P}) = \lambda x_1 \ldots \lambda x_m.y\mathcal{E}_m^h(\vec{P})\mathcal{E}_m^h([x_{n+1}^!]) \ldots \mathcal{E}_m^h([x_m^!])$$
$$\mathcal{E}_m^{h+1}([M_1, \ldots, M_k, \mathbb{M}^!]) = [\mathcal{E}_m^{h+1}(M_1), \ldots, \mathcal{E}_m^{h+1}(M_k), (\mathcal{E}_m^{h+1}(\mathbb{M}))^!]$$

Consider then $\mathbb{M}' = \mathcal{E}_k^k(\mathbb{M})$ for some $k > \max(\text{size}(\mathbb{M}), \text{size}(\mathbb{N}))$. $\qquad\qquad \square$

The next lemma will be the key ingredient for proving the resource Böhm's theorem (Theorem 2, below).

Lemma 7 (Main Lemma). *Let $M \in \Lambda_f^r$ be a finite β-nf and $\Gamma = \{x_1, \ldots, x_d\} \supseteq \text{FV}(M)$. Then, there exist a substitution σ and a sequence $\vec{R}$ of closed bags such that, for all β-normal $\mathbb{N} \in \mathbf{2}\langle\Lambda^r\rangle$ such that M is λ-wider than $\mathbb{N}$ and $\text{FV}(\mathbb{N}) \subseteq \Gamma$, we have:*

$$(1) \qquad\qquad \mathbb{N}\sigma\vec{R} \xrightarrow{\beta*} \begin{cases} \mathbf{I} & \text{if } \exists N' \in \mathbb{N}^\circ,\ M \xrightarrow{\varphi*} N', \\ 0 & \text{otherwise.} \end{cases}$$

Proof. The proof requires an induction loading, namely the fact that $\sigma = \{\mathbf{X}_{k_1}/x_1, \ldots, \mathbf{X}_{k_d}/x_d\}$ such that for all distinct $i, j \leq d$ we have $k_i, |k_i - k_j| > 2k$ for some fixed $k > \text{size}(M)$. Recall that $\mathbf{X}_n$ has been defined in section 1.

The proof is carried by induction on $\text{size}(M)$. Let

$$M = \lambda x_{d+1} \ldots \lambda x_{d+a}.x_h P_1 \ldots P_p$$

where $a, p \geq 0$, $h \leq d + a$ and, for every $i \leq p$, $P_i = [M_{i,1}, \ldots, M_{i,m_i}]$ with $m_i \geq 0$. Notice that, for every $j \leq m_i$, $\text{FV}(M_{i,j}) \subseteq \Gamma \cup \{x_{d+1}, \ldots, x_{d+a}\}$ and $k > \text{size}(M_{i,j})$. So, define $\sigma' = \sigma \cdot \{\mathbf{X}_{k_{d+1}}/x_{d+1}, \ldots, \mathbf{X}_{k_{d+a}}/x_{d+a}\}$ such that $k_i, |k_i - k_j| > 2k$, for every different $i, j \leq d + a$.

By induction hypothesis on $M_{i,j}$, we have a sequence $\vec{R}_{i,j}$ of closed bags satisfying condition (1) for every $\mathbb{N}'$ such that $M_{i,j}$ is λ-wider than $\mathbb{N}'$ and $\text{FV}(\mathbb{N}') \subseteq \Gamma \cup \{x_{d+1}, \ldots, x_{d+a}\}$.

We now build the sequence of bags $\vec{R}$ starting from such $\vec{R}_{i,j}$'s. First, we define a closed term H as follows (setting $m = \max\{k_1, \ldots, k_{d+a}\}$):

$$H := \lambda z_1 \ldots \lambda z_p \lambda w_1 \ldots \lambda w_{m+k_h - p}.\mathbf{I}[z_1\vec{R}_{1,1}] \ldots [z_1\vec{R}_{1,m_1}] \ldots [z_p\vec{R}_{p,1}] \ldots [z_p\vec{R}_{p,m_p}].$$

Then we set: $\vec{R} := [\mathbf{X}_{k_{d+1}}^!] \ldots [\mathbf{X}_{k_{d+a}}^!] \underbrace{1 \ldots 1}_{k_h - p \text{ times}} [H^!] \underbrace{1 \ldots 1}_{m \text{ times}}.$

Notice the base of induction is when $p = 0$ or for every $i \leq p$, $m_i = 0$. In these cases H will be of the form $\lambda\vec{z}\lambda\vec{w}.\mathbf{I}$.

We prove condition (1) for any $\mathbb{N}$ satisfying the hypothesis of the lemma. Note that we can restrict to the case $\mathbb{N}$ is a single term N, the general case follows by distributing the application $\mathbb{N}\sigma\vec{R}$ on every summand of $\mathbb{N}$. So, let

$$N = \lambda x_{d+1} \ldots \lambda x_{d+b}.x_{h'}Q_1 \ldots Q_q$$

and let us prove that $N\sigma\vec{R} \xrightarrow{\beta*} \mathbf{I}$ if there is $N' \in N^\circ, M \xrightarrow{\varphi*} N'$, otherwise $N\sigma\vec{R} \xrightarrow{\beta*} 0$. Since M is λ-wider than N, we have $a \geq b$. We then get, setting $\sigma'' = \sigma \cdot \{\mathbf{X}_{k_{d+1}}/x_{d+1}, \ldots, \mathbf{X}_{k_{d+b}}/x_{d+b}\}$:

$$(2) \qquad N\sigma\vec{R} \xrightarrow{\beta*} \mathbf{X}_{k_{h'}} Q_1 \ldots Q_q \sigma''[\mathbf{X}^!_{k_{d+b+1}}] \ldots [\mathbf{X}^!_{k_{d+a}}] \underbrace{1 \ldots 1}_{k_h - p} [H^!] \underbrace{1 \ldots 1}_{m}.$$

Indeed, since $a \geq b$, the variables $x_{d+b+1}, \ldots, x_{d+a}$ can be considered not occurring in $Q_1, \ldots, Q_q$, so that σ' acts on $Q_1, \ldots, Q_q$ exactly as σ''. Moreover, since $k > a$ and $k > q$, we have $k_{h'} \geq 2k > q + (a - b)$, so, setting $g = q + (a - b)$, we have (2) β-reduces to:

$$(3) \quad \left(\lambda y_{g+1} \ldots \lambda y_{k_{h'}}. \lambda y. y Q_1 \ldots Q_q [\mathbf{X}^!_{k_{d+b+1}}] \ldots [\mathbf{X}^!_{k_{d+a}}][y^!_{g+1}] \ldots [y^!_{k_{h'}}] \right) \sigma' \underbrace{1 \ldots 1}_{k_h - p} [H^!] \underbrace{1 \ldots 1}_{m}$$

We consider now three cases.

CASE I ($h' \neq h$). This means M and N differ on their head-variable, in particular, for every $N' \in N^\circ$, $M \xrightarrow{\varphi*} N'$. We prove then (3) $\xrightarrow{\beta*} 0$. By the hypothesis on $k_h, k_{h'}$, we have either $k_h > k_{h'} + 2k$ or $k_{h'} > k_h + 2k$. In the first case, we get (by the hypothesis on k and N) $k_h > k_{h'} + p + q + a + b > k_{h'} + p - g$, so that (3) $\xrightarrow{\beta*} 0$ since the head-variable y will get 0 from an empty bag of the bunch of the $k_h - p$ empty bags. In the second case, we get $m \geq k_{h'} - g > k_h - p$, so that (3) $\xrightarrow{\beta*} 0$ since the head-variable y will get 0 from an empty bag of the bunch of the m empty bags.

CASE II ($h' = h$ AND $p - a \neq q - b$). This means that, for every $N' \in N^\circ$, $M \xrightarrow{\varphi*} N'$ (in fact, note that $(\cdot)^\circ$ preserve the length of the head prefix of abstractions and that of the head sequence of applications, while $\xrightarrow{\varphi}$ preserves the difference between the two). We prove then (3) $\xrightarrow{\beta*} 0$. As before, we have two subcases. If $p - a < q - b$, then $k_{h'} - g = k_h - g < k_h - p$ so (3) $\xrightarrow{\beta*} 0$, the head variable y getting 0 from an empty bag of the bunch of the $k_h - p$ empty bags. Otherwise, $p - a > q - b$ implies $k_{h'} - g = k_h - g > k_h - p$ and so (3) $\xrightarrow{\beta*} 0$, the head variable y getting 0 from an empty bag of the bunch of the m empty bags.

CASE III ($h' = h$ AND $p - a = q - b$). In this case we have $k_{h'} - g = k_h - g = k_h - p$, so that (3) β-reduces to $H Q_1 \ldots Q_q [\mathbf{X}^!_{k_{d+b+1}}] \ldots [\mathbf{X}^!_{k_{d+a}}] \underbrace{1 \ldots 1}_{k_h - p} \underbrace{1 \ldots 1}_{m} \sigma'$. Notice that, by the definition of the substitution σ', we can rewrite this term as

$$(4) \qquad H Q_1 \ldots Q_q [x^!_{k_{d+b+1}}] \ldots [x^!_{k_{d+a}}] \underbrace{1 \ldots 1}_{k_h - p} \underbrace{1 \ldots 1}_{m} \sigma'.$$

Since N can be in $\Lambda^r - \Lambda^r_f$, the bags Q_i's may contain reusable resources. Hence, for every $i \leq q$, let us explicit $Q_i = [N_{i,1}, \ldots, N_{i,\ell_i}, (N_{i,\ell_i+1} + \cdots + N_{i,n_i})^!]$, where $n_i \geq \ell_i \geq 0$. We split into three subcases. Notice that $p \geq q$, indeed $a \geq b$ (as M is λ-wider than N) and we are considering the case $p - a = q - b$. Also, recall that m_i is the number of resources in P_i.

SUBCASE III.A ($\exists i \leq q, m_i < \ell_i$). In this case, for every $N' \in N^\circ$, we have $M \not\xrightarrow{\varphi*} N'$. In fact, any $N' \in N^\circ$ is of the form $\lambda x_{d+1} \ldots \lambda x_{d+b}.x_{h'}Q'_1 \ldots Q'_q$, with $Q'_j \in Q^\circ_j$ for every $j \leq q$. In particular, Q'_i has at least $\ell_i > m_i$ linear resources, hence $P_i \not\xrightarrow{\varphi*} Q'_i$, and hence $M \not\xrightarrow{\varphi*} N'$.

We have $(4) \xrightarrow{\beta*} 0$. Indeed, applying the β-reduction to (4) will eventually match the abstraction λz_i in H with the bag Q_i. The variable z_i has m_i linear occurrences in H and no reusable ones. This means there will be not enough occurrences of z_i to accommodate all the ℓ_i linear resources of Q_i, so giving $(4) \xrightarrow{\beta*} 0$.

SUBCASE III.B ($\exists i \leq q, m_i \neq \ell_i$ AND $n_i = \ell_i$). This case means that Q_i has no reusable resources and a number of linear resources different from P_i. Hence $M \not\xrightarrow{\varphi*} N'$, for every $N' \in N^\circ$. Also, $(4) \xrightarrow{\beta*} 0$ since the number of linear resources in Q_i does not match the number of linear occurrences of the variable z_i in H.

SUBCASE III.C ($\forall i \leq q, m_i \geq \ell_i$, AND $m_i > \ell_i$ ENTAILS $n_i > \ell_i$). The hypothesis of the case says that $\ell_i < m_i$ entails that Q_i has some reusable resources. Let $\mathcal{F}_i$ be the set of maps $s : \{1, \ldots, m_i\} \to \{1, \ldots, \ell_i, \ell_i + 1, \ldots, n_i\}$ such that

ℓ_i-**injectivity:** for every $j, h \leq m_i$, if $s(j) = s(h) \leq \ell_i$, then $j = h$,
ℓ_i-**surjectivity:** for every $h \leq \ell_i$, there is $j \leq m_i$, $s(j) = h$.

Intuitively, $\mathcal{F}_i$ describes the possible ways of replacing the m_i occurrences of the variable z_i in H by the n_i resources in Q_i: the two conditions say that each of the ℓ_i linear resources of Q_i must replace exactly one occurrence of z_i. Notice that, being under the hypothesis that $p - a = q - b$, we have $p - q = a - b$, hence (4) β-reduces to the following sum

$$(5) \quad \sum_{\substack{s_1 \in \mathcal{F}_1 \\ \vdots \\ s_q \in \mathcal{F}_q}} \mathbf{I}[N_{1,s_1(1)}\vec{R}_{1,1}]...[N_{1,s_1(m_1)}\vec{R}_{1,m_1}]...[N_{q,s_q(1)}\vec{R}_{q,1}]...[N_{q,s_q(m_q)}\vec{R}_{q,m_q}]$$
$$[x_{d+b+1}\vec{R}_{q+1,1}]...[x_{d+b+1}\vec{R}_{q+1,m_1}]...[x_{d+a}\vec{R}_{p,1}]...[x_{d+a}\vec{R}_{p,m_p}]\sigma'$$

Notice that for every $i \leq q$, $s_i \in \mathcal{F}_i$, $j \leq m_i$ the term $M_{i,j}$ is λ-wider than $N_{i,s_i(j)}$ and $FV(N_{i,s_i(j)}) \subseteq \Gamma \cup \{x_{d+1}, \ldots, x_{d+a}\}$, so by the induction hypothesis

$$(6) \qquad N_{i,s_i(j)}\sigma'\vec{R}_{i,j} \xrightarrow{\beta*} \begin{cases} \mathbf{I} & \text{if } \exists N' \in N^\circ_{i,s_i(j)}, M_{i,j} \xrightarrow{\varphi*} N', \\ 0 & \text{otherwise.} \end{cases}$$

Also, for every $i, 1 \leq i \leq p - q = a - b$, $j \leq m_{q+i}$, we have that $M_{q+i,j}$ is λ-wider than $x_{d+b+i} \in \Gamma \cup \{x_{d+1}, \ldots, x_{d+a}\}$, so by induction hypothesis

$$(7) \qquad x_{d+b+i}\sigma'\vec{R}_{q+i,j} \xrightarrow{\beta*} \begin{cases} \mathbf{I} & \text{if } M_{i,j} \xrightarrow{\varphi*} x_{d+b+i}, \\ 0 & \text{otherwise.} \end{cases}$$

From (6) and (7) we deduce that $(5) \xrightarrow{\beta*} \mathbf{I}$ if, and only if,

(i) for all $i \leq q$, there exists $s_i \in \mathcal{F}_i$ such that, for all $j \leq m_i$, $\exists N' \in N^\circ_{i,s_i(j)}$
   satisfying $M_{i,j} \xrightarrow{\varphi*} N'$, and

(ii) for all $i < p - q = a - b$ and for all $j \leq m_{q+i}$, we have $M_{q+i,j} \xrightarrow{\varphi*} x_{d+b+i}$.

Thanks to the conditions on the function s_i, item (i) is equivalent to say that for all $i \leq q$, $\exists Q'_i \in Q^\circ_i, P_i \xrightarrow{\varphi*} Q'_i$, while item (ii) is equivalent to say that for all $i < p - q = a - b$, $P_{q+i} \xrightarrow{\varphi*} [x^{m_{q+i}}_{d+b+i}]$. This means that $(5) \xrightarrow{\beta*} \mathbf{I}$ if, and only if,

$$M = \lambda x_{d+1} \ldots \lambda x_{d+a}.x_h P_1 \ldots P_p$$
$$\xrightarrow{\varphi*} \lambda x_{d+1} \ldots \lambda x_{d+b} \lambda x_{d+b+1} \ldots \lambda x_{d+a}.x_h Q'_1 \ldots Q'_q [x^{m_{q+1}}_{d+b+1}] \ldots [x^{m_p}_{d+a}]$$
$$\xrightarrow{\varphi*} \lambda x_{d+1} \ldots \lambda x_{d+b}.x_h Q'_1 \ldots Q'_q$$

where the last term is in N°. To sum up, $\mathbb{N}\sigma\vec{R}$ β-reduces to $\mathbf{I}$ if $\exists N' \in N^\circ$ such that $M \xrightarrow{\varphi*} N'$ and to 0, otherwise. We conclude that condition (1) holds.   □

4   A Resource Conscious Böhm's Theorem

In this section we will prove the main result of our paper, namely Böhm's theorem for the resource calculus. We first need the following technical lemma.

Lemma 8. *Let* $\mathbb{M}, \mathbb{N} \in \mathbf{2}\langle \Lambda^r \rangle$. *If* $\mathbb{M} \npreceq_\tau \mathbb{N}$ *then* $\mathbb{M}' \npreceq_\tau \mathbb{N}$ *for all* $\mathbb{M}' \xrightarrow{\eta*} \mathbb{M}$.

Proof. Suppose, by the way of contradiction, that there is an $\mathbb{M}' \xrightarrow{\eta*} \mathbb{M}$ such that $\mathbb{M}' \preceq_\tau \mathbb{N}$. Then, there are $\mathbb{M}'' \xrightarrow{\eta*} \mathbb{M}'$ and $\mathbb{N}' \xrightarrow{\eta*} \mathbb{N}$, such that $\mathbb{M}'' \sqsubseteq_\tau \mathbb{N}'$. By transitivity of $\xrightarrow{\eta*}$, $\mathbb{M}'' \xrightarrow{\eta*} \mathbb{M}$ holds, so we get $\mathbb{M} \preceq_\tau \mathbb{N}$ which is impossible.   □

We are now able to prove the main result of this paper.

Theorem 2 (Resource Böhm's Theorem). *Let* $\mathbb{M}, \mathbb{N} \in \mathbf{2}\langle \Lambda^r \rangle$ *be closed sums in* β-*nf. If* $\mathbb{M} \not\equiv_{\eta\tau} \mathbb{N}$ *then there is a sequence* $\vec{P}$ *of closed bags such that either* $\mathbb{M}\vec{P} \xrightarrow{\beta*} \mathbf{I}$ *and* $\mathbb{N}\vec{P} \xrightarrow{\beta*} 0$, *or vice versa.*

Proof. Let $\mathbb{M} \not\equiv_{\eta\tau} \mathbb{N}$, then $\mathbb{M} \npreceq_\tau \mathbb{N}$ or vice versa (Lemma 4): say $\mathbb{M} \npreceq_\tau \mathbb{N}$. Applying Lemma 8 we have $\mathbb{M}' \npreceq_\tau \mathbb{N}$ for all $\mathbb{M}' \xleftarrow{\eta*} \mathbb{M}$; in particular, by Lemma 6, $\mathbb{M}' \npreceq_\tau \mathbb{N}$ holds for an $\mathbb{M}'$ λ-wider than both $\mathbb{M}$ and $\mathbb{N}$. By Lemma 5 there is $M' \in (\mathbb{M}')^\circ$ such that for all $N \in \mathbb{N}^\circ$ we have $M' \xrightarrow{\varphi*}\hspace{-1.6em}/\hspace{1.0em} N$; such a term M' is in β-nf since $\mathbb{M}'$ is in β-nf and is λ-wider than both $\mathbb{M}$ and $\mathbb{N}$ by Remark 2.

From Lemma 7, recalling that $M', \mathbb{M}, \mathbb{N}$ are closed, there is a sequence $\vec{P}$ of closed bags such that: (*i*) $\mathbb{N}\vec{P} \xrightarrow{\beta*} 0$, since for all $N \in \mathbb{N}^\circ$ we have $M' \xrightarrow{\varphi*}\hspace{-1.6em}/\hspace{1.0em} N$, and (*ii*) $\mathbb{M}\vec{P} \xrightarrow{\beta*} \mathbf{I}$, since $\mathbb{M}' \xrightarrow{\eta*} \mathbb{M}$ and hence by Remark 1 there is an $M \in \mathbb{M}^\circ$ such that $M' \xrightarrow{\varphi*} M$. This concludes the proof of our main result.   □

Corollary 1. *Let* $\sim$ *be a congruence on* $\mathbf{2}\langle \Lambda^e \rangle$ *extending* β-*equivalence. If there are two closed* $\mathbb{M}, \mathbb{N} \in \mathbf{2}\langle \Lambda^r \rangle$ *in* β-*nf such that* $\mathbb{M} \not\equiv_{\eta\tau} \mathbb{N}$ *but* $\mathbb{M} \sim \mathbb{N}$, *then* $\sim$ *is trivial, i.e. for all sums* $\mathbb{L} \in \mathbf{2}\langle \Lambda^r \rangle$, $\mathbb{L} \sim 0$.

Proof. Suppose $\mathbb{M} \not\equiv_{\eta\tau} \mathbb{N}$ but $\mathbb{M} \sim \mathbb{N}$. From Theorem 2 there is $\vec{P}$ such that $\mathbb{M}\vec{P} \xrightarrow{\beta*} \mathbf{I}$ and $\mathbb{N}\vec{P} \xrightarrow{\beta*} 0$, or vice versa. By the congruence of $\sim$, we have $\mathbb{M}\vec{P} \sim \mathbb{N}\vec{P}$. By the hypothesis that $\sim$ extends β-equivalence, we get $\mathbf{I} \sim 0$. Now, take any term $\mathbb{L} \in 2\langle \Lambda^r \rangle$, we have $\mathbb{L} \equiv_\beta \mathbf{I}\mathbb{L} \sim 0\mathbb{L} = 0$. $\qquad\square$

Acknowlegements. We thank Stefano Guerrini and Henk Barendregt for stimulating discussions and remarks.

References

1. Böhm, C.: Alcune proprietà delle forme β-η-normali nel λ-K-calcolo. Pubbl. INAC 696 (1968)
2. Barendregt, H.: The lambda-calculus, its syntax and semantics, 2nd edn. Stud. Logic Found. Math., vol. 103. North-Holland, Amsterdam (1984)
3. Böhm, C., Dezani-Ciancaglini, M., Peretti, P., Ronchi Della Rocca, S.: A discrimination algorithm inside $\lambda\beta$-calculus. Theor. Comp. Sci. 8(3), 271–291 (1979)
4. Hyland, J.M.E.: A syntactic characterization of the equality in some models of the lambda calculus. J. London Math. Soc. 2(12), 361–370 (1976)
5. Girard, J.Y.: Linear logic. Th. Comp. Sc. 50, 1–102 (1987)
6. Carraro, A., Ehrhard, T., Salibra, A.: Exponentials with infinite multiplicities. In: Dawar, A., Veith, H. (eds.) CSL 2010. LNCS, vol. 6247, pp. 170–184. Springer, Heidelberg (2010)
7. Pagani, M., Ronchi Della Rocca, S.: Linearity, non-determinism and solvability. Fundamenta Informaticae 103(1-4), 173–202 (2010)
8. Boudol, G.: The lambda-calculus with multiplicities. INRIA Report 2025 (1993)
9. Boudol, G., Laneve, C.: The discriminating power of multiplicities in the lambda-calculus. Inf. Comput. 126(1), 83–102 (1996)
10. Boudol, G., Curien, P.L., Lavatelli, C.: A semantics for lambda calculi with resources. MSCS 9(4), 437–482 (1999)
11. Ehrhard, T., Regnier, L.: The differential lambda-calculus. Theor. Comput. Sci. 309(1), 1–41 (2003)
12. Tranquilli, P.: Intuitionistic differential nets and lambda-calculus. Theor. Comput. Sci. (2008) (to appear)
13. Bucciarelli, A., Carraro, A., Ehrhard, T., Manzonetto, G.: Full abstraction for resource calculus with tests (submitted), http://www.dsi.unive.it/~gmanzone/papers/bcem11.pdf
14. Ehrhard, T., Regnier, L.: Uniformity and the Taylor expansion of ordinary lambda-terms. Theor. Comput. Sci. 403(2-3), 347–372 (2008)
15. Pagani, M., Tranquilli, P.: Parallel reduction in resource lambda-calculus. In: Hu, Z. (ed.) APLAS 2009. LNCS, vol. 5904, pp. 226–242. Springer, Heidelberg (2009)
16. Manzonetto, G.: What is a categorical model of the differential and the resource λ-calculi? (submitted), http://www.dsi.unive.it/~gmanzone/papers/man10.pdf
17. Terese.: Term Rewriting Systems. Cambridge Tracts in Theoretical Computer Science, vol. 55. Cambridge University Press, Cambridge (2003)

Finite Combinatory Logic with Intersection Types

Jakob Rehof[1] and Paweł Urzyczyn[2],[*]

[1] Technical University of Dortmund
Department of Computer Science
jakob.rehof@cs.tu-dortmund.de
[2] University of Warsaw
Institute of Informatics
urzy@mimuw.edu.pl

Abstract. Combinatory logic is based on modus ponens and a schematic (polymorphic) interpretation of axioms. In this paper we propose to consider expressive combinatory logics under the restriction that axioms are not interpreted schematically but „literally", corresponding to a monomorphic interpretation of types. We thereby arrive at *finite combinatory logic*, which is strictly finitely axiomatisable and based solely on modus ponens. We show that the provability (inhabitation) problem for finite combinatory logic with intersection types is EXPTIME-complete with or without subtyping. This result contrasts with the general case, where inhabitation is known to be EXPSPACE-complete in rank 2 and undecidable for rank 3 and up. As a by-product of the considerations in the presence of subtyping, we show that standard intersection type subtyping is in PTIME. From an application standpoint, we can consider intersection types as an expressive specification formalism for which our results show that functional composition synthesis can be automated.

1 Introduction

Under the Curry-Howard isomorphism, combinatory terms correspond to Hilbert-style proofs, based on two principles of deduction: modus ponens and a schematic interpretation of axioms. The latter can be expressed as a meta-rule of instantiation or as a rule of substitution. The schematic interpretation corresponds to considering types of combinators as implicitly polymorphic.[1] Because of the schematic interpretation of axioms, such logics are not strictly finite.

In the present paper we propose to consider *finite combinatory logics*. Such logics arise naturally from combinatory logic by eliminating the schematic interpretation of axioms. This corresponds to a monomorphic interpretation of combinator types, and leaves modus ponens as the sole principle of deduction. We consider combinatory logic with intersection types [4]. In general, the provability problem for this logic is undecidable. We prove that provability in finite

---

[*] Partly supported by MNiSW grant N N206 355836.
[1] The schematic viewpoint is sometimes referred to as *typical ambiguity*.

L. Ong (Ed.): TLCA 2011, LNCS 6690, pp. 169–183, 2011.
© Springer-Verlag Berlin Heidelberg 2011

combinatory logic is decidable and EXPTIME-complete. We show, in fact, that all of the problems: emptiness, uniqueness and finiteness of inhabitation are in EXP-TIME by reduction to alternating tree automata. We then consider the provability problem in the presence of standard intersection type subtyping. We show that the subtyping relation is in PTIME (previous work only appears to establish an exponential algorithm). We prove that provability remains EXPTIME-complete in the presence of subtyping, and also that uniqueness and finiteness of inhabitation remain in EXPTIME. In contrast to the case without subtyping, which is solved by reduction to tree automata, the upper bounds are achieved using polynomial space bounded alternating Turing machines.

From an application standpoint, we are interested in finite combinatory logic with intersection types as an expressive specification logic for automatic function composition synthesis. Under the Curry-Howard isomorphism, validity is the question of inhabitation: Given a set Γ of typed combinators (function symbols), does there exist an applicative combination e of combinators in Γ having a given type τ? An algorithm for inhabitation in finite combinatory logic can be used to synthesize e from the specification given by Γ and τ.

Due to space limitations some proofs have been left out. All details can be found in the technical report [16] accompanying this paper.

Finite combinatory logic with intersection types. The strictly finite standpoint considered here is perhaps especially interesting for highly expressive logics for which validity is undecidable. Intersection type systems [3,15] belong to a class of propositional logics with enormous expressive power. Intersection types capture several semantic properties of lambda terms, and their type reconstruction problem has long been known to be undecidable, since they characterize exactly the set of strongly normalizing terms [15]. Intersection types have also been used more generally to capture semantic properties of programs, for example in the context of model checking [13,7]. The inhabitation problem for the λ-calculus with intersection types was shown to be undecidable in [18] (see also [17] for connections to λ-definability). More recently, the borderline between decidability and undecidability was clarified in [10,19] by rank restrictions (as defined in [11]), with inhabitation in rank 2 types being shown EXPSPACE-complete and undecidable from rank 3 and up.[2] Combinatory logic with intersections types was studied in [4], where it was shown that the combinatory system is complete in that it is logically equivalent to the λ-calculus by term (proof) translations in both directions. It follows that provability (inhabitation) for combinatory logic with intersections types is undecidable.

Application perspective. Finite combinatory logic can be used as a foundation for automatic composition synthesis via inhabitation: Given a set Γ of typed functions, does there exist a composition of those functions having a given type? Under this interpretation, the set Γ represents a library of functions, and the intersection type system provides a highly expressive specification logic for which our results show that the problem of typed function composition synthesis can be

---

[2] Other restrictions have also been considered, which arise by limiting the logic of intersection [9,20].

automated. Our approach provides an alternative to synthesis via proof counting (following [2]) for more complex term languages (e.g., [22]). Most notably, finite combinatory logic is naturally applicable to synthesis from libraries of reusable components which has been studied recently [12], leading to a 2EXPTIME-complete synthesis problem. Our setting is different, since our framework is functional (rather than reactive), it is based on a different specification format (types rather than temporal logic), so we obtain an EXPTIME-complete problem (using the identity APSPACE = EXPTIME). We gain algorithmic control over the solution space to the synthesis problem, since it follows from the reduction that not only the problem of inhabitation but also the problems of finiteness and uniqueness of inhabitation are solvable in EXPTIME. The incorporation of subtyping is important here, since it would allow synthesis specifications to take advantage of hierarchical structure, such as inheritance or taxonomies.

Although synthesis problems are by nature computationally complex, the exploitation of abstraction in component libraries and type structure allows new ways of scaling down problem size. An implementation of the algorithms described here is currently underway in the context of a component synthesis framework, and future work will address experimental evaluation.

2   Preliminaries

Trees: Let $\mathbf{A}$ and $\mathbf{L}$ be non-empty sets. A *labeled tree* over the *index set* $\mathbf{A}$ and the *alphabet of labels* $\mathbf{L}$ is understood as a partial function $T : \mathbf{A}^* \to \mathbf{L}$, such that the domain $dm(T)$ is non-empty and prefix-closed. The *size* of T, denoted $|T|$, is the number of nodes in (the domain of) T. The *depth* of a tree T, written $\|T\|$, is the maximal length of a member of $dm(T)$. Note that $\|T\| \leq |T|$. For $w \in dm(T)$, we write $T[w]$ for the subtree of T rooted at w, given by $T[w](w') = T(ww')$.

Terms: Applicative terms, ranged over by e etc., are defined by:

$$e ::= x \mid (e \, e'),$$

where x, y, z etc. range over a denumerable set of variables. We write $e \equiv e'$ for syntactic identity of terms. We abbreviate $(\ldots((e \, e_1) \, e_2) \ldots e_n)$ as $(e \, e_1 \, e_2 \ldots e_n)$ for $n \geq 0$ (for $n = 0$ we set $(e \, e_1 \ldots e_n) \equiv e$). Notice that any applicative term can be written uniquely as $(x \, e_1 \ldots e_n)$.

An alternative to the above standard understanding of expressions as words, is to represent them as finite labeled trees over the index set $\mathbb{N}$ and the alphabet $V \cup (\mathbb{N} - \{0\})$. A variable x is seen as a one-node tree where the only node is labeled x. An expression $(x \, e_1 \ldots e_n)$ is a tree, denoted $@_n(x, e_1, \ldots, e_n)$, with root labeled $@_n$ and $n + 1$ subtrees $x, e_1, \ldots, e_n$, which are addressed from left to right from the root by the indices $0, 1, \ldots, n$. That is, if $e = @_n(x, e_1, \ldots, e_n)$, we have $e(\varepsilon) = @_n$, $e(0) = x$ and $e(i) = e_i$ for $i = 1, \ldots, n$. In a term of the form $@_n(x, e_1, \ldots, e_n)$ we call the tree node x the *operator* of node $@_n$. We say that a variable x *occurs along an address* $w \in dm(e)$ if for some prefix u of w we have $u0 \in dm(e)$ with $e(u0) = x$.

Types: Type expressions, ranged over by τ, σ etc., are defined by

$$\tau ::= \alpha \mid \tau \to \tau \mid \tau \cap \tau$$

where α ranges over a denumerable set TV of type variables. We let T denote the set of type expressions. As usual, types are taken modulo idempotency ($\tau \cap \tau = \tau$), commutativity ($\tau \cap \sigma = \sigma \cap \tau$), and associativity (($\tau \cap \sigma) \cap \rho = \tau \cap (\sigma \cap \rho)$). A type *environment* Γ is a finite set of type assumptions of the form $x : \tau$, and we let $dm(\Gamma)$ and $rn(\Gamma)$ denote the domain and range of Γ.

A type $\tau \cap \sigma$ is said to have τ and σ as *components*. For an intersection of several components we sometimes write $\bigcap_{i=1}^{n} \tau_i$ or $\bigcap_{i \in I} \tau_i$. We stratify types by letting $A, B, C, \ldots$ range over types with no top level conjunction, that is

$$A ::= \alpha \mid \tau \to \tau$$

We can write any type τ as $\bigcap_{i=1}^{n} A_i$ for suitable $n \geq 1$ and A_i.

Types as trees: It is convenient to consider types as labeled trees. A *type tree* τ is a labeled tree over the index set $\mathbf{A} = \mathbb{N} \cup \{l, r\}$ and the alphabet of labels $\mathbf{L} = TV \cup \{\to, \cap\}$, satisfying the following conditions for $w \in \mathbf{A}^*$:

- if $\tau(w) \in TV$, then $\{a \in \mathbf{A} \mid wa \in dm(\tau)\} = \emptyset$;
- if $\tau(w) = \to$, then $\{a \in \mathbf{A} \mid wa \in dm(\tau)\} = \{l, r\}$;
- if $\tau(w) = \cap$, then $\{a \in \mathbf{A} \mid wa \in dm(\tau)\} = \{1, \ldots, n\}$, for some n.

Trees are identified with types in the obvious way. Observe that a type tree is LOGSPACE-computable from the corresponding type and vice versa. Recall that types are taken modulo idempotency, commutativity and associativity, so that a type has in principle several tree representations. In what follows we always assume a certain fixed way in which a type is interpreted. By representing A as $\bigcap_{i=1}^{1} A_i$ we can assume w.l.o.g. that types τ are *alternating*, i.e., that all $w \in dm(\tau)$ alternate between members of $\mathbb{N}$ and $\{l, r\}$ beginning and ending with members of $\mathbb{N}$, and we call such w *alternating addresses*. We define *right-addresses* to be members of $\mathbf{A}^*$ with no occurrence of l. For an alternating type τ, we define $\mathbf{R}_n(\tau)$ to be the set of alternating right-addresses in $dm(\tau)$ with n occurrences of r. Observe that alternating right-addresses have the form $n_0 r n_1 \ldots r n_k$ with $n_j \in \mathbb{N}$ for $0 \leq j \leq k$, in particular note that $\mathbf{R}_0(\tau) = \{1, \ldots, k\}$, for some k. For $w \in \mathbf{R}_n$, $n \geq 1$ and $1 \leq j \leq n$, we let $w \downarrow j$ denote the address that arises from chopping off w just before the j'th occurrence of r in w, and we define $L(j, w) = (w \downarrow j)l$. The idea is that alternating right-addresses in τ determine "target types" of τ, i.e., types that may be assigned to applications of a variable of type τ. For example, if $\tau = \cap\{\alpha_1 \cap \alpha_2 \to \beta_1 \cap \beta_2\}$, then we have $1r1, 1r2 \in \mathbf{R}_1(\tau)$ with $\tau[1r1] = \beta_1$ and $\tau[1r2] = \beta_2$. Moreover, $\tau[L(1, 1r1)] = \tau[1l] = \alpha_1 \cap \alpha_2$. Now suppose that $x : \tau$ and observe that an expression $(x\ e)$ can be assigned any of types β_1 and β_2, provided $e : \alpha_1 \cap \alpha_2$.

Definition 1 (*arg, tgt*). The j'th argument of τ at w is $arg(\tau, w, j) = \tau[L(j, w)]$, and the *target* of τ at w is $tgt(\tau, w) = \tau[w]$.

Definition 2 (Sizes). The *size* (resp. *depth*) of a term or type is the size (resp. depth) of the corresponding tree. The *size* of an environment Γ is the sum of the cardinality of the domain of Γ and the sizes of the types in Γ, i.e., $|\Gamma| = |dm(\Gamma)| + \sum_{x \in dm(\Gamma)} |\Gamma(x)|$. Also we let $\| \Gamma \|$ denote the maximal depth of a member of $rn(\Gamma)$.

Type assignment: We consider the applicative restriction of the standard intersection type system, which arises from that system by removing the rule $(\rightarrow I)$ for function type introduction. The system is shown in Figure 1 and is referred to as finite combinatory logic with intersection types, FCL($\cap$). We consider the following decision problem:

Inhabitation problem for FCL($\cap$): Given an environment Γ and a type τ, does there exist an applicative term e such that $\Gamma \vdash e : \tau$?

$$\frac{}{\Gamma, x : \tau \vdash x : \tau} \ (\text{var}) \qquad\qquad \frac{\Gamma \vdash e : \tau \rightarrow \tau' \quad \Gamma \vdash e' : \tau}{\Gamma \vdash (e\ e') : \tau'} \ (\rightarrow \text{E})$$

$$\frac{\Gamma \vdash e : \tau_1 \quad \Gamma \vdash e : \tau_2}{\Gamma \vdash e : \tau_1 \cap \tau_2} \ (\cap \text{I}) \qquad\qquad \frac{\Gamma \vdash e : \tau_1 \cap \tau_2}{\Gamma \vdash e : \tau_i} \ (\cap \text{E})$$

Fig. 1. Finite combinatory logic FCL($\cap$)

The difference between our type system and full combinatory logic based on intersection types [4,21] is that we do not admit a schematic (polymorphic) interpretation of the axioms (types in Γ) but take each of them monomorphically, "as is". Moreover, we do not consider a fixed base of combinators, since we are not here interested in combinatory completeness. The study of combinatory logics with intersection types usually contain, in addition, standard rules of subtyping which we consider in Section 4.

As has been shown [4], the standard translation of combinatory logic into the λ-calculus can be adapted to the intersection type system, preserving the main properties of the λ-calculus type system, including characterization of the strongly normalizing terms (see also [21, Section 2], which contains a compact summary of the situation). In particular, it is shown in [4, Theorem 3.11], that the combinatory system is logically equivalent to the λ-calculus with intersection types. By composing this result with undecidability of inhabitation in the λ-calculus [18], it immediately follows that inhabitation is undecidable for the full combinatory system.

We note that the system FCL($\cap$) is sufficiently expressive to uniquely specify any given expression: for every applicative term e there exists an environment Γ_e and a type τ_e such that e types uniquely in Γ_e at τ_e, i.e., $\Gamma_e \vdash e : \tau_e$ and for any term e' such that $\Gamma_e \vdash e' : \tau_e$ it is the case that $e \equiv e'$. This property is of interest for applications in synthesis. We refer to [16] for more details.

3   EXPTIME-completeness of Inhabitation

An EXPTIME-lower bound for applicative inhabitation follows easily from the known result (apparently going back to H. Friedman 1985 and published later by others, see [8]) that the composition problem for a set of functions over a finite set is EXPTIME-complete, because an intersection type can be used to represent a finite function. It may be instructive to observe that the problem (without subtyping) can be understood entirely within the theory of tree automata. In [16], we give an alternative, simple reduction from the intersection non-emptiness problem for tree automata. The proof works for types of rank 2 or more.

For the upper bound we prove that the inhabitation problem is reducible to the non-emptiness problem for polynomial-sized alternating top-down tree automata [5], and hence it is in APSPACE = EXPTIME. Representing the inhabitation problem by alternating tree automata, we can conclude that also finiteness and uniqueness of inhabitation are solvable in EXPTIME.

Let us point out that the presence of intersection makes an essential difference in complexity. It can easily be shown [16] that the inhabitation problem for the $\cap$-free fragment of FCL($\cap$) is solvable in polynomial time.

3.1   EXPTIME **Upper Bound**

We use the definition of alternating tree automata from [5]. We have adapted the definition accordingly, as follows. Below, $\mathcal{B}^+(S)$ denotes the set of positive propositional formulae over the set S, i.e., formulae built from members of S as propositional variables, conjunction, and disjunction.

Definition 3 (Alternating tree automaton). An *alternating tree automaton* is a tuple $\mathcal{A} = (Q, \mathcal{F}, q_0, \Delta)$, where q_0 is an initial state, $\mathcal{F}$ is a label alphabet, $I \subseteq Q$ is a set of initial states, and Δ is a mapping with $dm(\Delta) = Q \times \mathcal{F}$, such that $\Delta(q, @_n) \in \mathcal{B}^+(Q \times \{0, \ldots, n\})$ and $\Delta(q, x) \in \{true, false\}$.

We use the notation $\langle q, i \rangle$ for members of the set $Q \times \{0, \ldots, n\}$. For example, for a transition $\Delta(q, f) = \langle q', i \rangle \wedge \langle q'', i \rangle$, the automaton moves to both of the states q' and q'' along the i'th child of f (universal transition), whereas $\Delta(q, f) = \langle q', i \rangle \vee \langle q'', i \rangle$ is a nondeterministic (existential) transition. In the first case, it is necessary that both $\langle q', i \rangle$ and $\langle q'', i \rangle$ lead to accept states, in the second case it suffices that one of the successor states leads to an accept state. More generally, let $Accept(q, e)$ be true iff the automaton accepts a tree e starting from state q. Then $Accept(q, e)$ can be described as the truth value of $\Delta(q, f)$, where f is the label at the root of e, under the interpretation of each $\langle q', i \rangle$ as $Accept(q', e_i)$. We refer the reader to [5, Section 7.2] for full details on alternating tree automata.

Recognizing inhabitants. Given Γ and τ, we define a tree automaton $\mathcal{A}(\Gamma, \tau)$, assuming the alternating representation of types and term syntax based on the tree constructors $@_n$, as introduced in Section 2. The property of $\mathcal{A}(\Gamma, \tau)$ is:

$$\Gamma \vdash e : \tau \quad \text{if and only if} \quad e \in \mathbf{L}(\mathcal{A}(\Gamma, \tau)).$$

The definition is shown in Figure 2. The constructor alphabet of our automaton $\mathcal{A}(\Gamma, \tau)$ is $\Sigma = \{@_n\} \cup dm(\Gamma)$. The states of $\mathcal{A}(\Gamma, \tau)$ are either "search states" of the form $\mathrm{SRC}(\tau')$ or "check states" of the form $\mathrm{CHK}(A, B, x)$, where $x \in dm(\Gamma)$ and τ', A and B are subterms of types occurring in Γ (recall that types ranged over by A, B etc. do not have top level intersections, and τ are alternating types).

The initial state of $\mathcal{A}(\Gamma, \tau)$ is $\mathrm{SRC}(\tau)$. The transition relation of $\mathcal{A}(\Gamma, \tau)$ is given by the two transition rules shown in Figure 2, for $0 \leq n \leq \| \Gamma \|$.

$$\mathrm{SRC}(\textstyle\bigcap_{i=1}^{k} A_i) \xrightarrow{@_n} \bigvee_{x \in dm(\Gamma)} \bigwedge_{i=1}^{k} \bigvee_{w \in \mathbf{R}_n(\Gamma(x))} (\bigwedge_{j=1}^{n} \langle \mathrm{SRC}(arg(\Gamma(x), w, j)) , j \rangle$$
$$\wedge \langle \mathrm{CHK}(A_i, tgt(\Gamma(x), w), x) , 0 \rangle)$$

$$\mathrm{CHK}(A, B, x) \xrightarrow{y} \textit{true}, \text{ if } A = B \text{ and } x = y, \text{ otherwise } \textit{false}$$

Fig. 2. Alternating tree automaton for $\mathrm{FCL}(\cap)$

Intuitively, from search states of the form $\mathrm{SRC}(\tau)$ the automaton recognizes n-ary inhabitants (for $0 \leq n \leq \| \Gamma \|$) of τ by nondeterministically choosing an operator x and verifying that an expression with x as main operator inhabits all components of τ. Given the choice of x, a right-path w is guessed in every component, and the machine makes a universal move to inhabitation goals of the form $\mathrm{SRC}(arg(\Gamma(x), w, j))$, one for each of the argument types of τ as determined by the address w. In these goals, the type $arg(\Gamma(x), w, j)$ denotes the j'th such argument type, and the goal is coupled with the child index j, for $j = 1, \ldots, n$. Moreover, it is checked from the state $\mathrm{CHK}(A_i, tgt(\Gamma(x), w), x)$ that the target type of x at position w is identical to A_i.

Correctness. In order to prove correctness of our representation in alternating tree automata we need a special generation lemma (Lemma 4), in which we assume the alternating representation of general types τ.

Lemma 4 (Path Lemma). *The assertion* $\Gamma \vdash (x \, e_1 \ldots e_n) : A$ *holds if and only if there exists* $w \in \mathbf{R}_n(\Gamma(x))$ *such that*

1. $\Gamma \vdash e_j : arg(\Gamma(x), w, j)$, *for* $j = 1, \ldots, n$;
2. $A = tgt(\Gamma(x), w)$.

Proof. The proof from right to left is obvious. The proof from left to right is by induction with respect to $n \geq 0$. The omitted details can be found in [16]. □

Proposition 5. $\Gamma \vdash e : \tau$ *if and only if* $e \in \mathbf{L}(\mathcal{A}(\Gamma, \tau))$

Proof. For $e = @_n(x, e_1, \ldots, e_n)$ let $child_0(e)$ be x, and let $child_j(e)$ be e_j, when $j = 1, \ldots, n$. Let $\tau = \bigcap_{i=1}^{k} A_i$. By Lemma 4, we have $\Gamma \vdash e : \tau$ if and only if

$$\exists x \, \forall i \, \exists w \, \forall j \, [(\Gamma \vdash child_j(e) : arg(\Gamma(x), w, j)) \wedge A_i = tgt(\Gamma(x), w) \wedge x = child_0(e)].$$

The automaton $\mathcal{A}(\Gamma, \tau)$ is a direct implementation of this specification. □

As a consequence of the tree automata representation we immediately get decision procedures for emptiness, finiteness and uniqueness of inhabitation, via known results for the corresponding properties of tree automata.

Theorem 6. *In system* FCL($\cap$) *the following problems are in* EXPTIME.

1. *Emptiness of inhabitation;*
2. *Finiteness of inhabitation;*
3. *Uniqueness of inhabitation.*

In addition, problem (1) is EXPTIME-*complete.*

Proof. The size of $\mathcal{A}(\Gamma, \tau)$ is polynomial in the size of the problem, since its state set is composed of variables and subterms of types occurring in Γ and τ. Moreover, the number of transitions is bounded by $\|\Gamma\|$. The results now follow from Propostion 5 together with the polynomial time solvability of the respective problems for deterministic tree automata and the fact that alternation can be eliminated with an exponential blow-up in the size of the automaton [5]. □

4   Subtyping

In this section we prove that the inhabitation problem remains EXPTIME-complete in the presence of subtyping (Section 4.2), while the problems of uniqueness and finiteness remain in in EXPTIME. We also show (Section 4.1) that the subtyping relation itself is decidable in polynomial time.

We extend the type language with the constant ω and stratify types as follows. Type variables and ω are called *atoms* and are ranged over by a, b, etc. Function types $(A, B, \dots)$, and general types $(\tau, \sigma, \dots)$ are defined by

$$A ::= a \mid \tau \to \tau \qquad\qquad \tau ::= a \mid \tau \to \tau \mid \bigcap_{i \in I} \tau_i$$

Subtyping $\leq$ is the least preorder (reflexive and transitive relation), satisfying:

$$\sigma \leq \omega, \quad \omega \leq \omega \to \omega, \quad \sigma \cap \tau \leq \sigma, \quad \sigma \cap \tau \leq \tau, \quad \sigma \leq \sigma \cap \sigma;$$
$$(\sigma \to \tau) \cap (\sigma \to \rho) \leq \sigma \to \tau \cap \rho;$$

If $\sigma \leq \sigma'$ and $\tau \leq \tau'$ then $\sigma \cap \tau \leq \sigma' \cap \tau'$ and $\sigma' \to \tau \leq \sigma \to \tau'$.

We identify σ and τ when $\sigma \leq \tau$ and $\tau \leq \sigma$.

4.1   PTIME **Decision Procedure for Subtyping**

We show that the intersection type subtyping relation $\leq$ is decidable in polynomial time. It appears that very little has been published about algorithmic properties of the relation so far. The only result we have found is the decidability result of [6] from 1982. The key characterization in [6] (also to be found in [9, Lemma 2.1]) is that for *normalized* types A_i, B_j ($i \in I$, $j \in J$) one has

$$\bigcap_{i \in I} A_i \leq \bigcap_{j \in J} B_j \iff \forall j \exists i \; A_i \leq B_j.$$

Normalization, in turn, may incur an exponential blow-up in the size of types. Hence, a direct algorithm based on the above leads to an exponential procedure.

The PTIME algorithm uses the following property, probably first stated in [1], and often called *beta-soundness*. Note that the converse is trivially true.

Lemma 7. *Let a_j, for $j \in J$, be atoms.*

1. *If $\bigcap_{i \in I}(\sigma_i \to \tau_i) \cap \bigcap_{j \in J} a_j \leq \alpha$ then $\alpha = a_j$, for some $j \in J$.*
2. *If $\bigcap_{i \in I}(\sigma_i \to \tau_i) \cap \bigcap_{j \in J} a_j \leq \sigma \to \tau$, where $\sigma \to \tau \neq \omega$, then the set $\{i \in I \mid \sigma \leq \sigma_i\}$ is nonempty and $\bigcap\{\tau_i \mid \sigma \leq \sigma_i\} \leq \tau$.*

It is not too difficult to construct a PSPACE-algorithm by a direct recursive implementation of Lemma 7. We can however do better using memoization techniques in a combined top-down and bottom-up processing of type trees.

Theorem 8. *The subtyping relation $\leq$ is in* PTIME.

Proof. Given types τ_0 and σ_0, to decide if $\tau_0 \leq \sigma_0$, we describe a procedure for answering *simple questions* of the form $\tau \leq \sigma$ or $\tau \geq \sigma$ (where τ and σ are, respectively, subterms of τ_0 and σ_0) and *set questions* of the form $\bigcap_i \tau_i \leq \sigma$ or $\tau \geq \bigcap_i \sigma_i$ (where τ_i and σ_i, but not necesarily the intersections, are subterms of τ_0, σ_0). The problem is that the total number of set questions is exponential.

We introduce memoization, where it is assumed that when a question is asked, all simple questions (but not all set questions) concerning smaller types have been already answered and the answers have been recorded.

To answer a simple question $\tau \leq \sigma$, we must decide if $\tau \leq \theta$ for all components θ of the $\cap$-type σ. Let $\tau = \bigcap_{j \in J} a_j \cap \bigcap_{i \in I}(\varrho_i \to \mu_i)$. If θ is an atom different from ω, we only check if it occurs among the a_j's. Otherwise, $\theta = \xi \to \nu$ and we proceed as follows:

- Determine the set $K = \{i \in I \mid \xi \leq \varrho_i\}$ by inspecting the memoized answers;
- Recursively ask the set question $\bigcap_{i \in K} \mu_i \leq \nu$.

After all inequalities $\tau \leq \theta$ have been verified, the answer $\tau \leq \sigma$ or $\tau \not\leq \sigma$ is stored in memory.

For a set question $\bigcap_i \tau_i \leq \sigma$, write the $\bigcap_i \tau_i$ as $\bigcap_{j \in J} a_j \cap \bigcap_{i \in I}(\varrho_i \to \mu_i)$. Then we proceed as in the simple case, except that the result is not memoized. The algorithm is formalized in our report [16].

To answer a single question (simple or set) one makes as many recursive calls as there are components of the right-hand side. The right-hand sides ("targets") of the recursive calls are different subterms of the original target. Further recursive calls address smaller targets, and the principal observation is that *all recursive calls incurred by any fixed question have different targets*. Therefore the total number of calls to answer a single question is at most linear. Since linear time is enough to handle a single call, we have a quadratic bound to decide a simple question, and $\mathcal{O}(n^4)$ total time. $\qquad\square$

4.2   Inhabitation with Subtyping

We extend the system FCL($\cap$) with subtyping. The resulting system is denoted FCL($\cap, \leq$) and is shown in Figure 3. (The axiom $\Gamma \vdash e : \omega$ is left out for simplicity, since it has no interesting consequences for the theory developed in this paper.)

$$\frac{}{\Gamma, x : \tau \vdash x : \tau} \; (\text{var}) \qquad\qquad \frac{\Gamma \vdash e : \tau \to \tau' \quad \Gamma \vdash e' : \tau}{\Gamma \vdash (e\,e') : \tau'} \; (\to\text{E})$$

$$\frac{\Gamma \vdash e : \tau_1 \quad \Gamma \vdash e : \tau_2}{\Gamma \vdash e : \tau_1 \cap \tau_2} \; (\cap\text{I}) \qquad\qquad \frac{\Gamma \vdash e : \tau \quad \tau \leq \tau'}{\Gamma \vdash e : \tau'} \; (\leq)$$

Fig. 3. FCL($\cap, \leq$)

We now prove that inhabitation remains EXPTIME-complete in the presence of subtyping. We begin with a path characterization of typings.

Lemma 9. *For any τ and $w \in \mathbf{R}_n(\tau)$ one has:*

$$\tau \leq arg(\tau, w, 1) \to \cdots \to arg(\tau, w, n) \to tgt(\tau, w).$$

Proof. By induction on $n \geq 0$. Details omitted. $\square$

Lemma 10 (Path Lemma). *The assertion $\Gamma \vdash (x\, e_1 \ldots e_n) : \tau$ holds if and only if there exists a subset S of $\mathbf{R}_n(\Gamma(x))$ such that*

1. *$\Gamma \vdash e_j : \bigcap_{w \in S} arg(\Gamma(x), w, j)$, for $j = 1, \ldots, n$;*
2. *$\bigcap_{w \in S} tgt(\Gamma(x), w) \leq \tau$.*

Proof. Part ($\Leftarrow$) is an easy consequence of Lemma 9. Part ($\Rightarrow$) goes by induction with respect to the derivation of $\Gamma \vdash (x\, e_1 \ldots e_n) : \tau$. The omitted details can be found in the full report [16]. $\square$

Tree automaton. Using Lemma 10 it is possible to specify, as in the subtyping-free case, an alternating tree automaton $\mathcal{A}(\Gamma, \tau)$ of the following property (details are given in [16]):

$$\Gamma \vdash e : \tau \quad \text{if and only if} \quad e \in \mathbf{L}(\mathcal{A}(\Gamma, \tau)).$$

Unfortunately, Lemma 10 requires that we handle *subsets* S of right-addresses rather than single addresses, and hence the size of the modified automaton could be exponential in the size of the input. As a consequence, this solution would only imply a *doubly exponential* upper bound for the emptiness, finiteness and uniqueness problems of inhabitation (recall that an additional exponential blow-up is incurred by determinization, following the proof of Theorem 6). However, together with Theorem 6 this reduction to tree automata implies:

Corollary 11. *In system* FCL($\cap$) *and* FCL($\cap, \leq$) *the tree languages of inhabitants given by* $\mathbf{L}(\Gamma, \tau) = \{e \mid \Gamma \vdash e : \tau\}$ *are regular.*

Turing machine. We can achieve an exponential upper bound for FCL($\cap, \leq$) by a direct implementation of the algorithm implicit in Lemma 10 on a Turing machine. Notice that we cannot apply the same approach as with tree automata, namely define a machine recognizing inhabitants for a fixed pair Γ and τ, and then check if the language accepted by the machine is nonempty. Indeed, emptiness is undecidable even for LOGSPACE machines. But we can construct a polynomial space bounded alternating Turing machine $\mathcal{M}$ such that

$$\mathcal{M} \text{ accepts } (\Gamma, \tau) \quad \text{if and only if} \quad \Gamma \vdash e : \tau, \text{ for some } e.$$

The main point is that each choice of the set $S \subseteq \mathbf{R}_n(\Gamma(x))$ can be stored in linear space on the tape of the Turing machine.

The definition shown in Figure 4 specifies an alternating Turing machine $\mathcal{M}$ which accepts on input Γ and τ if and only if there exists an applicative inhabitant of τ in the environment Γ. Recall from e.g. [14, section 16.2] that the state set Q of an alternating Turing machine $\mathcal{M} = (Q, \Sigma, \Delta, s)$ is partitioned into two subsets, $Q = Q_\exists \cup Q_\forall$. States in $Q_\exists$ are referred to as existential states, and states in $Q_\forall$ are referred to as universal states. A configuration whose state is in $Q_\forall$ is accepting if and only if all its successor configurations are accepting, and a confuguration whose state is in $Q_\exists$ is accepting if and only if at least one of its successor configurations is accepting.

The alternating Turing machine defined in Figure 4 directly implements Lemma 10. We use shorthand notation for existential states (CHOOSE...) and universal states ($\bigwedge$...), where the branching factors depend on parameters of the input (Figure 4, lines 2–4, 5). For example, in line 2, the expression CHOOSE $x \in dm(\Gamma)$... denotes an existential state from which the machine transitions to a number of successor states that depends on Γ and writes a member of $dm(\Gamma)$ on a designated portion of the tape for later reference. We can treat the notation as shorthand for a loop, in standard fashion. Similarly, line 8 denotes a universal state ($\bigwedge_{j=1}^n$...) from which a loop is entered depending on n. In line 5, we call a decision procedure for subtyping, relying on Lemma 8.

The variable τ holds the value of the current inhabitation goal, processed as follows. First, values for x, n and S are existentially generated (lines 2–4). It is then tested (line 5) that the current goal type τ is a supertype of the target intersection and, if so, the machine transitions to the universal state (line 8). In every branch of computation created at step 8, the goal τ is set to the appropriate argument type, and the computation resumes again from line 2. The computation stops when there is no arguments to process (line 6).

Theorem 12. *The inhabitation problem in* FCL($\cap, \leq$) *is* EXPTIME-*complete.*

Proof. Inhabitation remains EXPTIME-hard in the presence of subtyping, and the proof remains essentially the same. Therefore it is omitted, see [16] for details.

Membership in EXPTIME follows from the fact that the Turing machine $\mathcal{M}$ shown in Figure 4 solves the inhabitation problem in alternating polynomial

$$
\begin{array}{ll}
& Input: \quad \Gamma, \tau \\[2mm]
1 & // \text{ loop} \\
2 & \quad \text{CHOOSE } x \in dm(\Gamma); \\
3 & \quad \text{CHOOSE } n \in \{0, \dots, \|\Gamma(x)\|\}; \\
4 & \quad \text{CHOOSE } S \subseteq \mathbf{R}_n(\Gamma(x)); \\[2mm]
5 & \quad \text{IF } (\bigcap_{w \in S} tgt(\Gamma(x), w) \leq \tau) \text{ THEN} \\
6 & \quad\quad \text{IF } (n = 0) \text{ THEN ACCEPT}; \\
7 & \quad\quad \text{ELSE} \\
8 & \quad\quad\quad \bigwedge_{j=1}^{n} \tau := \bigcap_{w \in S} arg(\Gamma(x), w, j); \\
9 & \quad\quad\quad \text{GOTO LINE 2}; \\
10 & \quad \text{ELSE REJECT};
\end{array}
$$

Fig. 4. Alternating Turing machine $\mathcal{M}$ deciding inhabitation for $\text{FCL}(\cap, \leq)$

space. Correctness (the machine accepts (Γ, τ) iff $\Gamma \vdash e : \tau$ for some e) follows easily from Lemma 10 (the machine is a direct implementation of the lemma). To see that the machine is polynomial space bounded, notice that the variable S holds a subset of $\mathbf{R}_n$ for $n \leq \|\Gamma\|$, hence it consumes memory linear in the size of the input. Moreover, Lemma 8 shows that the subtyping relation is decidable in polynomial time (and space). Finally, notice that the variable τ holds types that are intersections of distinct strict subterms of the type $\Gamma(x)$, and there is only a linear number of subterms.   □

Uniqueness of inhabitation with subtyping. In order to show that uniqueness of inhabitation in system $\text{FCL}(\cap, \leq)$ is in EXPTIME we modify the inhabitation algorithm using alternation to search for more than one inhabitant. Figure 5 shows an alternating polynomial space Turing machine accepting Γ and τ if and only if there exists more than one inhabitant e with $\Gamma \vdash e : \tau$ in system $\text{FCL}(\cap, \leq)$. In this machine, alternation is used in two different ways, to search for inhabitants of argument types (as was already done in Figure 4) as well as to search for two different inhabitants. The machine solves the *ambiguity problem* (given Γ, τ, does there exist *more than one* term e with $\Gamma \vdash e : \tau$?) by nondeterministically guessing an inhabitant (lines 3–5) and a point from which another inhabitant can be "split" off (lines 6–9). See [16] for full details.

Theorem 13. *In system $\text{FCL}(\cap, \leq)$, uniqueness of inhabitation is in* EXPTIME.

Finiteness of inhabitation with subtyping Lemma 10 (Path Lemma) yields an alternative proof system for $\text{FCL}(\cap, \leq)$, consisting only of the following rule schema (P), where $S \subseteq \mathbf{R}_n(\Gamma(x))$:

$$
\frac{\Gamma \vdash_P e_j : \bigcap_{w \in S} arg(\Gamma(x), w, j) \ (j = 1, \dots, n) \qquad \bigcap_{w \in S} tgt(\Gamma(x), w) \leq \tau}{\Gamma \vdash_P @_n(x, e_1, \dots, e_n) : \tau} \ (P)
$$

```
        Input :   Γ, τ

   1    do_split := TRUE;

   2    // loop
   3       CHOOSE x₁ ∈ dm(Γ);
   4       CHOOSE n₁ ∈ {0, ..., ‖Γ(x₁)‖};
   5       CHOOSE S₁ ⊆ Rₙ₁(Γ(x₁));

   6    IF (do_split) THEN
   7           CHOOSE x₂ ∈ dm(Γ);
   8           CHOOSE n₂ ∈ {0, ..., ‖Γ(x₂)‖};
   9           CHOOSE S₂ ⊆ Rₙ₂(Γ(x₂));
  10           IF (x₁ ≠ x₂ OR n₁ ≠ n₂) THEN
  11               do_split := FALSE;
  12               GOTO LINE 14   ∧   GOTO LINE 15;
  13           ELSE GOTO LINE 14;

  14    x := x₁;  n := n₁;  S := S₁;  GOTO LINE 16;
  15    x := x₂;  n := n₂;  S := S₂;

  16    IF (⋂_{w∈S} tgt(Γ(x), w) ≤ τ) THEN
  17       IF (n = 0) THEN
  18           IF (NOT do_split) THEN ACCEPT;
  19           ELSE REJECT;
  20       ELSE
  21           CHOOSE k ∈ {1, ..., n};
  22           ⋀ⁿ_{j=1}(IF (j = k) THEN do_split := TRUE; ELSE do_split := FALSE;
  23               τ := ⋂_{w∈S} arg(Γ(x), w, j);
  24               GOTO LINE 3)
  25    ELSE REJECT;
```

Fig. 5. Alternating Turing machine deciding ambiguity of inhabitation

Type judgements derivable by (P) are written as $\Gamma \vdash_P e : \tau$. Notice that rule (P) becomes an axiom scheme when $n = 0$.

Lemma 14. $\Gamma \vdash (x\, e_1 \ldots e_n) : \tau$ *if and only if* $\Gamma \vdash_P @_n(x, e_1, \ldots, e_n) : \tau$.

Proof. Immediate by induction using Lemma 10.   □

Clearly, a proof Π of $\Gamma \vdash_P e : \tau$ assigns a type to every subterm e' of e, applying rule (P), with some set S. If $e' = @_n(x, e_1, \ldots, e_n)$ then the triple (x, n, S) is called the Π-*stamp* at e. We say that a proof Π of $\Gamma \vdash_P e : \tau$ is *cyclic*, if there is a subterm e' of e and a proper subterm e'' of e' have the same Π-stamp.

Lemma 15. *Let* Π *be a proof of* $\Gamma \vdash_P e : \tau$. *If the depth of* e *exceeds* $|\Gamma| \cdot 2^{|\Gamma|} \cdot |dm(\Gamma)|$ *then the proof* Π *is cyclic.*

```
        Input : Γ, τ

   1    count := 0;
   2    inf := TRUE;

   3    CHOOSE y ∈ dm(Γ);
   4    CHOOSE m ∈ {0, . . . ‖Γ(y)‖};
   5    CHOOSE Q ⊆ Rₘ(Γ(y));

   6  // loop
   7      CHOOSE x ∈ dm(Γ);
   8      CHOOSE n ∈ {0, . . . , ‖Γ(x)‖};
   9      CHOOSE S ⊆ Rₙ(Γ(x));

  10      IF (inf AND (x, n, S) = (y, m, Q)) THEN count := count + 1;

  11      IF (⋂_{w∈S} tgt(Γ(x), w) ≤ τ) THEN
  12        IF (n = 0) THEN
  13          IF (inf) THEN
  14            IF (count = 2) THEN ACCEPT ELSE REJECT;
  15          ELSE ACCEPT
  16        ELSE
  17          CHOOSE k ∈ {1, . . . , n};
  18          ⋀ⁿⱼ₌₁ (IF (j = k) THEN inf := TRUE ELSE inf := FALSE;
  20            τ := ⋂_{w∈S} arg(Γ(x), w, j);
  21            GOTO LINE 7)
  22      ELSE REJECT;
```

Fig. 6. Alternating Turing machine deciding infiniteness of inhabitation

Lemma 16. *There are infinitely many inhabitants e with $\Gamma \vdash e : \tau$ if and only if there exists a cyclic proof of $\Gamma \vdash_P e' : \tau$ for some e'.*

Theorem 17. *In system* FCL$(\cap, \leq)$, *finiteness of inhabitation is in* EXPTIME.

Proof. Consider the algorithm shown in Figure 6. It is an alternating polynomial space Turing machine accepting Γ and σ if and only if there exists an inhabitant e with a cyclic proof of $\Gamma \vdash_P e : \sigma$. By Lemma 16, therefore, infiniteness of inhabitation is in EXPTIME for system FCL$(\cap, \leq)$, and hence finiteness of inhabitation is in EXPTIME. Full details can be found in [16]. □

Acknowledgement. The authors thank Boris Düdder and Moritz Martens for helpful comments.

References

1. Barendregt, H., Coppo, M., Dezani-Ciancaglini, M.: A filter lambda model and the completeness of type assignment. Journal of Symbolic Logic 48(4), 931–940 (1983)

2. Ben-Yelles, C.: Type Assignment in the Lambda-Calculus: Syntax and Semantics. PhD thesis, Department of Pure Mathematics, University College of Swansea (September 1979)
3. Coppo, M., Dezani-Ciancaglini, M.: An extension of basic functionality theory for lambda-calculus. Notre Dame Journal of Formal Logic 21, 685–693 (1980)
4. Dezani-Ciancaglini, M., Hindley, J.R.: Intersection types for combinatory logic. Theoretical Computer Science 100(2), 303–324 (1992)
5. Comon, H., et al.: Tree Automata Techniques and Applications (2008), http://tata.gforge.inria.fr
6. Hindley, J.R.: The simple semantics for Coppo-Dezani-Sallé types. In: Dezani-Ciancaglini, M., Montanari, U. (eds.) International Symposium on Programming. LNCS, vol. 137, pp. 212–226. Springer, Heidelberg (1982)
7. Kobayashi, N., Luke Ong, C.-H.: A type system equivalent to the modal mu-calculus model checking of higher-order recursion schemes. In: LICS, pp. 179–188. IEEE Computer Society, Los Alamitos (2009)
8. Kozik, M.: A finite set of functions with an EXPTIME-complete composition problem. Theoretical Computer Science 407, 330–341 (2008)
9. Kurata, T., Takahashi, M.: Decidable properties of intersection type systems. In: Dezani-Ciancaglini, M., Plotkin, G. (eds.) TLCA 1995. LNCS, vol. 902, pp. 297–311. Springer, Heidelberg (1995)
10. Kuśmierek, D.: The inhabitation problem for rank two intersection types. In: Ronchi Della Rocca, S. (ed.) TLCA 2007. LNCS, vol. 4583, pp. 240–254. Springer, Heidelberg (2007)
11. Leivant, D.: Polymorphic type inference. In: Proc. 10th ACM Symp. on Principles of Programming Languages, pp. 88–98. ACM, New York (1983)
12. Lustig, Y., Vardi, M.Y.: Synthesis from component libraries. In: de Alfaro, L. (ed.) FOSSACS 2009. LNCS, vol. 5504, pp. 395–409. Springer, Heidelberg (2009)
13. Naik, M., Palsberg, J.: A type system equivalent to a model checker. In: Sagiv, M. (ed.) ESOP 2005. LNCS, vol. 3444, pp. 374–388. Springer, Heidelberg (2005)
14. Papadimitriou, C.H.: Computational Complexity. Addison-Wesley, Reading (1994)
15. Pottinger, G.: A type assignment for the strongly normalizable lambda-terms. In: Hindley, J., Seldin, J. (eds.) To H. B. Curry: Essays on Combinatory Logic, Lambda Calculus and Formalism, pp. 561–577. Academic Press, London (1980)
16. Rehof, J., Urzyczyn, P.: Finite combinatory logic with intersection types. Technical Report 834, Dept. of Computer Science, Technical University of Dortmund (2011), http://ls14-www.cs.tu-dortmund.de/index.php/Datei:TR-834.pdf
17. Salvati, S.: Recognizability in the simply typed lambda-calculus. In: Ono, H., Kanazawa, M., de Queiroz, R. (eds.) WoLLIC 2009. LNCS, vol. 5514, pp. 48–60. Springer, Heidelberg (2009)
18. Urzyczyn, P.: The emptiness problem for intersection types. Journal of Symbolic Logic 64(3), 1195–1215 (1999)
19. Urzyczyn, P.: Inhabitation of low-rank intersection types. In: Curien, P.-L. (ed.) TLCA 2009. LNCS, vol. 5608, pp. 356–370. Springer, Heidelberg (2009)
20. Urzyczyn, P.: The logic of persistent intersection. Fundamenta Informaticae 103, 303–322 (2010)
21. Venneri, B.: Intersection types as logical formulae. Journal of Logic and Computation 4(2), 109–124 (1994)
22. Wells, J.B., Yakobowski, B.: Graph-based proof counting and enumeration with applications for program fragment synthesis. In: Etalle, S. (ed.) LOPSTR 2004. LNCS, vol. 3573, pp. 262–277. Springer, Heidelberg (2005)

Linear Lambda Calculus and Deep Inference

Luca Roversi

Dip. di Informatica, C.so Svizzera 185, 10149 Torino — Italy

Abstract. We introduce a deep inference logical system SBVr which extends SBV [6] with Rename, a self-dual atom-renaming operator. We prove that the cut free subsystem BVr of SBVr exists. We embed the terms of linear λ-calculus with explicit substitutions into formulas of SBVr. Our embedding recalls the one of full λ-calculus into π-calculus. The proof-search inside SBVr and BVr is complete with respect to the evaluation of linear λ-calculus with explicit substitutions. Instead, only soundness of proof-search in SBVr holds. Rename is crucial to let proof-search simulate the substitution of a linear λ-term for a variable in the course of linear β-reduction. Despite SBVr is a minimal extension of SBV its proof-search can compute all boolean functions, exactly like linear λ-calculus with explicit substitutions can do.

To Umberto Spina 1924 — 2000.

1 Introduction

We formalize unknown relations between basic functional computation and Deep inference (DI). We show that the computations of linear λ-calculus with explicit substitutions, at the core of functional programming, live as proof-search inside the logical system SBVr that we introduce, and which extends SBV, system at the core of DI [15].

Our motivation is to search how structural proof theory, based on DI methodology, can contribute to paradigmatic programming language design. We recall that logical systems designed under the prescriptions of DI allow inference rules to apply anywhere in a formula in the course of a derivation. This is in contrast to "shallow inference" of traditional structural proof theoretic formalisms, like natural deduction and sequent calculus, whose inference rules apply at the root of formulas and sequents only. Applying rules in depth implies that structural proof theory of a quite vast range of logics, i.e. classical, intuitionistic, linear and modal, has become very regular and modular. We expect that much regularity and modularity can highlight useful intrinsic properties and new primitives, or evaluation strategies, at the level of programs. The point is to look for the computational interpretation of derivations in DI style in the same vein as the one we are used to with shallow inference. We think that source of new programming primitives, or evaluation strategies, can be DI deductive systems whose inference rules only manipulate atoms of formulas, and for which new notions of proof normalization exist, in addition to cut elimination.

So far, [3] is the only work which focuses on the proposal of a Curry-Howard analogy for a DI system, and which we are aware of. It applies DI methodology to the natural

L. Ong (Ed.): TLCA 2011, LNCS 6690, pp. 184–197, 2011.
© Springer-Verlag Berlin Heidelberg 2011

deduction of the negative fragment of intuitionistic logic, and extracts an algebra of combinators from it where interpreting λ-terms.

Rather than intuitionistic logic, as in [3], we focus on SBV and its cut free subsystem BV. The point about BV is that, thanks to its "deepness", it naturally extends multiplicative linear logic (MLL) [5] with the non commutative binary operator Seq, denoted "$\triangleleft$". Second, from [4] we recall that Seq soundly and completely models the sequential communication of CCS concurrent and communicating systems [11], while Par ($\otimes$) models parallel composition. Third, we recall that λ-calculus β-reduction incorporates a sequential behavior. We mean that the overall computation the redex $(\lambda x.M)N$ embodies can complete only *after* executing the β-reduction $(\lambda x.M)N \rightarrow_\beta M\{^N/_x\}$ that substitutes N for every occurrence of x in M, if any. The sequential behavior of β-reduction closely recalls the one in CCS. The difference is that, after consuming a prefix that takes the form of a λ-abstraction, β-reduction also requires to perform a substitution. So, if we map a λ-variable x into a BV formula, by a map $(\!|\cdot|\!)$., as follows:

$$(\!|x|\!)_o = \langle x \triangleleft \overline{o} \rangle \tag{1}$$

we can start modeling the consumption of the λ-abstraction that takes place in the course of a β-reduction. Intuitively, x in (1) is the name of an input channel, to the left of Seq, eventually forwarded to the output channel o, associated to x by Seq. Definition (1) comes from the following embedding of sequents of *intuitionistic* multiplicative linear logic (IMLL) into SBV formulas:

$$(\alpha_1, \ldots, \alpha_m \vdash_{\mathsf{IMLL}} \beta)^\bullet = \langle (\overline{\alpha_1^\bullet} \otimes \cdots \otimes \overline{\alpha_m^\bullet}) \triangleleft \beta^\bullet \rangle \tag{2}$$

where Seq internalizes the meta-notation $\vdash_{\mathsf{IMLL}}$. We remark that (1) recalls:

$$[x]_o = x(\!\circ\!) . \overline{o}\langle \circ \rangle \tag{3}$$

the base clause of the, so called, *output-based embedding* of the *standard λ-calculus with explicit substitutions* into π-calculus [17]. In (3), "." is the sequential composition of the π-calculus, and $\circ$ a place-holder constant. In [8], (3) is called *forwarder*. So, SBV formulas contain the forwarder, the basic input/output communication flow that λ-variables realize, if we look at (1) as a process. However, we could not see how to obtain any *on-the-fly renaming* of channels able to model the substitution of a term for a bound variable, namely, β-reduction. So, we extended SBV to SBVr.

Contributions and paper organization

System SBVr. The system SBVr (Section 2) extends SBV by adding Rename, a binary renaming operator $\lceil \cdot \rfloor$.. Rename is self-dual, it binds atoms, and it is the inverse of α-rule. The prominent defining axiom of Rename is $R \approx \lceil R\{^a/_b\} \rfloor_a$ (Figure 3) where R is a formula of SBVr. The meta-notation $R\{^a/_b\}$ denotes the capture-free substitution of a for b, and of $\overline{a}$ for $\overline{b}$ in R. Roughly, we can think of $\lceil R\{^a/_b\} \rfloor_a$ as $\forall a.R\{^a/_b\}$. The idea is that we shall exploit Rename for renaming atoms of formulas which represent linear λ-terms with explicit substitutions. Like a universal quantifier, Rename sets the boundary where the name change can take place without altering the set of free names of SBVr of a formula.

Completeness of SBVr *and* BVr. First we recall that linear λ-calculus embodies the core of functional programming. The functions it represents use their arguments exactly once in the course of the evaluation. The set of functions we can express in it are quite limited, but "large" enough to let the decision about which is the normal form of its λ-terms a *polynomial time complete* problem [10], if we take the polynomial time Turing machines as computational model of reference. We shall focus on λ-calculus *with explicit substitutions* [1]. "Explicit" means that the operation that substitutes a λ-term for a λ-variable, in the course of a β-reduction, is not meta, but a term constructor.

Concerning the completeness of SBVr, we start defining an embedding $(\!|\cdot|\!)$. from λ-terms with explicit substitutions to formulas of SBVr (Figure 8, Section 5.) Then, we prove that, for every linear λ-term with explicit substitutions M, and every atom o, which plays the role of an output-channel, if M reduces to N, then there is a *derivation* of SBVr, with $(\!|M|\!)_o$ as conclusion, and $(\!|N|\!)_o$ as premise (Theorem 5.6.) Namely, completeness says that the evaluation of a linear λ-term with explicit substitutions M, viewed as a formula, is proof-search in SBVr. As a corollary, we show that BVr, the cut free subsystem of SBVr, is complete with respect to linear λ-calculus with explicit substitutions as well. We mean that a computation from M to N in BVr becomes a process of annihilation between the formula $(\!|M|\!)_o$ that represents M, and the *negation of* $(\!|N|\!)_o$, representing N (Corollary 5.7.)

It is worth remarking that the above embedding $(\!|\cdot|\!)$. strongly recalls *output-based* embedding of *standard* λ-calculus into π-calculus [17]. The reason why we think this is relevant is simple. We recall that the traditional embeddings of *standard* λ-calculus into π-calculus follow, instead, the *input-based* pattern of [12]. Input-based embeddings are quite restrictive. The only β-reduction strategy they allow to simulate by π-calculus processes is the *lazy* one. Instead, output-based embedding extends the simulation to *spine* evaluation strategy, which strictly includes the lazy one. So, extending SBVr with logical operators that model replication of λ-terms will possibly extend completeness we now have for SBVr, relatively to linear λ-calculus with explicit substitutions, to wider subsets of full λ-calculus.

Cut elimination for SBVr. The above mentioned completeness of BVr follows from proving that the cut elimination holds inside SBVr (Theorem 3.6.) The main steps to prove cut elimination are *shallow splitting*, *context reduction*, *splitting*, and *admissibility of the up fragment*, following [6]. The cut elimination for SBVr says that its theorems can be proved by means of the rules of its subsystem BVr free of cut rule.

Soundness of SBVr. It says that proof-search inside SBVr is an interpreter of linear λ-terms with explicit substitutions. We show that, if a derivation of SBVr proves that $(\!|M|\!)_o$ derives from $(\!|N|\!)_o$ in SBVr, then M reduces N (Theorem 5.6.) However, the derivation of $(\!|M|\!)_o$ from $(\!|N|\!)_o$ must be under a specific proof-search strategy. This, might limit efficiency. We mean that all the freedom we might gain thanks to the deep application of the logical rules, in the course of a proof-search, can be lost by sticking to the specific strategy we are referring to and that we shall see.

Expressiveness of SBVr *and* BVr. Linear λ-calculus can compute all boolean functions [10]. Proof-search of SBVr, and BVr can do the same thanks to the above

$$R ::= a \mid \overline{a} \mid \circ \mid [R \,\mathbin{\bindnasrepma}\, R] \mid (R \otimes R) \mid \langle R \blacktriangleleft R \rangle \mid \lceil R \rfloor_a$$

Fig. 1. Structures

completeness. So, the extension of SBV to SBVr is not trivial. Moreover, since BV is NPtime-complete [9], so is BVr.

Extended version. This work is an extended abstract and contains a single full detailed proof. The technical report [13] contains all proofs of the coming statements, and a richer bibliography.

2   Systems SBVr and BVr

Structures. Let $a, b, c, \ldots$ denote the elements of a countable set of *positive propositional variables*. Instead, $\overline{a}, \overline{b}, \overline{c}, \ldots$ denote the elements of a countable set of *negative propositional variables*. The set of *atoms* contains both positive and negative propositional variables, and nothing else. Let $\circ$ be a *constant* different from any atom. The grammar in Figure 1 gives the set of *structures*, the standard name for formulas in SBV. We shall range over structures with R, P, T, U, and V. Par $[R \,\mathbin{\bindnasrepma}\, R]$, CoPar $(R \otimes R)$, and Seq $\langle R \blacktriangleleft R \rangle$ come from SBV. Rename $\lceil R \rfloor_a$ is new and comes with the proviso that a must be a positive atom. Namely, $\lceil R \rfloor_{\overline{a}}$ is not in the syntax. Rename implies the definition of the *free names* FN(R) of R as in Figure 2.

Size of the structures. The *size* $|R|$ of R sums the number of occurrences of atoms in R and the number of occurrences of renaming operators $\lceil T \rfloor_a$ inside R whose bound variable a belongs to FN(T). For example, $|\lceil [b \,\mathbin{\bindnasrepma}\, \overline{b}] \rfloor_a| = 2$, while $|\lceil [a \,\mathbin{\bindnasrepma}\, \overline{a}] \rfloor_a| = 3$.

Equivalence on structures. Structures are equivalent up to the smallest congruence defined by the set of axioms in Figure 3 where Rename is a self-dual operator. By $R\{^a/_b\}$ we denote the capture-free substitution of a for b, and of $\overline{a}$ for $\overline{b}$ in R. The intuitive reason why Rename is self-dual follows. Since $R\{^a/_b\}$ denotes R where every free occurrence of the atom a, and its dual $\overline{a}$, replaces b, and $\overline{b}$, respectively, nothing changes when operating on $\overline{R}$ in place of R. Every occurrence of a in $\overline{R}$ corresponds to one of $\overline{a}$ of R, and

$$\{a\} = \text{FN}(a) \cup \text{FN}(\overline{a})$$
$$a \in \text{FN}(\langle R \blacktriangleleft T \rangle) \text{ if } a \in \text{FN}(R) \cup \text{FN}(T)$$
$$a \in \text{FN}((R \otimes T)) \text{ if } a \in \text{FN}(R) \cup \text{FN}(T)$$
$$a \in \text{FN}([R \,\mathbin{\bindnasrepma}\, T]) \text{ if } a \in \text{FN}(R) \cup \text{FN}(T)$$
$$a \in \text{FN}(\lceil R \rfloor_b) \text{ if } a \neq b \text{ and } a \in \text{FN}(R)$$

Fig. 2. Free names of structures

Negation

$$\overline{o} \approx o$$

$$\overline{[R \,\bindnasrepma\, T]} \approx (\overline{R} \otimes \overline{T})$$

$$\overline{(R \otimes T)} \approx [\overline{R} \,\bindnasrepma\, \overline{T}]$$

$$\overline{\langle R \triangleleft T \rangle} \approx \langle \overline{R} \triangleleft \overline{T} \rangle$$

$$\overline{\lceil R \rfloor_a} \approx \lceil \overline{R} \rfloor_a$$

Renaming

$$R \approx \lceil R\{^a/_b\} \rfloor_a \text{ if } a \notin \mathrm{FN}(R)$$

$$o\{^a/_b\} \approx o$$

$$b\{^a/_b\} \approx a$$

$$\overline{b}\{^a/_b\} \approx \overline{a}$$

$$c\{^a/_b\} \approx c$$

$$\overline{c}\{^a/_b\} \approx \overline{c}$$

$$[R \,\bindnasrepma\, T]\{^a/_b\} \approx [R\{^a/_b\} \,\bindnasrepma\, T\{^a/_b\}]$$

$$(R \otimes T)\{^a/_b\} \approx (R\{^a/_b\} \otimes T\{^a/_b\})$$

$$\langle R \triangleleft T \rangle\{^a/_b\} \approx \langle R\{^a/_b\} \triangleleft T\{^a/_b\} \rangle$$

$$\lceil R \rfloor_b\{^a/_b\} \approx R$$

$$\lceil R \rfloor_c\{^a/_b\} \approx \lceil R\{^a/_b\} \rfloor_c$$

Contextual Closure

$$S\{R\} \approx S\{T\} \text{ if } R \approx T$$

Unit

$$R \approx [o \,\bindnasrepma\, R] \approx (o \otimes R)$$

$$R \approx \langle o \triangleleft R \rangle \approx \langle R \triangleleft o \rangle$$

Associativity

$$[[R \,\bindnasrepma\, T] \,\bindnasrepma\, V] \approx [R \,\bindnasrepma\, [T \,\bindnasrepma\, V]]$$

$$((R \otimes T) \otimes V) \approx (R \otimes (T \otimes V))$$

$$\langle \langle R \triangleleft T \rangle \triangleleft V \rangle \approx \langle R \triangleleft \langle T \triangleleft V \rangle \rangle$$

Symmetry

$$[R \,\bindnasrepma\, T] \approx [T \,\bindnasrepma\, R]$$

$$(R \otimes T) \approx (T \otimes R)$$

Distributivity

$$\lceil \langle R \triangleleft U \rangle \rfloor_a \approx \langle \lceil R \rfloor_a \triangleleft \lceil U \rfloor_a \rangle$$

Exchange

$$\lceil \lceil R \rfloor_b \rfloor_a \approx \lceil \lceil R \rfloor_a \rfloor_b$$

Fig. 3. Equivalence $\approx$ on structures

vice-versa. We observe that $o \approx \lceil o \rfloor_a$ is as particular case of $R \approx \lceil R\{^a/_a\} \rfloor_a \approx \lceil R \rfloor_a$, which holds whenever $a \notin \mathrm{FN}(R)$. Finally, Distributivity in Figure 3 must be understood in relation to the deductive rules of **SBVr**, so we postpone the discussion.

(Structure) Contexts. They are $S\{\ \}$, i.e. a structure with a single hole $\{\ \}$ in it. If $S\{R\}$, then R is a *substructure* of S. For example, we shall tend to shorten $S\{[R \,\bindnasrepma\, U]\}$ as $S[R \,\bindnasrepma\, U]$ when $[R \,\bindnasrepma\, U]$ fills the hole $\{\ \}$ of $S\{\ \}$ exactly.

The system **SBVr**. It contains the set of inference rules in Figure 4 with form $\rho \, \dfrac{T}{R}$, name ρ, premise T, and *conclusion* R. The typical use of an inference rules is $\rho \, \dfrac{S\{T\}}{S\{R\}}$. It specifies that if a structure U matches R in a context $S\{\ \}$, it can be rewritten to $S\{T\}$. Since rules apply in any context, and we use them as rewriting rules, R is the *redex* of ρ, and T its *reduct*.

BVr is the *down fragment* $\{\mathsf{ai}{\downarrow}, \mathsf{s}, \mathsf{q}{\downarrow}, \mathsf{r}{\downarrow}\}$ of **SBVr**. The *up fragment* of **SBVr** is $\{\mathsf{ai}{\uparrow}, \mathsf{s}, \mathsf{q}{\uparrow}, \mathsf{r}{\uparrow}\}$. So s belongs to both. The rules that formalize Rename are $\mathsf{r}{\downarrow}$, and $\mathsf{r}{\uparrow}$. The former can be viewed as the restriction to a self-dual quantifier of the rule $\mathsf{u}{\downarrow}$ which, in [14], formalizes the standard universal quantifier.

$$\text{ai}{\downarrow}\ \frac{\circ}{[a \mathbin{⅋} \bar{a}]} \qquad\qquad \text{ai}{\uparrow}\ \frac{(a \otimes \bar{a})}{\circ}$$

$$\text{q}{\downarrow}\ \frac{\langle [R \mathbin{⅋} U] \triangleleft [T \mathbin{⅋} V]\rangle}{[\langle R \triangleleft T\rangle \mathbin{⅋} \langle U \triangleleft V\rangle]} \qquad \text{s}\ \frac{([R \mathbin{⅋} T] \otimes U)}{[(R \otimes U) \mathbin{⅋} T]} \qquad \text{q}{\uparrow}\ \frac{(\langle R \triangleleft T\rangle \otimes \langle U \triangleleft V\rangle)}{\langle (R \otimes U) \triangleleft (T \otimes V)\rangle}$$

$$\text{r}{\downarrow}\ \frac{\lceil [R \mathbin{⅋} U]\rceil_a}{[\lceil R\rceil_a \mathbin{⅋} \lceil U\rceil_a]} \qquad\qquad \text{r}{\uparrow}\ \frac{(\lceil R\rceil_a \otimes \lceil U\rceil_a)}{\lceil (R \otimes U)\rceil_a}$$

Fig. 4. System SBVr

Derivation and proof. A *derivation* in SBVr is either a structure or an instance of the above rules or a sequence of two derivations. The topmost structure in a derivation is its *premise*. The bottommost is its *conclusion*. The *length* $|\mathscr{D}|$ of a derivation $\mathscr{D}$ is the number of rule instances in $\mathscr{D}$. A derivation $\mathscr{D}$ of a structure R in SBVr from a structure T in SBVr, only using a subset $\mathsf{B} \subseteq$ SBVr is $\mathscr{D}\|\mathsf{B}$ with $\frac{T}{R}$. The equivalent *space-saving* form we shall tend to use is $\mathscr{D} : T \vdash_\mathsf{B} R$. The derivation $\mathscr{P}\|\mathsf{B}$ is a *proof* whenever $T \approx \circ$. We denote it as $\overset{\circ}{\underset{R}{\mathscr{P}\|\mathsf{B}}}$, or $\mathscr{P}\|\mathsf{B}$, or $\mathscr{P} : \vdash_\mathsf{B} R$. In general, we shall drop B when clear from the context. We shall write $\frac{T}{R}$ to mean $R \approx T$.

We are in the position to give an account of Distributivity in Figure 3. It morally embodies two rules, one flipping the other:

$$\frac{\lceil \langle R \triangleleft U\rangle\rceil_a}{\langle \lceil R\rceil_a \triangleleft \lceil U\rceil_a\rangle} \qquad\qquad \frac{\langle \lceil R\rceil_a \triangleleft \lceil U\rceil_a\rangle}{\lceil \langle R \triangleleft U\rangle\rceil_a}$$

Replacing these two rules by a single equivalence axiom is coherent to the intuitive interpretation of every Seq structure $\langle R \triangleleft U\rangle$ in [6]. Whatever the derivation $\mathscr{D}$ that contains $\langle R \triangleleft U\rangle$ is, neither going upward in $\mathscr{D}$, nor moving downward in it, any atom of R will ever interact with any atom of U. So, Rename can indifferently operate locally to R, and U or globally on $\langle R \triangleleft U\rangle$.

The following proposition shows when two structures R, T can be "moved" inside a context so that they are one aside the other and may eventually communicate going upward in a derivation.

Proposition 2.1 (*Context extrusion*.) $S[R \mathbin{⅋} T] \vdash_{\{\text{q}{\downarrow},\text{s},\text{r}{\downarrow}\}} [S\{R\} \mathbin{⅋} T]$, *for every* S, R, T.

Equivalence of systems. A subset $\mathsf{B} \subseteq$ SBVr *proves* R if $\mathscr{P} : \vdash_\mathsf{B} R$, for some proof $\mathscr{P}$. Two subsets B and B' of SBVr are *strongly equivalent* if, for every derivation $\mathscr{D} : T \vdash_\mathsf{B} R$, there exists a derivation $\mathscr{D}' : T \vdash_{\mathsf{B}'} R$, and vice versa. Two *systems are equivalent* if they prove the same structures.

$$i\downarrow \frac{\circ}{[R \,\bar{\otimes}\, \overline{R}]} \qquad i\uparrow \frac{(R \otimes \overline{R})}{\circ} \qquad \text{mixp} \frac{(R \otimes T)}{\langle R \triangleleft T \rangle}$$

$$\text{def}\downarrow \frac{\langle R \triangleleft T \rangle}{[\langle R \triangleleft \bar{a} \rangle \,\bar{\otimes}\, (a \otimes T)]} \qquad \text{def}\uparrow \frac{(\langle R \triangleleft \bar{a} \rangle \otimes [a \,\bar{\otimes}\, T])}{\langle R \triangleleft T \rangle} \qquad \text{pmix} \frac{\langle R \triangleleft T \rangle}{[R \,\bar{\otimes}\, T]}$$

Fig. 5. A core-set of rules derivable in SBVr

Derivable rules. A rule ρ is *derivable* in B $\subseteq$ SBVr if $\rho \notin$ B and, for every instance $\rho \frac{T}{R}$, there exists a derivation $\mathscr{D}$ in B such that $\mathscr{D} : T \vdash_B R$. Figure 5 recalls a core set of rules derivable in SBV, hence in SBVr. *General interaction down* and *up* rules are i↓, and i↑, respectively. The rule def↓ uses $\bar{a}$ as a place-holder, and a as name for T. Building the derivation upward, we literally replace T for $\bar{a}$. Symmetrically for def↑. The rules mixp, and pmix show a hierarchy between the connectives, where $\bar{\otimes}$ is the lowermost, $\triangleleft$ lies in the middle, and $\otimes$ on top. General interaction up is derivable in {ai↑, s, q↑, r↑}, while def↓ is derivable in {ai↓, s, q↓}, and mixp is derivable in {q↑}:

$$r\uparrow \frac{\dfrac{\dfrac{([R]_a \otimes \overline{[R]}_a)}{([R]_a \otimes [\overline{R}]_a)}}{[(R \otimes \overline{R})]_a}}{i\uparrow \dfrac{[(R \otimes \overline{R})]_a}{[\circ]_a}} \text{ ind. hypothesis}$$
$$\circ$$

$$\text{ai}\downarrow \frac{\dfrac{\dfrac{\langle R \triangleleft T \rangle}{\langle R \triangleleft (\circ \otimes T) \rangle}}{\langle R \triangleleft ([a \,\bar{\otimes}\, \bar{a}] \otimes T) \rangle}}{s} $$
$$q\downarrow \frac{\dfrac{\langle [R \,\bar{\otimes}\, \circ] \triangleleft [a \,\bar{\otimes}\, (\bar{a} \otimes T)] \rangle}{[\langle R \triangleleft a \rangle \,\bar{\otimes}\, \langle \circ \triangleleft (\bar{a} \otimes T) \rangle]}}{[\langle R \triangleleft a \rangle \,\bar{\otimes}\, (\bar{a} \otimes T)]}$$

$$q\uparrow \frac{\dfrac{\dfrac{(R \otimes T)}{(\langle R \triangleleft \circ \rangle \otimes \langle \circ \triangleleft T \rangle)}}{(\langle R \otimes \circ \rangle \triangleleft (\circ \otimes T))}}{\langle R \triangleleft T \rangle}$$

The leftmost derivation sketches how to derive general interaction up in the case new to SBVr, as compared to SBV. Symmetrically, *general interaction down* is derivable in {ai↓, s, q↓, r↓}, while def↑ in {ai↑, s, q↑}, and pmix in {q↓}.

3   Cut Elimination of SBVr

The goal is to prove that SBVr, and BVr are equivalent. Proving the equivalence, amounts to proving that every up rule is admissible in BVr or, equivalently, that we can eliminate them from any derivation of SBVr. Splitting theorem for BVr, which extends the namesake theorem for BV [6], is the effective tool we prove to exist for showing that the up fragment of SBVr is admissible for BVr.

Proposition 3.1 (BVr *is affine.*) *In every* $\mathscr{D} : T \vdash R$, *we have* $|R| \geq |T|$.

Proposition 3.2 here below says that the components of Seq, and CoPar can be proved independently, and that Rename is a universal quantifier on atoms.

Proposition 3.2 (*Derivability of structures in* BVr.) *For all* R, T, *and* a:

1. $\mathscr{P} : \vdash \langle R \triangleleft T \rangle$ *iff* $\mathscr{P}_1 : \vdash R$ *and* $\mathscr{P}_2 : \vdash T$.
2. $\mathscr{P} : \vdash (R \otimes T)$ *iff* $\mathscr{P}_1 : \vdash R$ *and* $\mathscr{P}_2 : \vdash T$.
3. $\mathscr{P} : \vdash [R]_a$ *iff* $\mathscr{P}' : \vdash R\{^b/_a\}$, *for every atom* b.

Proposition 3.3 here below says that part of the structure in the conclusion of every proof of BVr, which in the proposition will be P, is the context that, suitably decomposed, allows to annihilate each component of the remaining part of the conclusion.

Proposition 3.3 (*Shallow Splitting in* BVr.) *For all R, T, P, and a:*

1. *If $\mathscr{P} : \vdash [\langle R \triangleleft T \rangle \,\bindnasrepma\, P]$, then $\langle P_1 \triangleleft P_2 \rangle \vdash P$, and $\vdash [R \,\bindnasrepma\, P_1]$, and $\vdash [T \,\bindnasrepma\, P_2]$, for some P_1, P_2.*
2. *If $\mathscr{P} : \vdash [(R \otimes T) \,\bindnasrepma\, P]$, then $[P_1 \,\bindnasrepma\, P_2] \vdash P$, and $\vdash [R \,\bindnasrepma\, P_1]$, and $\vdash [T \,\bindnasrepma\, P_2]$, for some P_1, P_2.*
3. *If $\mathscr{P} : \vdash [a \,\bindnasrepma\, P]$, then $\mathscr{P}' : \bar{a} \vdash P$.*
4. *If $\mathscr{P} : \vdash [\lceil R \rceil_a \,\bindnasrepma\, P]$, then $\lceil P' \rceil_a \vdash P$, and $\vdash [R \,\bindnasrepma\, P']$, for some P'.*

Proposition 3.4 here below says that in the conclusion of every proof of BVr we can always focus on a specific structure R so that, upward in the proof, a suitable structure U allows to annihilate R. If some Rename gets in the way in between R, and U, we can always move it outward so that it does not interfere.

Proposition 3.4 (*Context Reduction in* BVr.) *For all R and contexts $S\{\ \}$ such that $\mathscr{P} : \vdash S\{R\}$, there are U, a such that $\mathscr{D} : \lceil [\{\ \} \,\bindnasrepma\, U] \rceil_a \vdash S\{\ \}$, and $\vdash [R \,\bindnasrepma\, U]$.*

Namely, here above, $S\{\ \}$ supplies the "context" U, required for annihilating R, no matter which structure fills the hole of $S\{\ \}$, and no matter the existence of Rename enclosing R and U.

Theorem 3.5 here below results from composing Context Reduction (Proposition 3.4), and Shallow Splitting (Proposition 3.3) in this order. It says that in every proof of BVr we can always focus on any structure inside the conclusion and, moving upward in the proof, we can annihilate it, the final goal being obtaining only ∘ as premise of the whole proof.

Theorem 3.5 (*Splitting in* BVr.) *For all R, T, and contexts $S\{\ \}$:*

1. *If $\mathscr{P} : \vdash S\langle R \triangleleft T \rangle$, then $\lceil [\{\ \} \,\bindnasrepma\, \langle K_1 \triangleleft K_2 \rangle] \rceil_a \vdash S\{\ \}$, and $\vdash [R \,\bindnasrepma\, K_1]$, and $\vdash [T \,\bindnasrepma\, K_2]$, for some K_1, K_2, a.*
2. *If $\mathscr{P} : \vdash S(R \otimes T)$, then $\lceil [\{\ \} \,\bindnasrepma\, [K_1 \,\bindnasrepma\, K_2]] \rceil_a \vdash S\{\ \}$, and $\vdash [R \,\bindnasrepma\, K_1]$, and $\vdash [T \,\bindnasrepma\, K_2]$, for some K_1, K_2, a.*
3. *If $\mathscr{P} : \vdash S\lceil R \rceil_a$, then $\lceil [\{\ \} \,\bindnasrepma\, K] \rceil_a \vdash S\{\ \}$, and $\vdash [R \,\bindnasrepma\, K]$, for some K, a.*

Theorem 3.6 here below results from composing Splitting (Theorem 3.5), and Shallow Splitting (Proposition 3.3). Just to give an informal idea about its meaning, in its proof we can show that every derivation of SBVr that contains an instance of r ↑ can be rewritten as a derivation with the same conclusion, without the above occurrence of r↑, but with a couple of new instances of both r↓, and s.

Theorem 3.6 (*Admissibility of the up fragment for* BVr.) *The set of rules $\{\mathsf{ai}{\uparrow}, \mathsf{q}{\uparrow}, \mathsf{r}{\uparrow}\}$ of SBVr is admissible for BVr.*

Theorem 3.6 here above directly implies:

Corollary 3.7 *The cut elimination holds for SBVr.*

$$(\lambda x.M)N \to (M)\{x = N\}$$
$$(x)\{x = P\} \to P$$
$$(\lambda y.M)\{x = P\} \to \lambda y.(M)\{x = P\}$$
$$((M)N)\{x = P\} \to (M)(N)\{x = P\} \qquad \text{if } x \in \mathrm{fv}(N)$$
$$((M)N)\{x = P\} \to ((M)\{x = P\})N \qquad \text{if } x \in \mathrm{fv}(M)$$

Fig. 6. β-reduction $\to\, \subseteq \Lambda \times \Lambda$ with explicit substitution

4   Linear λ-Calculus with Explicit Substitutions

Linear λ-calculus with explicit substitutions is a pair with a set of linear λ-terms, and an operational semantics on them. The operational semantics looks at substitution as explicit syntactic component and not as meta-operation.

The linear λ-terms. Let $\mathscr{V}$ be a countable set of variable names we range over by x, y, w, z. We call $\mathscr{V}$ the *set of λ-variables*. The set of *linear λ-terms with explicit substitutions* is $\Lambda = \bigcup_{X \subset \mathscr{V}} \Lambda_X$ we range over by M, N, P, Q. For every $X \subset \mathscr{V}$, the set Λ_X contains the *linear λ-terms with explicit substitutions whose free variables are in* X, and which we define as follows: (i) $x \in \Lambda_{\{x\}}$; (ii) $\lambda x.M \in \Lambda_X$ if $M \in \Lambda_{X \cup \{x\}}$; (iii) $(M)N \in \Lambda_{X \cup Y}$ if $M \in \Lambda_X$, $N \in \Lambda_Y$, and $X \cap Y = \emptyset$; (iv) $(M)\{x = P\} \in \Lambda_{X \cup Y}$ if $M \in \Lambda_{X \cup \{x\}}$, $P \in \Lambda_Y$, and $X \cap Y = \emptyset$.

β-reduction on linear λ-terms with explicit substitutions. It is the relation $\to$ in Figure 6. It is the core of the very simple, indeed, computations the syntax of the terms in Λ allow to develop. The point, however, is that the core computational mechanism that replaces a term for a variable is there, and we aim at modeling it inside **SBVr**. Moreover, despite Λ is so simple, it can model all the boolean functions [10].

Operational semantics on linear λ-terms with explicit substitutions. It is the relation $\Rightarrow$ in Figure 7, i.e. the reflexive, contextual, and transitive closure of the above β-reduction with explicit substitution. The number of instances of rules in Figure 7 that yield $M \Rightarrow N$ is $|M \Rightarrow N|$.

$$\mathrm{rfl}\ \frac{}{M \Rightarrow M} \qquad \mathrm{lft}\ \frac{M \to N}{M \Rightarrow N} \qquad \mathrm{tra}\ \frac{M \Rightarrow P \quad P \Rightarrow N}{M \Rightarrow N}$$

$$\mathrm{f}\ \frac{M \Rightarrow N}{\lambda x.M \Rightarrow \lambda x.N} \qquad @\mathrm{l}\ \frac{M \Rightarrow N}{(M)P \Rightarrow (N)P} \qquad @\mathrm{r}\ \frac{M \Rightarrow N}{(P)M \Rightarrow (P)N}$$

$$\sigma\mathrm{l}\ \frac{M \Rightarrow N}{(M)\{x = P\} \Rightarrow (N)\{x = P\}} \qquad \sigma\mathrm{r}\ \frac{M \Rightarrow N}{(P)\{x = M\} \Rightarrow (P)\{x = N\}}$$

Fig. 7. Rewriting relation $\Rightarrow\, \subseteq \Lambda \times \Lambda$

5   Completeness and Soundness of SBVr and BVr

We relate functional and proof-theoretic worlds. First we map terms of Λ into structures of SBVr. Then, we show the *completeness* of SBVr and BVr, i.e. that the computations of Λ correspond to proof-search inside the two systems. Finally, we prove *soundness* of SBVr with respect to the computations of λ-calculus with explicit substitutions under a specific proof-search strategy. This means that we can use SBVr to compute any term which any given M reduces to.

The map $(\!|\cdot|\!)..$ We start with the following "fake" map from Λ to SBVr:

$$(\!|x|\!)_o = \langle x \cdot \overline{o} \rangle \tag{4}$$

$$(\!|\lambda x.M|\!)_o = \forall x.\exists p.[(\!|M|\!)_p \,\mathbin{⅋}\, (p \otimes \overline{o})] \tag{5}$$

$$(\!|(M)N|\!)_o = \exists p.[(\!|M|\!)_p \,\mathbin{⅋}\, \exists q.(\!|N|\!)_q \,\mathbin{⅋}\, (p \otimes \overline{o})] \tag{6}$$

$$(\!|(M)\{x = P\}|\!)_o = \forall x.[(\!|M|\!)_o \,\mathbin{⅋}\, (\!|P|\!)_x] \tag{7}$$

We use it only to intuitively illustrate how we shall effectively represent terms of Λ as structures of SBVr. The map here above translates M into $(\!|M|\!)_o$ where o is a unique output channel, while the whole expression depends on a set of free input channels, each for every free variable of M. Clause (4) associates the input channel x to the fresh output channel $\overline{o}$, under the intuition that x is *forwarded* to o, using the terminology of [8]. Clause (5) assumes $(\!|M|\!)_p$ has p as output and (at least) x as input. It renames p, hidden by $\exists$, as $\overline{o}$ thanks to $(p \otimes \overline{o})$. This must work for every input x. For this reason we hide x by means of $\forall$. Clause (6) makes the output channels of both $(\!|M|\!)_p$ and $(\!|N|\!)_q$ local, while renaming p to $\overline{o}$ thanks to $(p \otimes \overline{o})$. If $(\!|M|\!)_p$ will result in the translation of a λ-abstraction $\lambda z.P$, then the existential quantifier immediately preceding $(\!|N|\!)_q$ will interact with the universal quantifier in front of $(\!|M|\!)_p$. The result will be an on-the-fly channel name renaming. Clause (7) identifies the output of $(\!|P|\!)_x$ with one of the existing free names of $(\!|M|\!)_o$. The identification becomes local thanks to the universal quantifier.

In a setting where the second order quantifiers $\forall$, and $\exists$ only operate on atoms, distinguishing between the two is meaningless. So, the renaming can be self-dual and the true map $(\!|\cdot|\!)$. which adheres to the above intuition is in Figure 8.

Remark 5.1   We keep stressing that $(\!|\cdot|\!)$. strongly recalls output-based embedding of *standard* λ-calculus with explicit substitutions into π-calculus [17]. In principle, this

$$(\!|x|\!)_o = \langle x \cdot \overline{o} \rangle$$

$$(\!|\lambda x.M|\!)_o = \lceil\lceil[(\!|M|\!)_p \,\mathbin{⅋}\, (p \otimes \overline{o})]\rfloor_p\rfloor_x$$

$$(\!|(M)N|\!)_o = \lceil[(\!|M|\!)_p \,\mathbin{⅋}\, \lceil(\!|N|\!)_q\rfloor_q \,\mathbin{⅋}\, (p \otimes \overline{o})]\rfloor_p$$

$$(\!|(M)\{x = P\}|\!)_o = \lceil[(\!|M|\!)_o \,\mathbin{⅋}\, (\!|P|\!)_x]\rfloor_x$$

Fig. 8. Map $(\!|\cdot|\!)$ from Λ to structures

$$\text{o-ren} \frac{(\!|M|\!)_o}{[(\!|M|\!)_p \mathbin{\rotatebox[origin=c]{180}{\&}} (p \otimes \overline{o})]} \qquad \text{s-var} \frac{(\!|P|\!)_o}{(\!|(x)\{x = P\}|\!)_o}$$

$$\text{s-intro} \frac{(\!|(M)\{x = N\}|\!)_o}{(\!|(\lambda x.M)N|\!)_o} \qquad \text{s-}\lambda \frac{(\!|\lambda y.(M)\{x = P\}|\!)_o}{(\!|(\lambda y.M)\{x = P\}|\!)_o}$$

$$\text{s-@l} \frac{(\!|((M)\{x = P\})N|\!)_o \quad x \in \text{fv}(M)}{(\!|((M)N)\{x = P\}|\!)_o} \quad \text{s-@r} \frac{(\!|(M)(N)\{x = P\}|\!)_o \quad x \in \text{fv}(N)}{(\!|((M)N)\{x = P\}|\!)_o}$$

Fig. 9. Derivable rules that simulate β-reduction with explicit substitutions

means that extending SBVr with the right logical operators able to duplicate atoms, and consequently upgrading $(\!|\cdot|\!).$, we could model full β-reduction as proof-search.

Lemma 5.2 (Output names are linear). *Every output name of* $(\!|M|\!)_o$ *occurs once in it.*

Lemma 5.3 (Output renaming). *For every* $M, o,$ *and* p *we can derive* o$-$ren *in* BVr.

Lemma 5.4 (Simulating $\rightarrow$). *For every* $M, N, P, o, p,$ *and* $q,$ *we can derive:*

1. s$-$intro, s$-\lambda$, s$-$@l, *and* s$-$@r *in* BVr, *and*
2. s$-$var *in* BVr $\cup$ $\{q\!\uparrow\}$.

Appendix A shows how proving Lemma 5.4 here above, by detailing out the derivations of SBVr that derive the rules in Figure 9.

Remark 5.5. Were the clause "$(y)\{x = P\} \rightarrow y$" in the definition of $\rightarrow$ we could not prove Lemma 5.4 because $(\!|P|\!)_o \vdash (\!|(y)\{x = P\}|\!)_o$ would not exist in SBVr. The reason is that, given $(\!|(y)\{x = P\}|\!)_o$, it is not evident which logical tool can erase any translation of P as directly as happens in $(y)\{x = P\} \rightarrow y$. The only erasure mechanism existing in BVr is atom annihilation through the rules ai$\downarrow$, and ai$\uparrow$, indeed.

Theorem 5.6 here below says that to every β-reduction sequence from M to N in linear λ-calculus with explicit substitutions corresponds at least a proof-search process inside SBVr that builds a derivation with $(\!|N|\!)_o$ as premise and $(\!|M|\!)_o$ as conclusion.

Theorem 5.6 (Completeness of SBVr). *For every* $M,$ *and* $o,$ *if* $M \Rightarrow N,$ *then there is* $\mathscr{D} : (\!|N|\!)_o \vdash (\!|M|\!)_o$ *in* SBVr, *where* $q\uparrow$ *is the only rule of the up-fragment in* SBVr *that can occur in* $\mathscr{D}$.

We remark that the strategy to derive $(\!|M|\!)_o$ from $(\!|N|\!)_o$ in SBVr has no connection to the one we might use for rewriting M to N. Corollary 5.7 here below follows from Theorem 5.6 and cut elimination of SBVr (Corollary 3.7). It reinterprets logically the meaning of rewriting M to N. It says that M transforms to N if $\overline{(\!|N|\!)_o}$ logically annihilate the components of $(\!|M|\!)_o$.

Corollary 5.7 (Completeness of BVr). *For every* $M, N,$ *and* $o,$ *if* $M \Rightarrow N,$ *then, in* BVr, *we have* $\vdash [(\!|M|\!)_o \mathbin{\rotatebox[origin=c]{180}{\&}} \overline{(\!|N|\!)_o}].$

Theorem 5.8 here below says that proof-search inside SBVr can be used as an interpreter of linear λ-calculus with explicit substitutions. We say it is *Weak* Soundness because the interpreter follows a specific strategy of proof-search. The strategy is made of somewhat rigid steps represented by the rules in Figure 9 that simulate β-reduction.

Theorem 5.8 (*Weak Soundness of* SBVr). *For every M, N, and o, let $\mathscr{D} : (\!(N)\!)_o \vdash (\!(M)\!)_o$ in SBVr be derived by composing a, possibly empty, sequence of rules in Figure 9. Then $M \Rightarrow N$.*

Conjecture 5.9 (*Soundness of* SBVr *and* BVr). *As a referee suggested, Theorem 5.8 above should hold by dropping the requirement that, for building $\mathscr{D}$ in SBVr, we forcefully have to compose rules in Figure 9. The proof of this conjecture could take advantage of the method that works for proving that* BV, *extended with exponentials of linear logic, is undecidable [16].*

A corollary of the conjecture would say that, for every M, N, and o, if $\mathscr{P} : \vdash [(\!(M)\!)_o \, \mathbin{⅋} \, \overline{(\!(N)\!)_o}]$ in BVr, *then $M \Rightarrow N$. Namely, we could use the process that searches the shortest cut free proof of $[(\!(M)\!)_o \, \mathbin{⅋} \, \overline{(\!(N)\!)_o}]$ in* BVr *as an interpreter of linear λ-calculus with explicit substitutions. Of course, the possible relevance of this would be evident in a further extension of* BVr *where full β-reduction could be simulated as proof-search of cut free proofs.*

6   Conclusions and Future Work

We define an extension SBVr of SBV by introducing an atom renaming operator Rename which is a self-dual limited version of universal and existential quantifiers. Rename and Seq model the evaluation of linear λ-terms with explicit substitutions as proof-search in SBVr. So, we do not apply DI methodology to reformulate an existing logical system we already know to enjoy Curry-Howard correspondence with λ-calculus. Instead, we use logical operators at the core of DI, slightly extended, to get a computational behavior we could not obtain otherwise.

We conclude with a possible list of natural developments.

Of course proving Conjecture 5.9 is one of the obvious goals. Proving it would justify the relevance of looking for further extensions of SBVr whose unconstrained proof-search strategies could supply sound interpreters of *full* λ-calculus (with explicit substitutions). Starting points could be [16,7,15].

Also, we can think of extending SBVr by an operator that models non-deterministic choice. One reason would be proving the following statement. Let us assume we know that a λ-term M can only reduce to one of the normal forms $N_1, \ldots, N_m$. Let us assume the following statement can hold in the hypothetical extension of SBVr:

If $[\overline{(\!(P_1)\!)_o} \oplus \cdots \oplus \overline{(\!(P_{i-1})\!)_o} \oplus \overline{(\!(P_{i+1})\!)_o} \oplus \cdots \oplus \overline{(\!(P_m)\!)_o}] \vdash [(\!(M)\!)_o \, \mathbin{⅋} \, \overline{[(\!(N_1)\!)_o \oplus \cdots \oplus (\!(N_m)\!)_o]}]$,
then M reduces to N_i, for some $P_1, \ldots, P_m$.

This way we would represent the evaluation space of any linear λ-term with explicit substitutions as a non-deterministic process searching for normal forms. Candidate rules for non-deterministic choice to extend SBVr could be[1]:

---

[1] The conjecture about the existence of the two rules p $\downarrow$, and p $\uparrow$, that model non-deterministic choice, results from discussions with Alessio Guglielmi.

$$\mathsf{p}\!\downarrow \frac{[[R \,\invamp\, T] \oplus [U \,\invamp\, T]]}{[[R \oplus U] \,\invamp\, T]} \qquad \mathsf{p}\!\uparrow \frac{([R \oplus U] \otimes T)}{[(R \otimes T) \oplus (U \otimes T)]}$$

A further reason to extend **SBVr** with non-deterministic choice is to keep developing the programme started in [4], aiming at a purely logical characterization of *full* CCS.

Finally, we could explore if any relation between linear λ-calculus with explicit substitutions, as we embed it in **SBVr** using a calculus-of-process style, and the evolution of quantum systems as proofs of **BV** [2], exists. Exploration makes sense if we observe that modeling a λ-variable x as a forwarder $\langle x \triangleleft \overline{o} \rangle$ is, essentially, looking at x as a sub-case of $\langle (x_1 \otimes \cdots \otimes x_k) \triangleleft [\overline{o}_1 \,\invamp\, \cdots \,\invamp\, \overline{o}_l] \rangle$, the representation of edges in DAGs that model quantum systems evolution in [2].

References

1. Abadi, M., Cardelli, L., Curien, P.-L., Lévy, J.-J.: Explicit substitutions. JFP 1(4), 375–416 (1991)
2. Blute, R.F., Guglielmi, A., Ivanov, I.T., Panangaden, P., Straßburger, L.: A Logical Basis for Quantum Evolution and Entanglement. Private Communication
3. Brünnler, K., McKinley, R.: An algorithmic interpretation of a deep inference system. In: Cervesato, I., Veith, H., Voronkov, A. (eds.) LPAR 2008. LNCS (LNAI), vol. 5330, pp. 482–496. Springer, Heidelberg (2008)
4. Bruscoli, P.: A purely logical account of sequentiality in proof search. In: Stuckey, P.J. (ed.) ICLP 2002. LNCS, vol. 2401, pp. 302–316. Springer, Heidelberg (2002)
5. Girard, J.-Y., Taylor, P., Lafont, Y.: Proofs and Types. CUP (1989)
6. Guglielmi, A.: A system of interaction and structure. ToCL 8(1), 1–64 (2007)
7. Guglielmi, A., Straßburger, L.: A system of interaction and structure V: The exponentials and splitting. To appear on MSCS (2009)
8. Honda, K., Yoshida, N.: On the Reduction-based Process Semantics. TCS (151), 437–486 (1995)
9. Kahramanoğulları, O.: System BV is NP-complete. APAL 152(1-3), 107–121 (2007)
10. Mairson, H.G.: Linear lambda calculus and ptime-completeness. JFP 14(6), 623–633 (2004)
11. Milner, R.: Communication and Concurrency. International Series in Computer Science. Prentice Hall, Englewood Cliffs (1989)
12. Milner, R.: Functions as processes. MSCS 2(2), 119–141 (1992)
13. Roversi, L.: Linear lambda calculus with explicit substitutions as proof-search in Deep Inference (November 2010), http://arxiv.org/abs/1011.3668
14. Straßburger, L.: Some observations on the proof theory of second order propositional multiplicative linear logic. In: Curien, P.-L. (ed.) TLCA 2009. LNCS, vol. 5608, pp. 309–324. Springer, Heidelberg (2009)
15. Straßburger, L., Guglielmi, A.: A system of interaction and structure IV: The exponentials and decomposition. ToCL (2010) (in press)
16. Straßburger, L.: System NEL is undecidable. In: De Queiroz, R., Pimentel, E., Figueiredo, L. (eds.) WoLLIC 2003. ENTCS, vol. 84. Elsevier, Amsterdam (2003)
17. van Bakel, S., Vigliotti, M.G.: A logical interpretation of the λ-calculus into the π-calculus, preserving spine reduction and types. In: Bravetti, M., Zavattaro, G. (eds.) CONCUR 2009. LNCS, vol. 5710, pp. 84–98. Springer, Heidelberg (2009)

A   Proof of Lemma 5.4

We simulate s−var with the following derivation:

$$
\cfrac{
e_0 \cfrac{
\cfrac{
\text{mixp} \cfrac{
\mathscr{D} \|
\cfrac{(\!|P|\!)_o}{\lceil [(\!|P|\!)_x \,\otimes\, (x \otimes \overline{o})] \rfloor_x}
}{\lceil [(\!|P|\!)_x \,\otimes\, \langle x \triangleleft \overline{o}\rangle] \rfloor_x}
}{}
}{(\!|(x)\{x = P\}|\!)_o \equiv \lceil [(\!|x|\!)_o \,\otimes\, (\!|P|\!)_x] \rfloor_x \equiv \lceil [\langle x \triangleleft \overline{o}\rangle \,\otimes\, (\!|P|\!)_x] \rfloor_x}
$$

where Lemma 5.3 implies the existence of $\mathscr{D}$. The here above derivation requires q ↑ because mixp is derivable in {q↑}.

We simulate s−intro with the following derivation:

$$
\cfrac{
e_0 \cfrac{
\text{r↓} \cfrac{
\mathscr{D} \|
\cfrac{
e_1 \cfrac{
\mathscr{D}' \|
\cfrac{
e_2 \cfrac{(\!|(M)\{x = N\}|\!)_o}{\lceil (\!|(M)\{x = N\}|\!)_o \rfloor_p}
}{
\lceil [\lceil [(\!|M|\!)_p \,\otimes\, (\!|N|\!)_x] \rfloor_x \,\otimes\, (p \otimes \overline{o})] \rfloor_p \equiv \lceil [(\!|(M)\{x = N\}|\!)_p \,\otimes\, (p \otimes \overline{o})] \rfloor_p
}
}{\lceil [\lceil [\lceil (\!|M|\!)_p \rfloor_{p'} \,\otimes\, (\!|N|\!)_x] \rfloor_x \,\otimes\, (p \otimes \overline{o})] \rfloor_p}
}{\lceil [\lceil [\lceil (\!|M|\!)_{p'} \,\otimes\, (p' \otimes \overline{p})] \rfloor_{p'} \,\otimes\, (\!|N|\!)_x] \rfloor_x \,\otimes\, (p \otimes \overline{o})] \rfloor_p}
}{\lceil [\lceil [\lceil (\!|M|\!)_{p'} \,\otimes\, (p' \otimes \overline{p})] \rfloor_{p'} \rfloor_x \,\otimes\, \lceil (\!|N|\!)_x \rfloor_x \,\otimes\, (p \otimes \overline{o})] \rfloor_p}
}{(\!|(\lambda x.M) N|\!)_o \equiv \lceil [\lceil [\lceil (\!|M|\!)_{p'} \,\otimes\, (p' \otimes \overline{p})] \rfloor_{p'} \rfloor_x \,\otimes\, \lceil (\!|N|\!)_q \rfloor_q \,\otimes\, (p \otimes \overline{o})] \rfloor_p}
$$

where:

- Lemma 5.3 implies the existence of both $\mathscr{D}, \mathscr{D}'$;
- $\lceil (\!|N|\!)_q \rfloor_q$ in the conclusion of e_0 becomes $\lceil (\!|N|\!)_q\{^x/_q\} \rfloor_x \approx \lceil (\!|N|\!)_x \rfloor_x$ in its premise because q only occurs as output channel name in a pair $(p'' \otimes q)$, for some p'', and nowherelse;
- in the conclusion of e_1, the channel p' has disappeared from $(\!|M|\!)_p$;
- in the conclusion of e_2, the channel p has disappeared from $(\!|(M)\{x = N\}|\!)_o$.

We simulate s−λ with the following derivation:

$$
\cfrac{
e_0 \cfrac{
\text{r↓,r↓} \cfrac{
\cfrac{\lceil [\lceil [(\!|M|\!)_p \,\otimes\, (\!|P|\!)_x] \rfloor_x \,\otimes\, (p \otimes \overline{o})] \rfloor_p \rfloor_y \equiv \lceil [\lceil [(\!|(M)\{x = P\}|\!)_p \,\otimes\, (p \otimes \overline{o})] \rfloor_p \rfloor_y \equiv (\!|\lambda y.(M)\{x = P\}|\!)_o}{\lceil [\lceil [(\!|M|\!)_p \,\otimes\, (p \otimes \overline{o}) \,\otimes\, (\!|P|\!)_x] \rfloor_p \rfloor_y \rfloor_x}
}{\lceil [\lceil [(\!|M|\!)_p \,\otimes\, (p \otimes \overline{o})] \rfloor_p \rfloor_y \,\otimes\, \lceil [(\!|P|\!)_x \rfloor_p \rfloor_y] \rfloor_x}
}{(\!|(\lambda y.M)\{x = P\}|\!)_o \equiv \lceil [\lceil [(\!|M|\!)_p \,\otimes\, (p \otimes \overline{o})] \rfloor_p \rfloor_y \,\otimes\, (\!|P|\!)_x] \rfloor_x}
$$

where e_0 applies three of the axioms in Figure 3.

We simulate s − @l with the following derivation:

$$
\cfrac{
\text{r↓} \cfrac{
\cfrac{
\cfrac{
\begin{array}{c}(\!|((M)\{x = P\}) N|\!)_o \\ \equiv \lceil [(\!|(M)\{x = P\}|\!)_p \,\otimes\, \lceil (\!|N|\!)_q \rfloor_q \,\otimes\, (p \otimes \overline{o})] \rfloor_p \\ \equiv \lceil [\lceil [(\!|M|\!)_p \,\otimes\, (\!|P|\!)_x] \rfloor_x \,\otimes\, \lceil (\!|N|\!)_q \rfloor_q \,\otimes\, (p \otimes \overline{o})] \rfloor_p\end{array}
}{\lceil [\lceil [(\!|M|\!)_p \,\otimes\, \lceil (\!|N|\!)_q \rfloor_q \,\otimes\, (p \otimes \overline{o}) \,\otimes\, (\!|P|\!)_x] \rfloor_x \rfloor_p}
}{\lceil [\lceil [(\!|M|\!)_p \,\otimes\, \lceil (\!|N|\!)_q \rfloor_q \,\otimes\, (p \otimes \overline{o}) \,\otimes\, (\!|P|\!)_x] \rfloor_p \rfloor_x}
}{\lceil [\lceil [(\!|M|\!)_p \,\otimes\, \lceil (\!|N|\!)_q \rfloor_q \,\otimes\, (p \otimes \overline{o})] \rfloor_p \,\otimes\, \lceil (\!|P|\!)_x \rfloor_p] \rfloor_x}
}{(\!|((M) N)\{x = P\}|\!)_o \equiv \lceil [(\!|(M) N|\!)_o \,\otimes\, (\!|P|\!)_x] \rfloor_x \equiv \lceil [\lceil [(\!|M|\!)_p \,\otimes\, \lceil (\!|N|\!)_q \rfloor_q \,\otimes\, (p \otimes \overline{o})] \rfloor_p \,\otimes\, (\!|P|\!)_x] \rfloor_x}
$$

The simulation of s − @r works like the one of s − @l.

Partiality, State and Dependent Types[*]

Kasper Svendsen[1], Lars Birkedal[1], and Aleksandar Nanevski[2]

[1] IT University of Copenhagen
{kasv,birkedal}@itu.dk
[2] IMDEA Software
aleks.nanevski@imdea.org

Abstract. Partial type theories allow reasoning about recursively-defined computations using fixed-point induction. However, fixed-point induction is only sound for admissible types and not all types are admissible in sufficiently expressive dependent type theories.

Previous solutions have either introduced explicit admissibility conditions on the use of fixed points, or limited the underlying type theory. In this paper we propose a third approach, which supports Hoare-style partial correctness reasoning, without admissibility conditions, but at a tradeoff that one cannot reason equationally about effectful computations. The resulting system is still quite expressive and useful in practice, which we confirm by an implementation as an extension of Coq.

1 Introduction

Dependent type theories such as the Calculus of Inductive Constructions [2] provide powerful languages for integrated programming, specification, and verification. However, to maintain soundness, they typically require all computations to be pure and terminating, severely limiting their use as general purpose programming languages.

Constable and Smith [9] proposed adding partiality by introducing a type $\bigcirc\tau$ of potentially non-terminating computations of type τ, along with the following fixed point principle for typing recursively defined computations:

$$\text{if } M : \bigcirc\tau \to \bigcirc\tau \quad \text{then} \quad \textit{fix}(M) : \bigcirc\tau$$

Unfortunately, in sufficiently expressive dependent type theories, there exists types τ for which the above fixed point principle is unsound [10]. For instance, in type theories with subset-types, the fixed point principle allows reasoning by a form of fixed point induction, which is only sound for admissible predicates (a predicate is admissible if it holds for the limit whenever it holds for all finite approximations). Previous type theories based on the idea of partial types which admit fixed points have approached the admissibility issue in roughly two different ways:

[*] This research has been partially supported by MICINN Project TIN2010-20639 Paran10; AMAROUT grant PCOFUND-GA-2008-229599; and Ramon y Cajal grant RYC-2010-0743.

L. Ong (Ed.): TLCA 2011, LNCS 6690, pp. 198–212, 2011.

1. The notion of admissibility is axiomatized in the type theory and explicit admissibility conditions are required in order to use *fix*. This approach has, e.g., been investigated by Crary in the context of Nuprl [10]. The resulting type theory is expressive, but admissibility conditions lead to significant proof obligations, in particular, when using Σ types.

2. The underlying dependent type theory is restricted in such a way that one can only form types that are trivially admissible. This approach has, e.g., been explored in recent work on Hoare Type Theory (HTT) [21]. The restrictions exclude usage of subset types and Σ types, which are often used for expressing properties of computations and for modularity. Another problem with this approach is that since it limits the underlying dependent type theory one cannot easily implement it as a simple extension of existing implementations.

In this paper we explore a third approach, which ensures that all types are admissible, not by limiting the underlying standard dependent type theory, but by limiting only the partial types. The limitation on partial types consists of equating all effectful computations at a given type: if M and N are both of type $\bigcirc \tau$, then they are propositionally equal. Thus, with this approach, the only way to reason about effectful computations is through their type, rather than via equality or predicates. With sufficiently expressive types, the type of an effectful computation can serve as a partial correctness specification of the computation. Our hypothesis is that this approach allows us to restrict attention to a subset of admissible types, which is closed under the standard dependent type formers and which suffices for reasoning about partial correctness.

To demonstrate that this approach scales to expressive type theories and to effects beyond partiality, we extend the Calculus of Inductive Constructions (CIC) [2] with stateful and potentially non-terminating computations. Since reasoning about these effectful computations is limited to their type, our partial types are further refined into a Hoare-style partial correctness specifications, and have the form $ST \tau (P, Q)$, standing for computations with pre-condition P, post-condition Q, that diverge or terminate with a value of type τ.

The resulting type theory is an impredicative variant of Hoare Type Theory [17], which differs from previous work on Hoare Type Theory in the scope of features considered and the semantic approach. In particular, this paper is the first to clarify semantically the issue of admissibility in Hoare Type Theory.

Impredicative Hoare Type Theory (iHTT) features the universes of propositions (*prop*), small types (*set*), and large types (*type*), with *prop* included in *set*, *set* included in *type*, and axioms *prop:type* and *set:type*. The *prop* and *set* universes are impredicative, while *type* is predicative. There are two main challenges in building a model to justify the soundness of iHTT: (1) achieving that Hoare types are small ($ST\ \tau\ s : set$), which enables *higher-order store*; that is, storing side-effectful computations into the heap, and (2) supporting arbitrary Σ types, and more generally, inductive types. In this respect iHTT differs from the previous work on Hoare Type Theory, which either lacks higher-order store [19], lacks strong Σ types [21], or whose soundness has been justified using specific syntactic methods that do not scale to fully general inductive definitions [17,18].

The model is based on a standard realizability model of partial equivalence relations (PERs) and assemblies over a combinatory algebra A. These give rise to a model of the Calculus of Constructions [14], with *set* modelled using PERs. Restricting PERs to complete PERs (i.e., PERs closed under limits of ω-chains) over a suitable universal domain, allows one to model recursion in a simply-typed setting [4], or in a dependently-typed setting, but without strong Σ types [21].

Our contribution is in identifying a set of complete *monotone* PERs that are closed under Σ types and Hoare types. Complete PERs do not model Σ types because, given a chain of dependent pairs, in general, due to dependency, the second components of the chain are elements of *distinct* complete PERs. To apply completeness, we need a fixed single complete PER. Monotonicity will equate the first components of the chain and give us the needed single complete PER for the second components. Monotonicity further forces a trivial equality on Hoare types, equating all effectful computations satisfying a given specification. However, it does not influence the equality on the total, purely functional, fragment of iHTT, ensuring that we still model CIC. This is sufficient for very expressive Hoare-style reasoning, and avoids admissibility conditions on the use of *fix*.

As iHTT is an extension of CIC, we have implemented iHTT as an axiomatic extension of Coq [1], available at: http://www.itu.dk/people/kasv/ihtt.tgz. The implementation is carried out in Ssreflect [12] (a recent extension of Coq), based on the previous implementation of predicative Hoare Type Theory [19].

Details and proofs can be found in the accompanying technical report [23].

2   Hoare Types by Example

To illustrate Hoare types, we sketch a specification of a library for arrays in iHTT. We assume that array indexes range over a finite type $\iota{:}set_{fin}$, that the elements of ι can be enumerated as $\iota_0, \iota_1, \ldots, \iota_n$, and that equality between these elements can be decided by a function $={:}\ \iota \to \iota \to bool$.

Each array is implemented as a contiguous block of locations, each location storing a value from the range type $\tau{:}set$. The space occupied by the array is uniquely determined by ι, τ, and the pointer to the first element, justifying that the *array* type be defined as this first pointer.

$$array : set_{fin} \to set \to set = \lambda\iota.\,\lambda\tau.\,ptr.$$

Here, *ptr* is the type of pointers, which we assume isomorphic to *nat*. Each array is essentially a stateful implementation of some finite function $f{:}\iota\to\tau$. To capture this, we define a predicate indexed by f, that describes the layout of an array in the heap.

$$\mathsf{shape} : (array\ \iota\ \tau) \to (\iota \to \tau) \to heap \to prop = \\ \lambda a.\,\lambda f.\,\lambda h.\,h = a \mapsto f\,\iota_o \bullet a{+}1 \mapsto f\,\iota_1 \bullet \cdots \bullet a{+}n \mapsto f\,\iota_n.$$

In other words, h stores an array a, representing a finite function f, if shape $a\ f\ h$ holds, that is, if h consists of $n{+}1$ consecutive locations $a, a{+}1, \ldots, a{+}n$, storing $f\,\iota_0, f\,\iota_1, \ldots, f\,\iota_n$, respectively. The property is stated in terms of singleton heaps

$a+k \mapsto f\,\iota_k$, connected by the operator $\bullet$ for *disjoint* heap union. Later in the text, we will also require a constant **empty** denoting the empty heap.

The type of arrays comes equipped with several methods for accessing and manipulating the array elements. For example, the method for reading the value at index $k{:}\iota$ can be given the following type.

$$\text{read} : \Pi a{:}\boldsymbol{array}\ \iota\ \tau.\ \Pi k{:}\iota.$$
$$\boldsymbol{ST}\ \tau\ (\ \lambda h.\ \exists f.\ \mathsf{shape}\ a\ f\ h,$$
$$\lambda r.\ \lambda h.\ \lambda m.\ \forall f.\ \mathsf{shape}\ a\ f\ h \to r = f\,k \wedge m = h)$$

Informally, read $a\ k$ is specified as a stateful computation whose precondition permits the execution only in a heap h which stores a valid array at address a ($\exists f.\ \mathsf{shape}\ a\ f\ h$). The postcondition, on the other hand, specifies the result of executing read $a\ k$ as a relation between output result $r{:}\tau$, input heap h and output heap m. In particular, the result r is indeed the array value at index k ($r = f\,k$), and the input heap is unchanged ($m = h$).

Unlike in ordinary Hoare logic, but similar to VDM [6], our postcondition is parametrized wrt. both input and the output heaps in order to directly express the relationship between the two. In particular, when this relationship depends on some specification-level value, such as f above, the dependency can be expressed by an ordinary propositional quantification.

Hoare types employ *small footprint* specifications, as in separation logic [20], whereby the specifications only describe the parts of the heap that the computation traverses. The untraversed parts are by default invariant. To illustrate, consider the type for the method new that generates a fresh array, indexed by ι, and populated by the value $x{:}\tau$.

$$\text{new} : \Pi x{:}\tau.\ \boldsymbol{ST}\ (\boldsymbol{array}\ \iota\ \tau)\ (\lambda h.\ h = \mathsf{empty}, \lambda a.\ \lambda h.\ \lambda m.\ \mathsf{shape}\ a\ (\lambda z.\,x)\ m)$$

The type states *not* that new x can only run in an empty heap, but that new x *changes* the empty subheap of the current heap into a heap m containing an array rooted at a and storing all x's. In other words, new is *adding* fresh pointers, and the resulting array a itself is fresh. On the other hand, unlike in separation logic, we allow that the specifications can directly use and quantify over variables of type $\boldsymbol{heap}$. For completeness, we next simply list without discussion the types of the other methods for arrays.

$\text{new\_from\_fun} : \Pi f{:}\iota{\to}\tau.\ \boldsymbol{ST}\ (\boldsymbol{array}\ \iota\ \tau)\ (\lambda h.\ h = \mathsf{empty}, \lambda a\,h\,m.\ \mathsf{shape}\ a\ f\ m)$

$\text{free} : \Pi a{:}\boldsymbol{array}\ \iota\ \tau.\ \boldsymbol{ST}\ unit\ (\lambda h.\ \exists f.\ \mathsf{shape}\ a\ f\ h, \lambda r\,h\,m.\ m = \mathsf{empty})$

$\text{write} : \Pi a{:}\boldsymbol{array}\ \iota\ \tau.\ \Pi k{:}\iota.\ \Pi x{:}\tau.$
$$\boldsymbol{ST}\ unit\ (\ \lambda h.\ \exists f.\ \mathsf{shape}\ a\ f\ h, \lambda r\,h\,m.\ \forall f.\ \mathsf{shape}\ a\ f\ h \to$$
$$\mathsf{shape}\ a\ (\lambda z.\ \text{if}\ z == k\ \text{then}\ x\ \text{else}\ f(z))\ m)$$

At this point, we emphasize that various type theoretic abstractions are quite essential for practical work with Hoare types. The usefulness of Π types and the propositional quantifiers is apparent from the specification of the array methods. But the ability to structure specification is important too. For example, we can pair pre- and postconditions into a type **spec** $\tau = (\textbf{heap} \rightarrow \textbf{prop}) \times (\tau \rightarrow \textbf{heap} \rightarrow \textbf{heap} \rightarrow \textbf{prop})$, which is then used to specify the fixpoint combinator.

fix : $\Pi\alpha{:}\textbf{set}.\,\Pi\beta{:}\alpha{\rightarrow}\textbf{set}.\,\Pi s{:}\Pi x.\,\textbf{spec}\,(\beta\ x).$
$\quad (\Pi x.\,\textbf{ST}\,(\beta\ x)\,(s\ x) \rightarrow \Pi x.\,\textbf{ST}\,(\beta\ x)\,(s\ x)) \rightarrow \Pi x.\,\textbf{ST}\,(\beta\ x)\,(s\ x).$

Structuring proofs and specifications with programs is also necessary, and is achieved using dependent records (i.e., Σ types), which we illustrate next.

The first example of a dependent record is the $\textbf{set}_{fin}$ type. This is an algebraic structure containing the carrier type σ, the operation == for deciding equality on σ, and a list enumerating σ's elements. Additionally, $\textbf{set}_{fin}$ needs proofs that == indeed decides equality, and that the enumeration list contains each element exactly once. Using the record notation $[x_1{:}\tau_1,\ldots x_n{:}\tau_n]$ instead of the more cumbersome $\Sigma x_1{:}\tau_1 \ldots \Sigma x_n{:}\tau_n.\,1$, the $\textbf{set}_{fin}$ type is defined as follows.

$$\begin{aligned}
\textbf{set}_{fin} = [\,&\sigma : \textbf{set}, \mathsf{enum} : \textbf{list}\,\sigma, == : \sigma \rightarrow \sigma \rightarrow \mathsf{bool}, \\
&\mathsf{eqp} : \forall x\,y{:}\sigma.\,x == y = \mathsf{true} \iff x = y, \\
&\mathsf{enump} : \forall x{:}\sigma.\,\mathsf{count}\,x\,\mathsf{enum} = 1\,]
\end{aligned}$$

The above dependent record refines a type. In practice, we will also use records that refine *values*. For example, in programming, arrays are often indexed by the type of bounded integers $I_n = [x : \textbf{nat}, \mathsf{boundp} : x \leq n]$. I_n can be extended with appropriate fields, to satisfy the specification for $\textbf{set}_{fin}$, but the important point here is that the elements of I_n are dependent records containing a number x and a proof that $x \leq n$. Of course, during actual execution, this proof can be ignored (proofs are computationally irrelevant), but it is clearly important statically, during verification.

Finally, the library of arrays itself can be ascribed a signature which will serve as an interface to the client programs. This signature too is a dependent record, providing types for all the array methods. Just as in the case of $\textbf{set}_{fin}$ and I_n, the signature may also include properties, similar to object invariants [13,15]. For example, we have found it useful in practice to hide from the clients the definitions of the array type and the array shape predicate, but expose that two arrays in stable states; that is, between two method calls, stored in compatible heaps must be equal (i.e., that the shape predicate is "functional"):

functional : $\Pi\mathsf{shape}.\,\forall a_1\ a_2\ f_1\ f_2\ h_1\ h_2.\,\mathsf{shape}\,a_1\ f_1\ h_1 \rightarrow \mathsf{shape}\,a_2\ f_2\ h_2 \rightarrow$
$\quad (\exists j_1\ j_2.\,h_1 \bullet j_1 = h_2 \bullet j_2) \rightarrow a_1 = a_2 \wedge f_1 = f_2 \wedge h_1 = h_2.$

Then the signature for arrays indexed by ι, containing values of type τ, is provided by the following dependent record parametrized by ι and τ.

$ArraySig = \Pi\iota{:}\textbf{set}_{fin}.\,\Pi\tau{:}\textbf{set}.$
$\quad\quad\quad\quad [\,\textbf{array} : \textbf{set},\ \mathsf{shape} : \textbf{array} \rightarrow (\iota \rightarrow \tau) \rightarrow \textbf{heap} \rightarrow \textbf{prop},$
$\quad\quad\quad\quad \mathsf{funcp} : \mathsf{functional}\,\mathsf{shape},\ \mathsf{read} : \Pi a{:}\textbf{array}.\,\Pi k{:}\iota.\ \ldots].$

Therefore, Σ types are central for building verified libraries of programs, specifications and proofs.

3   Semantics

In presenting the model, we first focus on the *set*-universe, and then scale up to cover all of iHTT. Since the purely-function fragment of iHTT is terminating, we take our universe of realizers to be a universal *pre-domain* with a suitable sub-domain for modelling stateful and potentially non-terminating computations.

Definition 1. *Let* $\mathbb{V}$ *denote a pre-domain satisfying the following recursive pre-domain equation:*

$$\mathbb{V} \cong 1 + \mathbb{N} + (\mathbb{V} \times \mathbb{V}) + (\mathbb{V} \to_c \mathbb{V}_\perp) + T(\mathbb{V}) + H(\mathbb{V})$$

where $\to_c$ *is the space of continuous functions and*

$$T(\mathbb{V}) \stackrel{def}{=} H(\mathbb{V}) \to_c ((\mathbb{V} \times H(\mathbb{V})) + 1)_\perp$$

$$H(\mathbb{V}) \stackrel{def}{=} \{h : \mathbf{ptr} \to_c \mathbb{V}_\perp \mid supp(h) \text{ finite} \wedge h(\mathbf{null}) = \perp\}$$

$$supp(h : \mathbf{ptr} \to \mathbb{V}_\perp) \stackrel{def}{=} \{l \in \mathbf{ptr} \mid h(l) \neq \perp\}$$

The first four summands of $\mathbb{V}$ *model the underlying dependent type theory, and* $T(\mathbb{V})$ *and* $H(\mathbb{V})$ *model computations and heaps, respectively. The ordering on* $T(\mathbb{V})$ *is the standard pointwise order and the ordering on* $H(\mathbb{V})$ *is as follows:*

$$h_1 \leq h_2 \quad \text{iff} \quad supp(h_1) = supp(h_2) \wedge \forall n \in supp(h_1). \ h_1(n) \leq h_2(n)$$

Let $in_1, in_\mathbb{N}, in_\times, in_\to, in_T,$ *and* in_H *denote injections into* $\mathbb{V}$ *corresponding to each of the above summands.*

$\mathbb{V}$ defines a partial combinatory algebra with the following partial application operator:

Definition 2. *Let* $\cdot : \mathbb{V} \times \mathbb{V} \rightharpoonup \mathbb{V}$ *denote the function,*

$$a \cdot b = \begin{cases} f(b) & \text{if } a = in_\to(f) \wedge f(b) \neq \perp \\ \mathbf{undef} & \text{otherwise} \end{cases}$$

We recall some notation and definitions. If $R \subseteq A \times A$ is a PER then its domain, denoted $|R|$, is $\{x \in A \mid (x, x) \in R\}$. If $R, S \subseteq A \times A$ are PERs, then $R \to S$ is the PER $\{(\alpha, \beta) \in A \times A \mid \forall x, y \in A. \ (x, y) \in R \Rightarrow (\alpha \cdot x, \beta \cdot y) \in S\}$. If $R \subseteq A \times A$ is a PER and $f : A \to B$ then $f(R)$ denotes the PER $\{(f(x), f(y)) \mid (x, y) \in R\} \subseteq B \times B$. For a subset $X \subseteq A$, we use $\Delta(X)$ to denote the PER $\{(x, y) \mid x \in X \wedge y \in X\}$. Lastly, if $R \subseteq A \times A$ is a PER, we use $[R]$ to denote the set of R equivalence classes.

Definition 3 (Per(A)). *The category of PERs,* $\mathrm{Per}(A)$, *over a partial combinatory algebra* $(A, \cdot)$, *has PERs over* A *as objects. Morphisms from* R *to* S *are set-theoretic functions* $f : [R] \to [S]$, *such that there exists a realizer* $\alpha \in A$ *such that,*

$$\forall e \in |R|. \ [\alpha \cdot e]_S = f([e]_R)$$

$\mathrm{Per}(\mathbb{V})$ is cartesian closed and thus models simple type theory. To model recursion, note that a realized set-theoretic function is completely determined by its realizers (i.e., $\mathrm{Per}(\mathbb{V})(R, S) \cong [R \to S]$) and that we have the standard least fixed-point operator on the sub-domain of computations of $\mathbb{V}$. This lifts to a least fixed-point operator on those PERs that are admissible on the sub-domain of computations:

Definition 4

1. *A PER $R \subseteq A \times A$ on a pre-domain A is* complete *if, for all chains $(c_i)_{i \in \mathbb{N}}$ and $(d_i)_{i \in \mathbb{N}}$ such that $(c_i, d_i) \in R$ for all i, also $(\bigsqcup_i c_i, \bigsqcup_i d_i) \in R$.*
2. *A PER $R \subseteq A \times A$ on a domain A is* admissible *if it is complete and $\bot \in |R|$.*

Let $\mathrm{CPer}(A)$ and $\mathrm{AdmPer}(A)$ denote the full sub-categories of $\mathrm{Per}(A)$ consisting of complete PERs and admissible PERs, respectively.

Definition 5. *Define $u : \mathbb{V} \to_c (T(\mathbb{V}) \to_c T(\mathbb{V}))$ as follows,*

$$u(x)(y) \stackrel{def}{=} \begin{cases} z & \text{if } x \cdot in_T(y) = in_T(z) \\ \bot & \text{otherwise} \end{cases}$$

and let lfp denote the realizer $in_\to(\lambda x.\ [in_T(\bigsqcup_n(u(x))^n)])$.

Lemma 1. *Let $R \in \mathrm{AdmPer}(T(\mathbb{V}))$, then $lfp \in |(in_T(R) \to in_T(R)) \to in_T(R)|$ and for all $\alpha \in |in_T(R) \to in_T(R)|$, $\alpha \cdot (lfp \cdot \alpha) = lfp \cdot \alpha$.*

The above development is standard and suffices to model fixed points over partial types in a non-stateful, simply-typed setting. However, it does not extend directly to a stateful dependently-typed setting: Assume $\vdash \tau : \textbf{type}$ and $x : \tau \vdash \sigma : \textbf{type}$. Then τ is interpreted as a PER $R \in \mathrm{Per}(\mathbb{V})$, and σ as an $[R]$-indexed family of PERs $S \in [R] \to \mathrm{Per}(\mathbb{V})$, and $\vdash \Sigma x : \tau.\sigma : \textbf{type}$ as the PER $\Sigma_R(S)$:

$$\Sigma_R(S) = \{(in_\times(a_1, b_1), in_\times(a_2, b_2)) \mid a_1\ R\ a_2 \wedge b_1\ S([a_1]_R)\ b_2\}.$$

In general, this PER is not chain-complete even if R and each S_x is: given a chain $(a_i, b_i)_{i \in \mathbb{N}}$, we do not know in general that $[a_i]_R = [a_j]_R$ and hence cannot apply the completeness of S_x to the chain $(b_i)_{i \in \mathbb{N}}$ for any $x \in [R]$.

To rectify this problem we impose the following monotonicity condition on the PERs, which ensures exactly that $a_i\ R\ a_j$, for all $i, j \in \mathbb{N}$, and hence that $\Sigma_R(S)$ is chain-complete.

Definition 6 (CMPer(A)). *A PER $R \subseteq A \times A$ on a pre-domain A is* monotone *if, for all $x, y \in |R|$ such that $x \leq y$, we have $(x, y) \in R$. Let $\mathrm{CMPer}(A)$ denote the full sub-category of $\mathrm{Per}(A)$ consisting of complete monotone PERs.*

Restricting to complete monotone PERs forces a trivial equality on any particular Hoare type, as all of the elements of the type have to be equal to the diverging computation. However, it does not trivialize the totality of Hoare types, as we can still interpret each distinct Hoare type, $\textbf{ST}\ \tau\ s$, as a distinct PER R, containing the computations that satisfy the specification s.

Restricting to complete monotone PERs does not collapse the equality on types in the purely functional fragment of iHTT, as these types are all interpreted as PERs without a bottom element in their domain. In particular, Π types, which are modelled as elements of $\mathbb{V} \to \mathbb{V}_\perp$, picks out only those elements that map to non-bottom on their domain.

We shall see later that the monotonicity condition is also used to interpret partial types with a post-condition, which induces a dependency similar to that of Σ types, in the semantics of partial types.

3.1   iHTT

So far, we have informally introduced a PER model of a single dependent type universe extended with partial types. The next step is to scale the ideas to a model of all of iHTT, and to prove that we do indeed get a model of iHTT. We start by showing that CMPERs and assemblies form a model of the underlying dependent type theory. Next, we sketch the interpretation of iHTT specific features such as heaps and computations in the model. Lastly, we show that the model features W-types at both the *set* and ***type*** universe.

Underlying DTT. We begin by defining a general class of models for the dependent type theory underlying iHTT, and then present a concrete instance based on complete monotone PERs. To simplify the presentation and exploit existing categorical descriptions of models of the Calculus of Constructions, the model will be presented using the fibred approach of [14]. To simplify the definition of the interpretation function, we consider a split presentation of the model (i.e., with canonical choice of all fibred structure, preserved on-the-nose).

Definition 7 (Split iHTT structure). *A split iHTT structure is a structure*

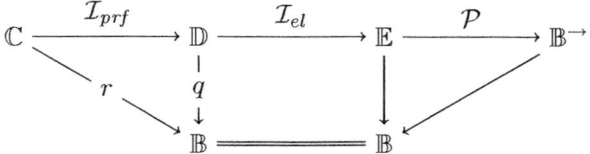

such that (1) $\mathcal{P}$ is a split closed comprehension category, (2) $\mathcal{I}_{el}$ and $\mathcal{I}_{prf}$ are split fibred reflections, (3) the coproducts induced by the $\mathcal{I}_{el}$ reflection are strong (i.e., $\mathcal{P} \circ \mathcal{I}_{el}$ is a split closed comprehension category), and (4) there exists objects $\Omega_{el}, \Omega_{prf} \in \mathbb{E}_1$ such that $\{\Omega_{el}\}$ is a split generic object for the q fibration and $\{\Omega_{prf}\}$ is a split generic object for the r fibration, where $\{-\} = \mathrm{Dom} \circ \mathcal{P} : \mathbb{E} \to \mathbb{B}$.

The idea is to model contexts in $\mathbb{B}$, and the three universes, ***prop***, ***set***, and ***type*** in fibres of $\mathbb{C}$, $\mathbb{D}$ and $\mathbb{E}$, respectively. The split closed comprehension category structure models unit, Π and Σ types in the ***type*** universe. The split fibred reflections models the inclusion of ***prop*** into ***set*** and ***set*** into ***type*** and induces unit, Π and weak Σ types in ***prop*** and ***set***. Lastly, the split generic objects models the axioms ***prop*** : ***type*** and ***set*** : ***type***, respectively.

The concrete model we have in mind is mostly standard: the contexts and the **type** universe will be modelled with assemblies, the **set** universe with complete monotone PERs, and the **prop** universe with regular subobjects of assemblies. We begin by defining a category of uniform families of complete monotone PERs. Uniformity refers to the fact that each morphism is realized by a single $\alpha \in \mathbb{V}$.

Definition 8 (UFam(CMPer(A)))

- *Objects are pairs* $(I, (S_i)_{i \in |I|})$ *where* $I \in \mathrm{Asm}(\mathbb{V})$ *and each* $S_i \in \mathrm{CMPer}(\mathbb{V})$.
- *Morphisms from* (I, S_i) *to* (J, T_j) *are pairs* $(u, (f_i)_{i \in |I|})$ *where*

$$u : I \to J \in \mathrm{Asm}(\mathbb{V}) \qquad and \qquad f_i : \mathbb{V}/S_i \to \mathbb{V}/T_{u(i)}$$

such that there exists an $\alpha \in \mathbb{V}$ *satisfying*

$$\forall i \in |I|. \; \forall e_i \in E_I(i). \; \forall e_v \in |S_i|. \; \alpha \cdot e_i \cdot e_v \in f_i([e_v]_{S_i})$$

Recall [14] that the standard category UFam(Per(A)) is defined in the same manner, using all PERs instead of only the complete monotone ones.

Let RegSub(Asm($\mathbb{V}$)) denote the standard category of regular subobjects of assemblies and UFam(Asm($\mathbb{V}$)) the standard category of uniform families of assemblies (see [14] for a definition). There is a standard split fibred reflection of UFam(Per(A)) into UFam(Asm(A)): the inclusion views a PER R as the assembly $([R], id_{[R]})$ [14]. This extends to a split fibred reflection of UFam(CMPer(A)) into UFam(Asm(V)) by composing with the following reflection of UFam(CMPer(A)) into UFam(Per(A)).

Lemma 2. *The inclusion* $\mathcal{I} : \mathrm{UFam}(\mathrm{CMPer}(A)) \to \mathrm{UFam}(\mathrm{Per}(A))$ *is a split fibred reflection.*

Proof (Sketch). We show that CMPer(A) is a reflective sub-category of Per(A); the same construction applies to uniform families, by a point-wise lifting. The left-ajoint, $\mathcal{R} : \mathrm{Per}(A) \to \mathrm{CMPer}(A)$ is given by monotone completion:

$$\mathcal{R}(S) = \overline{S} \qquad\qquad \mathcal{R}([\alpha]_{R \to S}) = [\alpha]_{\overline{R} \to \overline{S}}$$

where $\overline{R} \overset{\mathrm{def}}{=} \bigcap \{S \in \mathrm{CMPer}(\mathbb{V}) \mid R \subseteq S\}$ for a PER $R \in \mathrm{Per}(\mathbb{V})$. Since the underlying realizers are continuous functions, for all $R, S \in \mathrm{Per}(A)$, $\overline{R} \to \overline{S} = R \to \overline{S}$, which induces the adjoint-isomorphism:

$$\mathrm{CMPer}(A)(\mathcal{R}(S), T) \cong [\overline{S} \to T] = [S \to T] \cong \mathrm{Per}(A)(S, \mathcal{I}(T))$$

for $S \in \mathrm{Per}(A)$ and $T \in \mathrm{CMPer}(A)$.

Lemma 3. *The coproducts induced by the* $\mathcal{I}_{el} : \mathrm{UFam}(\mathrm{CMPer}(\mathbb{V})) \to \mathrm{UFam}(\mathrm{Asm}(\mathbb{V}))$ *reflection are strong.*

Proof. The coproducts induced by $\mathcal{I}_{el}$ coincide with those induced by the UFam (Per($\mathbb{V}$)) $\to$ UFam(Asm($\mathbb{V}$)) reflection, which are strong [14, Section 10.5.8].

Theorem 1. *The diagram below forms a split iHTT structure.*

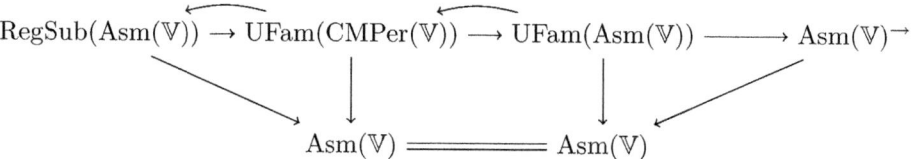

Heaps. Pre- and post-conditions in iHTT are expressed as predicates over a type *heap*, of heaps, which supports basic operations such as update, which stores a value at a given location, and peek, which reads the value at a given location. *heap* is a large type (i.e., *heap* : *type*) and the types of update and peek are as follows.

$$\textit{upd} : \Pi\alpha : \textit{set. heap} \to \textit{ptr} \to \alpha \to \textit{heap}$$
$$\textit{peek} : \textit{heap} \to \textit{ptr} \to 1 + \Sigma\alpha : \textit{set. } \alpha$$

We use $h[m \mapsto_\tau v]$ as shorthand for *upd* τ h m v. These operations can be used to define the points-to ($\mapsto$) and disjoint union ($\bullet$) operations used in Section 2.

We would like to model the values of *heap* as elements of $H(\mathbb{V})$. However, with this interpretation of *heap* we cannot interpret the update operation, $h[m \mapsto_\tau v]$, as the definition would not be independent of the choice of realizer for v. Rather, we introduce a notion of a world, which gives a notion of heap equivalence and interpret a heap as a pair consisting of a world and an equivalence class of heaps in the given world. A world is a finite map from locations to small semantic types:

$$\mathbb{W} \stackrel{\text{def}}{=} \textit{ptr} \stackrel{\text{fin}}{\rightharpoonup} \text{CMPer}(\mathbb{V})$$

and two heaps $h_1, h_2 \in H(\mathbb{V})$ are considered equivalent in a world $w \in \mathbb{W}$ iff their support equals the domain of the world and their values are point-wise related by the world:

$$h_1 \sim_w h_2 \quad \text{iff} \quad supp(h_1) = supp(h_2) = \text{Dom}(w) \land \forall l \in \text{Dom}(w). \, h_1(l) \, w(l) \, h_2(l)$$

A typed heap is then a pair consisting of a world and an equivalence class of domain-theoretic heaps,

$$\mathbb{H}_t \stackrel{\text{def}}{=} \coprod_{w \in \mathbb{W}} [\sim_w]$$

and *heap* is interpreted as the set of typed heaps with the underlying domain-theoretic heaps as realizers:

$$[\![\Gamma \vdash \textit{heap}]\!] = (\mathbb{H}_t, (w, U) \mapsto in_H(U))_{i \in |I|}$$

for $I = [\![\Gamma]\!]$. That is, for each i, we have the assembly with underlying set $\mathbb{H}_t$ and with realizability map $\mathbb{H}_t \to P(\mathbb{V})$ given by $(w, U) \mapsto in_H(U)$. The realizers themselves do not have to contain any typing information, as we interpret small types with trivial realizability information (i.e., $[\![\Gamma \vdash \textbf{\textit{set}}]\!] = \nabla(\mathrm{CMPer}(\mathbb{V})))$. We now sketch the interpretation of update and peek in the empty context. Let $(w, [h]) = [\![\vdash M]\!]$, $in_{\mathrm{N}}(n) \in [\![\vdash N_1]\!]$, $R = [\![\vdash \tau]\!]$, and $v \in [\![\vdash N_2]\!]$. Then,

$$[\![\vdash M[N_1 \mapsto_\tau N_2]]\!] = (w[n \mapsto R], [h[n \mapsto v]]_{w[n \mapsto R]})$$

$$[\![\vdash \textbf{\textit{peek}}\ M\ N_1]\!] = \begin{cases} inr(w(n), [h(n)]_{w(n)}) & \text{if } w(n) \text{ defined} \\ inl(*) & \text{otherwise} \end{cases}$$

In the interpretation of update, the world is used to ensure that the interpretation is independent of the choice of realizer v for N_2. Since the underlying domain-theoretic heaps do not contain any typing information, the world is also used in the interpretation of peek, to determine the type of the value stored at the given location.

Note that iHTT has "strong" update and that the world is modified to contain the new type (semantically, the per R) upon update. Thus our notion and use of worlds is different from the use of worlds in models of "weak" ML-like reference types, e.g. [5]; in particular, note that we do not index every type by a world, but only use worlds to interpret the type of heaps and the operations thereon (see also the next subsection for further discussion).

Computations. We are now ready to sketch the interpretation of Hoare-types. The idea is to interpret Hoare-types as PERs on elements of $T(\mathbb{V})$ that satisfy the given specification, but with a trivial equality. Specifically, given a partial correctness specification, we define an admissible subset $X \subseteq T(\mathbb{V})$ of computations satisfying the specification, and interpret the associated Hoare type as the PER,

$$R = in_T(\Delta(X)) = \{(in_T(f), in_T(g)) \mid f \in X \wedge g \in X\}$$

The trivial equality ensures that R is trivially monotone and admissibility on the sub-domain of computations follows from admissibility of X.

Assume a semantic pre-condition $P \in \mathbb{H}_t \to 2$, a small semantic type $R \in \mathrm{CMPer}(\mathbb{V})$, and a semantic post-condition $Q \in [R] \times \mathbb{H}_t \times \mathbb{H}_t \to 2$. As explained in the previous section, the pre- and post-condition is expressed in terms of a typed heaps, $\mathbb{H}_t$, instead of the underlying domain theoretic heaps. The subset of computations satisfying the specification is thus defined using the usual "forall initial worlds, there exists a terminal world"-formulation, known from models of ML-like references [5]. However, as iHTT supports strong update and dealloca-tion, the terminal world is not required to be an extension of the initial world. Specifically, define $hoare(R, P, Q)$ as the following subset of $T(\mathbb{V})$:

$$\{f \in T(\mathbb{V}) \mid \forall w \in \mathbb{W}.\ \forall h \in |\sim_w|.\ P([h]_w) = \top \Rightarrow$$
$$(f(h) = \bot\ \vee \exists v', h'.\ f(h) = (v', h') \wedge v' \in |R| \wedge$$
$$h' \in \overline{\{h' \in H(\mathbb{V}) \mid \exists w' \in \mathbb{W}.\ Q([v']_R)([h]_w)([h']_{w'}) = \top\}})\}$$

where $\overline{(-)}$ denotes the chain-completion operator on $T(\mathbb{V})$. The explicit chain-completion of the post-condition is required because of the existential quan-tification over worlds. Furthermore, since the post-condition is indexed by the return value $[v']_R$, monotonicity is used to collapse a chain of domain-theoretic values $v_1 \leq v_2 \leq \cdots$ into a single type-theoretic value $[v_1]_R$, when proving that $hoare(R, P, Q)$ is an admissible subset of $T(\mathbb{V})$.

A Hoare-type in the empty context is now interpreted as follows:

$$[\![\vdash \textbf{\textit{st}}_\tau \ (\mathrm{P}, \mathrm{Q})]\!] = in_T(\Delta(hoare([\![\vdash \tau]\!], [\![\vdash \mathrm{P}]\!], [\![\vdash \mathrm{Q}]\!])))$$

The previous model of iHTT [21] featured a non-trivial computational equality. However, the previous model lacked the worlds introduced in the previous section and with it a useful notion of heap equivalence. As a result, the computational equality in the previous model was very strict, rendering the structural rule for existentials in Hoare logic unsound. The new model validates all the usual structural rules of Hoare-logic.

Theorem 2. *The interpretation of computations is sound, i.e., well-typed computations satisfy their specifications.*

W-types. The presentation of (co)-inductive types in CIC is based on an intricate syntactic scheme of inductive families. So far, this presentation of (co)-inductive types has eluded a categorical semantics. Martin-Löf type theory features an alternative presentation, based on W-types (a type-theoretic formalization of well-founded trees), which is strong enough to represent a wide range of predicative inductive types in extensional models (such as ours) [11,3]. Since W-types in addition have a simple categorical semantics, we have chosen to show that iHTT models inductive types by showing that it models W-types. Specifically, we show that it models W-types at both the ***type*** and ***set*** universe, and that the W-types at the ***set*** universe supports elimination over large types.

Semantically, in the setting of locally cartesian closed categories, W-types are modelled as initial algebras of polynomial functors [16]. In the setting of split closed comprehension categories we define:

Definition 9. *A split closed comprehension category* $\mathcal{P} : \mathbb{E} \to \mathbb{B}^{\to}$ *has split W-types, if for every* $I \in \mathbb{B}$, $X \in \mathbb{E}_I$ *and* $Y \in \mathbb{E}_{\{X\}}$ *the endo-functor,*

$$P_{X,Y} = \Sigma_X \circ \Pi_Y \circ (\pi_X \circ \pi_Y)^* : \mathbb{E}_I \to \mathbb{E}_I$$

has a chosen initial algebra $\alpha_{X,Y} : P_{X,Y}(W_{X,Y}) \to W_{X,Y} \in \mathbb{E}_I$, *which is preserved on-the-nose by re-indexing functors.*

As is well-known, $\mathrm{Asm}(\mathbb{V})$ is locally cartesian closed and all polynomial functors on $\mathrm{Asm}(\mathbb{V})$ have initial algebras. This yields initial algebras for functors $P_{X,Y} :$ $\mathrm{UFam}(\mathrm{Asm}(\mathbb{V}))_1 \to \mathrm{UFam}(\mathrm{Asm}(\mathbb{V}))_1$ for $X \in \mathbb{E}_1$ and $Y \in \mathbb{E}_{\{X\}}$. This lifts to an arbitrary context $I \in \mathrm{Asm}(\mathbb{V})$ by a point-wise construction, which yields split W-types, as re-indexing in $\mathrm{UFam}(\mathrm{Asm}(\mathbb{V}))$ is by composition:

Lemma 4. *The split ccomp* $\mathcal{P} : \mathrm{UFam}(\mathrm{Asm}(\mathbb{V})) \to \mathrm{Asm}(\mathbb{V})^{\to}$ *has split W-types.*

This models W-types in the ***type***-universe. In addition, iHTT features small W-types over small types, with elimination over large types. The idea is to model these small W-types by forming the W-type in UFam(Asm($\mathbb{V}$)) and mapping it back into UFam(CMPer($\mathbb{V}$)) using the reflection. This yields a small type, and by the following lemma, this small type is still a W-type in UFam(Asm($\mathbb{V}$)), and thus models elimination over large types:

Lemma 5. *For every* $I \in \mathbb{B}$, $X \in$ UFam(CMPer($\mathbb{V}$))$_I$, *and* $Y \in$ UFam (CMPer($\mathbb{V}$))$_{\{X\}}$, *there is a chosen isomorphism,* $W_{\mathcal{I}_{el}(X),\mathcal{I}_{el}(Y)} \cong \mathcal{I}_{el}(\mathcal{R}_{el} (W_{\mathcal{I}_{el}(X),\mathcal{I}_{el}(Y)}))$, *which is preserved on-the-nose by reindexing functors.*

Proof (Sketch). The reflection $\mathcal{R}_{el}$ collapses an assembly into a PER by equating values with overlapping sets of realizers and into a complete monotone PER by taking the monotone completion of said PER. The result follows by proving that $W_{X,Y}$ is a modest set (i.e., has no overlapping sets of realizers), and that the induced PER is already monotone complete, for X and Y in the image of $\mathcal{I}_{el}$.

4   Implementation

One of the advantages of weakening Hoare types instead of the underlying dependent type theory – as in the case of [21] – is that it can simplify implementation of the resulting type theory. In our case, presenting iHTT as an extension of CIC allows for easy implementation as an axiomatic extension of the Coq proof assistant.

Our implementation is based on the Coq infrastructure developed by Nanevski et. al. [19], to support efficient reasoning about stateful computations in Coq. This infrastructure defines a new Hoare type on top of the built-in Hoare type with support for more efficient reasoning, based on ideas from separation logic. Specifically, as illustrated in Section 2, this new Hoare type features (1) small footprint specifications and (2) efficient reasoning about heaps. Efficient reasoning about heaps is achieved by reasoning using the partial commutative monoid $(1 + \boldsymbol{heap}, \bullet)$ instead of the total non-commutative monoid $(\boldsymbol{heap}, \bullet)$, where $\boldsymbol{heap}$ is the actual type of heaps and $\bullet$ is heap union [19].

Compared to Nanevski et. al.'s implementation of predicative Hoare Type Theory (pHTT), this new implementation features higher-order store and impredicative quantification, but the predicative hierarchy lacks Hoare-types. The new implementation is almost source compatible with verifications in pHTT that do not exploit the predicative hierarchy.

Compared to the Ynot implementation of pHTT [18], in addition to impredicativity the main difference lies in the treatment of ghost variables. Post conditions in Ynot are unary and thus employ ghost-variables to relate the pre- and postcondition. Ynot expresses ghost variables of a specification as computationally irrelevant arguments to the computation. As Coq lacks support for computationally irrelevant variables, Ynot extends Coq with an injectivity axiom, which gives an embedding of **set** into the computationally irrelevant **prop**-universe. This axiom is inconsistent with a proof irrelevant **prop**-universe and thus in particular unsound in our model. Additionally, this limits Ynot's ghost variables to small types, whereas iHTT supports ghost variables of large types.

5   Related Work

Our approach to partiality is based on the idea of partial types, as introduced by Constable and Smith [9]. We have already discussed its relation to the work on admissibility by Crary [10] in the introduction. Below we first discuss related work on partiality, followed by related work on partial correctness reasoning.

Bove and Capretta [7] proposed representing a partial function $f : A \rightharpoonup B$ as a total function $\overline{f} : \Pi a : A.\ P(a) \rightarrow B$, defined by recursion over an inductively defined predicative $P : A \rightarrow \textbf{\textit{prop}}$, expressing the domain of the partial function. This allows the definition of partial computations by general recursion, but does not model non-termination, as $\overline{f}$ can only be applied to arguments on which it terminates. Capretta [8] proposed an alternative co-inductive representation, which does model non-termination, representing a partial function $f : A \rightharpoonup B$ as a total function $\overline{f} : A \rightarrow B^v$, where B^v is co-inductive type of partial elements of type B. This representation yields a least fixed point operator on finitary (continuous) endo-functions on $A \rightarrow B^v$. Capretta does not provide a fixed point induction principle, but we believe such a principle would require admissibility proofs.

Another alternative approach to partiality is to give a model of a language featuring general recursion inside a dependent type theory. This allows one to model and reason about partial computations inside the type theory, but does not extend the type theory itself with partial computations. This approach has for instance been studied and implemented by Reus [22], who formalized Synthetic Domain Theory in the Lego proof checker. The resulting type theory can be seen as a very expressive version of LCF. The synthetic approach alleviates the need for continuity proofs, but still requires admissibility proofs when reasoning by fixed point induction.

Hoare-style specification logics is another line of closely related work. With a collapsed computational equality, reasoning about a partial computation in iHTT is limited to Hoare-style partial correctness reasoning, as in a specification logic. With the modularity features provided by the underlying dependent type theory (i.e., Σ types), iHTT can thus be seen as a modular specification logic for a higher-order programming language.

6   Conclusion

We have presented a new approach for extending dependent type theory with potentially non-terminating computations, without weakening the underlying dependent type theory or adding restrictions on the use of fixed points in defining partial computations. We have shown that it scales to very expressive dependent type theories and effects beyond partiality, by extending the Calculus of Inductive Constructions with stateful and potentially non-terminating computations. We have further demonstrated that this approach is practical, by implementing this extension of CIC as an axiomatic extension of the Coq proof assistant.

To justify the soundness of our extension of CIC, we have presented a realizability model of the theory. For lack of space, we have limited the presentation to a single predicative universe, but the model can be extended to the whole predicative hierarchy $\textbf{\textit{type}} \subseteq \textbf{\textit{type}}_1 \subseteq \ldots$ of CIC.

References

1. The Coq Proof Assistant, http://coq.inria.fr/
2. Coq Reference Manual, Version 8.3
3. Abbott, M., Altenkirch, T., Ghani, N.: Representing nested inductive types using W-types. In: Díaz, J., Karhumäki, J., Lepistö, A., Sannella, D. (eds.) ICALP 2004. LNCS, vol. 3142, pp. 59–71. Springer, Heidelberg (2004)
4. Amadio, R.M.: Recursion over Realizability Structures. Information and Computation 91, 55–85 (1991)
5. Birkedal, L., Støvring, K., Thamsborg, J.: Realisability semantics of parametric polymorphism, general references and recursive types. Math. Struct. Comp. Sci. 20(4), 655–703 (2010)
6. Bjorner, D., Jones, C.B. (eds.): The Vienna Development Method: The Meta-Language. LNCS, vol. 61. Springer, Heidelberg (1978)
7. Bove, A.: Simple General Recursion in Type Theory. Nordic Journal of Computing 8 (2000)
8. Capretta, V.: General Recursion via Coinductive Types. Logical Methods in Computer Science 1(2), 1–28 (2005)
9. Constable, R.L., Smith, S.F.: Partial Objects in Constructive Type Theory. In: Proceedings of Second IEEE Symposium on Logic in Computer Science (1987)
10. Crary, K.: Admissibility of Fixpoint Induction over Partial Types. In: Automated Deduction - CADE-15 (1998)
11. Dybjer, P.: Representing inductively defined sets by wellorderings in Martin-Löf's type theory. Theor. Comput. Sci. 176(1-2), 329–335 (1997)
12. Gonthier, G., Mahboubi, A.: A Small Scale Reflection Extension for the Coq system. Technical report, INRIA (2007)
13. Hoare, C.A.R.: Proof of correctness of data representations. Acta Informatica 1, 271–281 (1972)
14. Jacobs, B.: Categorical Logic and Type Theory. Elsevier Science, Amsterdam (1999)
15. Meyer, B.: Object-oriented software construction. Prentice Hall, Englewood Cliffs (1997)
16. Moerdijk, I., Palmgren, E.: Wellfounded trees in categories. Annals of Pure and Applied Logic 104(1-3), 189–218 (2000)
17. Nanevski, A., Morrisett, G., Birkedal, L.: Hoare Type Theory, Polymorphism and Separation. Journal of Functional Programming 18(5-6), 865–911 (2008)
18. Nanevski, A., Morrisett, G., Shinnar, A., Govereau, P., Birkedal, L.: Ynot: Dependent Types for Imperative Programs. In: Proceedings of ICFP 2008, pp. 229–240 (2008)
19. Nanevski, A., Vafeiadis, V., Berdine, J.: Structuring the Verification of Heap-Manipulating Programs. In: Proceedings of POPL 2010 (2010)
20. O'Hearn, P., Reynolds, J., Yang, H.: Local reasoning about programs that alter data structures. In: Fribourg, L. (ed.) CSL 2001 and EACSL 2001. LNCS, vol. 2142, pp. 1–19. Springer, Heidelberg (2001)
21. Petersen, R.L., Birkedal, L., Nanevski, A., Morrisett, G.: A realizability model for impredicative hoare type theory. In: Gairing, M. (ed.) ESOP 2008. LNCS, vol. 4960, pp. 337–352. Springer, Heidelberg (2008)
22. Reus, B.: Synthetic Domain Theory in Type Theory: Another Logic of Computable Functions. In: von Wright, J., Harrison, J., Grundy, J. (eds.) TPHOLs 1996. LNCS, vol. 1125. Springer, Heidelberg (1996)
23. Svendsen, K., Birkedal, L., Nanevski, A.: Partiality, State and Dependent Types. Technical report, IT University of Copenhagen (2011), http://www.itu.dk/people/kasv/ihtt-adm-tr.pdf

A Filter Model for the $\lambda\mu$-Calculus

(Extended Abstract)

Steffen van Bakel[1], Franco Barbanera[2], and Ugo de'Liguoro[3]

[1] Department of Computing, Imperial College London,
180 Queen's Gate, London SW7 2BZ, UK
svb@doc.ic.ac.uk
[2] Dipartimento di Matematica e Informatica, Università degli Studi di Catania,
Viale A. Doria 6, 95125 Catania, Italia
barba@dmi.unict.it
[3] Dipartimento di Informatica, Università degli Studi di Torino,
Corso Svizzera 185, 10149 Torino, Italy
deliguoro@di.unito.it

Abstract. We introduce an intersection type assignment system for the pure $\lambda\mu$-calculus, which is invariant under subject reduction and expansion. The system is obtained by describing Streicher and Reus's denotational model of continuations in the category of ω-algebraic lattices via Abramsky's domain logic approach. This provides a tool for showing the completeness of the type assignment system with respect to the continuation models via a filter model construction. We also show that typed $\lambda\mu$-terms in Parigot's system have a non-trivial intersection typing in our system.

1 Introduction

The $\lambda\mu$-calculus is a pure calculus introduced by Parigot [27] to denote classical proofs and to compute with them. It is an extension of the proofs-as-programs paradigm where types can be understood as classical formulas and (closed) terms inhabiting a type as the respective proofs in a variant of Gentzen's natural deduction calculus for classical logic [16]. Since the early days, the study of the syntactic properties of the $\lambda\mu$-calculus has been challenging, motivating the introduction of variants of term syntax, of reduction rules, and of type assignment, as is the case of de Groote's variant of the $\lambda\mu$-calculus [18]. These changes have an impact on the deep nature of the calculus which emerges both in the typed and in the untyped setting [12,29,19].

Types are of great help in understanding the computational properties of terms in an abstract way. Although in [7] Barendregt treats the theory of the pure λ-calculus without a reference to types, most of the fundamental results of the theory can be exposed in a quite elegant way by using the Coppo-Dezani intersection type system [10]. This is used by Krivine [22], where the treatment of the pure λ-calculus relies on intersection type assignment systems called $\mathcal{D}$ and $\mathcal{D}\Omega$.

The quest for more expressive notions of type assignment for $\lambda\mu$ is part of an ongoing investigation into calculi for classical logic. In order to come to a characterisation of strong normalisation for Curien and Herbelin's (untyped) sequent calculus $\overline{\lambda}\mu\tilde{\mu}$ [11],

L. Ong (Ed.): TLCA 2011, LNCS 6690, pp. 213–228, 2011.

Dougherty, Ghilezan and Lescanne presented System $\mathcal{M}^{\cap\cup}$ [14,15], that defines a notion of intersection and union typing for that calculus. However, in [4] van Bakel showed that this system is not closed under conversion, an essential property of Coppo-Dezani systems; in fact, it is shown that it is *impossible* to define a notion of type assignment for $\overline{\lambda}\mu\tilde{\mu}$ that satisfies that property. In [5] van Bakel brought intersection (and union) types to the context of the (untyped) $\lambda\mu$-calculus, and showed type preservation with respect to $\lambda\mu$-conversion. However, union types are no longer dual to intersection types and play only a marginal role, as was also the intention of [15]. In particular, the normal $(\cup I)$ and $(\cup E)$ rules as used in [6], which are known to create the same soundness problem in the context of the λ-calculus, are not allowed. Moreover, although one can link intersection types with the logical connector *and*, the union types used in [5] bear *no* relation with *or*; one could argue that therefore *union* might perhaps not be the right name to use for this type constructor.

In the view of the above mentioned failure, the result of [5] came as a surprise, and led automatically to the question we answer here: does a filter semantics for $\lambda\mu$ exist?

Building on Girard's ideas [23] and Ong and Stewart's work [25], in [30,21] the Streicher and Reus proposed a model of both typed and untyped λ-calculi embodying some idea of continuations, including a version of pure $\lambda\mu$. Their model is based on the solution of the domain equations $D = C{\rightarrow}R$ and $C = D \times C$, where R is an arbitrary domain of 'results' (we call the triple (R, D, C) a $\lambda\mu$-model). With respect to this, here we adapt the term-interpretation map of Streicher and Reus [30] to Parigot's original calculus (it is not difficult to do the same for de Groote's $\Lambda\mu$); we deviate from the variant studied in [30], where also continuation terms are included in the syntax.

Following Abramski's approach [1], we reconstruct the initial/final solution of these equations in the category of ω-algebraic lattices by describing compact points and the ordering of these domains as Lindenbaum-Tarski algebras of certain intersection type theories. Types are generated from type variables, the trivial type ω, the connective $\wedge$, plus the domain specific type constructors $\times$ and $\rightarrow$. This way, we obtain an extension of the type theory used in [8] which is a *natural equated* intersection type theory in terms of [3] and hence isomorphic to the inverse limit construction of a D_∞ λ-model (as an aside, we observe that this perfectly matches with Theorem 3.1 in [30]). The thus obtained type theory and term interpretation guide the definition of an intersection type assignment system. We prove this to be invariant under conversion, and sound and complete with respect to the validity of type judgements in any $\lambda\mu$-model; this is shown through the filter model which, together with the system, is the main contribution of this paper.

At this point we have to stress the quite different meaning of our types and type assignment system with respect to the system originally presented by Parigot. The types we use here are *not* logical formulas, and the $\wedge$ connective is not conjunction. As is the case with ordinary intersection types, $\wedge$ is not even the left adjoint of $\rightarrow$, which instead is the case for the $\times$ connective. Nonetheless, we show there exists a strong connection between Parigot's type assignment and ours. In fact, we show that any typed term in Parigot's system has a non-trivial type (i.e. one that cannot be equated to ω) in our system, and that this is true of all the subjects in its derivation. We interpret this result as evidence that terms that actually represent logical proofs do have a computational meaning in a precise sense. Assuming the model captures relevant computational properties, this might provide a characterisation of strong normalisation for $\lambda\mu$.

Due to space restrictions, proofs are omitted from this paper.

2   The $\lambda\mu$ Calculus

In this section we briefly recall Parigot's pure $\lambda\mu$-calculus introduced in [27], slightly changing the notation.

Definition 2.1 (Term Syntax [27]).   The sets Trm of *terms* and Cmd of *commands* are defined inductively by the following grammar (where $x \in Var$, a set of *term variables*, and $\alpha \in Name$, a set of *names*, that are both denumerable):

$$M, N ::= x \mid \lambda x.M \mid MN \mid \mu\alpha.Q \qquad \text{(terms)}$$
$$Q ::= [\alpha]M \qquad \text{(commands)}$$

As usual, we consider λ and μ to be binders; we adopt Barendregt's convention on terms, and will assume that free and bound variables are different.

In [27] terms and commands are called unnamed and named terms respectively. In the same place names are called μ-variables, while they are better understood as *continuation variables* (see [30]), but this would be a direct link to the interpretation; we prefer a more neutral terminology.

Definition 2.2 (Substitution [27]).   Substitution takes three forms:

term substitution:	$M[N/x]$ (N is substituted for x in M, avoiding capture)
renaming:	$Q[\alpha/\beta]$ (every free occurrence of β in Q is replaced by α)
structural substitution:	$T[\alpha \Leftarrow L]$ (every subterm $[\alpha]N$ of M is replaced by $[\alpha]NL$)

where $M, N, L \in$ Trm, $Q \in$ Cmd and $T \in$ Trm $\cup$ Cmd. More precisely, $T[\alpha \Leftarrow L]$ is defined by:

$$([\alpha]M)[\alpha \Leftarrow L] \equiv [\alpha](M[\alpha \Leftarrow L])L$$

whereas in all the other cases it is defined by:

$$x[\alpha \Leftarrow L] \equiv x$$
$$(\lambda x.M)[\alpha \Leftarrow L] \equiv \lambda x.M[\alpha \Leftarrow L]$$
$$(MN)[\alpha \Leftarrow L] \equiv (M[\alpha \Leftarrow L])(N[\alpha \Leftarrow L])$$
$$(\mu\beta.Q)[\alpha \Leftarrow L] \equiv \mu\beta.Q[\alpha \Leftarrow L]$$
$$([\beta]M)[\alpha \Leftarrow L] \equiv [\beta]M[\alpha \Leftarrow L]$$

Definition 2.3 (Reduction [27]).   The reduction relation $T \rightarrow_\mu S$, where $(T, S) \subseteq$ (Trm $\times$ Trm) $\cup$ (Cmd $\times$ Cmd) is defined as the compatible closure of the following rules :

$$\begin{aligned}
(\beta): &\quad (\lambda x.M)N \rightarrow M[N/x] \\
(\mu): &\quad (\mu\beta.Q)N \rightarrow \mu\beta.Q[\beta \Leftarrow N] \\
(ren): &\quad [\alpha]\mu\beta.Q \rightarrow Q[\alpha/\beta] \\
(\mu\eta): &\quad \mu\alpha.[\alpha]M \rightarrow M \qquad\qquad \text{if } \alpha \notin fn(M)
\end{aligned}$$

Parigot's original paper [27] just mentions rule $(\mu\eta)$, while proving confluence of the reduction relation axiomatised by rules (β), (μ), (ren) only. The full reduction relation in Def 2.3 has been proved confluent by Py [28].

Theorem 2.4 (Confluence of $\rightarrow_\mu$ [28]). The reduction relation $\rightarrow_\mu$ is confluent.

Because of Thm. 2.4 the convertibility relation $=_\mu$ determined by $\rightarrow_\mu$ is consistent in the usual sense that different normal forms are not equated. If we add the rule

$$(\eta): \quad \lambda x.Mx \rightarrow M \quad \text{if } x \notin fv(M)$$

of the λ-calculus, we obtain a non-confluent reduction relation, that we call $\rightarrow_{\mu\eta}$; see [28]§2.1.6 for an example of non-confluence and possible repairs. However, the convertibility relation $=_{\mu\eta}$ induced by $\rightarrow_{\mu\eta}$ is consistent (namely non-trivial) by a semantic argument (see Sect. 3). The theory of $=_{\mu\eta}$ is interesting because it validates the untyped version of Ong's equation (ζ) in [26], which has the following form:

$$(\zeta): \quad \mu\alpha.Q = \lambda x.\mu\alpha.Q[\alpha \Leftarrow x]$$

where $x \notin fv(Q)$. Indeed, for a fresh x we have:

$$\mu\alpha.Q \quad_\eta\!\leftarrow \quad \lambda x.(\mu\alpha.Q)x \quad\rightarrow_\mu\quad \lambda x.\mu\alpha.Q[\alpha \Leftarrow x]$$

So it is possible to define more reduction rules, but Parigot refrained from that since he aimed at defining a confluent reduction system.

With $\lambda\mu$ Parigot created a multi-conclusion typing system which corresponds to classical logic; the derivable statements have the shape $\Gamma \vdash M : A \mid \Delta$, where A is the main conclusion of the statement, expressed as the *active* conclusion, and Δ contains the alternative conclusions. The reduction rules for the terms that represent the proofs correspond to proof contractions.

Parigot's type assignment for $\lambda\mu$ is defined by the following natural deduction system; there is a *main*, or *active*, conclusion, labelled by a term of this calculus, and the alternative conclusions are labelled names.

Definition 2.5 (Typing rules for $\lambda\mu$ [27]). Types are those of the simply typed λ-calculus, extended with the type constant $\perp$ (essentially added to express negation), *i.e.*:

$$A, B ::= \varphi \mid \perp \mid A{\rightarrow}B \quad (A \neq \perp)$$

The type assignment rules are:

$$(Ax): \frac{}{\Gamma \vdash x : A \mid \Delta} \; (x{:}A \in \Gamma) \quad (\perp): \frac{\Gamma \vdash M : B \mid \beta{:}B, \Delta}{\Gamma \vdash [\beta]M : \perp \mid \beta{:}B, \Delta} \quad (\mu): \frac{\Gamma \vdash Q : \perp \mid \alpha{:}A, \Delta}{\Gamma \vdash \mu\alpha.Q : A \mid \Delta}$$

$$(\rightarrow I): \frac{\Gamma, x{:}A \vdash M : B \mid \Delta}{\Gamma \vdash \lambda x.M : A{\rightarrow}B \mid \Delta} \quad (\rightarrow E): \frac{\Gamma \vdash M : A{\rightarrow}B \mid \Delta \quad \Gamma \vdash N : A \mid \Delta}{\Gamma \vdash MN : B \mid \Delta}$$

We write $\Gamma \vdash_P N : A \mid \Delta$ for statements derivable using these rules.

We can think of $[\alpha]M$ as storing the type of M amongst the alternative conclusions by giving it the name α.

Example 2.6. As an example illustrating the fact that this system is more powerful than the system for the λ-calculus, Fig. 2 contains a proof of Peirce's Law.

$$\cfrac{\cfrac{\cfrac{\cfrac{x{:}(A{\to}B){\to}A, y{:}A \vdash y : A \mid \alpha{:}A, \beta{:}B}{x{:}(A{\to}B){\to}A, y{:}A \vdash [\alpha]y : \perp \mid \alpha{:}A} \;(\perp)}{x{:}(A{\to}B){\to}A, y{:}A \vdash \mu\beta.[\alpha]y : B \mid \alpha{:}A} \;(\mu)}{x{:}(A{\to}B){\to}A \vdash \lambda y.\mu\beta.[\alpha]y : A{\to}B \mid \alpha{:}A} \;(\to I)}{}$$

$$x{:}(A{\to}B){\to}A \vdash x : (A{\to}B){\to}A \mid \alpha{:}A \qquad x{:}(A{\to}B){\to}A \vdash \lambda y.\mu\beta.[\alpha]y : A{\to}B \mid \alpha{:}A \;(\to E)$$

$$\cfrac{x{:}(A{\to}B){\to}A \vdash x(\lambda y.\mu\beta.[\alpha]y) : A \mid \alpha{:}A}{\cfrac{x{:}(A{\to}B){\to}A \vdash [\alpha](x(\lambda y.\mu\beta.[\alpha]y)) : \perp \mid \alpha{:}A}{\cfrac{x{:}(A{\to}B){\to}A \vdash \mu\alpha.[\alpha](x(\lambda y.\mu\beta.[\alpha]y)) : A \mid}{\vdash \lambda x.\mu\alpha.[\alpha](x(\lambda y.\mu\beta.[\alpha]y)) : ((A{\to}B){\to}A){\to}A \mid} \;(\to I)} \;(\mu)} \;(\perp)}$$

Fig. 1. A proof of Peirce's Law (due to Ong and Stewart [25])

3   Semantics

The semantics considered here is due to Streicher and Reus [30]. Their idea is to work in the category $\mathcal{N}_R$ of 'negated' domains of the shape $A{\to}R$, where R is a parameter for the domain of results. In such a category, continuations are directly modelled and treated as the fundamental concept, providing a semantics both to Felleisen's λC-calculus and to a variant of $\lambda\mu$, with three sorts of terms, instead of two.

In this section we adapt such a semantics to Parigot's original $\lambda\mu$, which does not have continuation terms $M_1 :: \dots :: M_k$. We rephrase the model definition in the setting of ordinary categories of domains, getting something similar to the Hindley-Longo 'syntactical models', but without pretending to achieve the general definition of what a $\lambda\mu$-model is, an issue which is dealt with in [26,20].

Definition 3.1 ($\lambda\mu$-Model).   We say that a triple (R, D, C) is a $\lambda\mu$-model in a category of domains $\mathcal{D}$ if $R \in \mathcal{D}$ is a fixed domain of *results* and D, C (called domains of *denotations* and of *continuations* respectively), are solutions in $\mathcal{D}$ of the equations:

$$\begin{cases} D = C \to R \\ C = D \times C \end{cases} \tag{1}$$

In the terminology of [30] elements of D are *denotations*, while those of C are *continuations*. We refer to these equations as the *continuation domain equations*.

Definition 3.2 (Term Interpretation).   Let (R, D, C) be a $\lambda\mu$-model, and $\mathsf{Env} = (Var \to D) + (Name \to C)$. The interpretation mappings $\llbracket \cdot \rrbracket^D : \mathsf{Trm} \to \mathsf{Env} \to D$ and $\llbracket \cdot \rrbracket^C : \mathsf{Cmd} \to \mathsf{Env} \to C$ are mutually defined by the following equations, where $e \in \mathsf{Env}$ and $k \in C$:

$$\llbracket x \rrbracket^D e\, k = e\, x\, k$$
$$\llbracket \lambda x.M \rrbracket^D e\, k = \llbracket M \rrbracket^D e[x := d]\, k' \quad \text{where } k = \langle d, k' \rangle$$
$$\llbracket MN \rrbracket^D e\, k = \llbracket M \rrbracket^D e\, \langle \llbracket N \rrbracket^D e, k \rangle$$
$$\llbracket \mu\alpha.Q \rrbracket^D e\, k = d\, k' \qquad\qquad \text{where } \langle d, k' \rangle = \llbracket Q \rrbracket^C e[\alpha := k]$$
$$\llbracket [\alpha]M \rrbracket^C e\ = \langle \llbracket M \rrbracket^D e, e\, \alpha \rangle$$

This definition has (of course), a strong similarity with Bierman's interpretation of $\lambda\mu$ [9]; however, he considers a *typed* version.

In the second equation the assumption $k = \langle d, k' \rangle$ is not restrictive: in particular, if $k = \perp_C = \langle \perp_D, \perp_C \rangle$ then $d = \perp_D$ and $k' = k = \perp_C$. We shall omit the superscripts in $\llbracket \cdot \rrbracket^C$ and $\llbracket \cdot \rrbracket^D$ when clear from the context.

Remark 3.3. Let us recall $\Lambda\mu$, the variant of $\lambda\mu$ deviced by de Groote [18] (see also [19]), where there is no distinction between terms and commands:

$$\Lambda\mu\text{-Trm}: \quad M, N ::= x \mid \lambda x.M \mid MN \mid \mu\alpha.M \mid [\alpha]M$$

Then, given a solution D, C of the continuation domain equations, it is possible to define a similar interpretation map $\llbracket \cdot \rrbracket^D : \Lambda\mu\text{-Trm} \to \text{Env} \to D$, where the first three clauses are the same as in Def. 3.2, while the last two become:

$$\llbracket \mu\alpha.M \rrbracket^D e\, k = \llbracket M \rrbracket^D e[\alpha := k]\, k$$
$$\llbracket [\alpha]M \rrbracket^D e\, k = \llbracket M \rrbracket^D e\, (e\, \alpha)$$

The distinctive feature of $\Lambda\mu$ is that in $\mu\alpha.M$ the subterm M need not be a named term (of the shape $[\alpha]L$), and that there exist terms of the form $\lambda x.[\alpha]M$, which play a key role in [18] to inhabit the type $\neg\neg A \to A$ with a closed term.

The analogous of Thm. 3.7 below can be established for $\Lambda\mu$ and the above semantics, with the proviso that the renaming reduction rule:

$$[\alpha]\mu\beta.M \to M[\alpha/\beta]$$

is unsound for arbitrary M. On the contrary, renaming is sound in the expected contexts:

$$[\alpha]\mu\beta.[\gamma]M \to ([\gamma]M)[\alpha/\beta] \quad \text{and} \quad \mu\alpha[\alpha]\mu\beta.M \to \mu\alpha.M[\alpha/\beta]$$

where the second one is essentially the same as Parigot's. The form of the renaming rule is a delicate point, for which we refer the reader to [19].

Since Def. 3.2 does not coincide exactly with Streicher and Reus's, we check below that it actually models $\lambda\mu$ convertibility.

Proposition 3.4. 1. $\llbracket M[N/x] \rrbracket e = \llbracket M \rrbracket e[x := \llbracket N \rrbracket e]$
   2. $\llbracket M[\alpha/\beta] \rrbracket e = \llbracket M \rrbracket e[\beta := e\, \alpha]$

The semantics satisfies the following "swapping continuations" equation[1]:

Lemma 3.5. $\llbracket \mu\alpha.[\beta]M \rrbracket e\, k = \llbracket M \rrbracket e[\alpha := k]\, (e[\alpha := k]\beta)$.

We can now show:

Lemma 3.6. $\llbracket M[\alpha \Leftarrow N] \rrbracket e\, k = \llbracket M \rrbracket e[\alpha := \langle \llbracket N \rrbracket e, e\, \alpha \rangle]\, k$.

Theorem 3.7 (Soundness of $\lambda\mu$). $\vdash_{\lambda\mu} M = N \Rightarrow \llbracket M \rrbracket = \llbracket N \rrbracket$

The soundness property holds also in case rule (η) is taken into account. The proof of Thm. 3.7 can in fact be easily extended with a further induction case.

Proposition 3.8 (Soundness of $\lambda\mu\eta$). The (η) rule is sound with respect to the semantics, i.e. $\llbracket \lambda x.Mx \rrbracket = \llbracket M \rrbracket$ if $x \notin fv(M)$.

---

[1] The equation in [30] is actually $\llbracket \mu\alpha.[\beta]M \rrbracket e\, k = \llbracket M \rrbracket e[\alpha := k]\, (e\, \beta)$, but this is a typo.

4   The Filter Domain

We look for a solution of the continuation domain equations (1) in the category **ALG** of ω-algebraic lattices and Scott-continuous functions (see the classic [17]), described as a domain of filters generated by a suitable intersection type theory (below, a *domain* is an object of **ALG**). Filter models appeared first in the theory of intersection type assignment in [8] (see also [2]). A general theory of Stone duality for domains is treated in [1]; more recent contributions for filter models and recursive domain equations can be found in [13,3].

Definition 4.1 (Intersection Type Language and Theory). An *intersection type language* is a countable set of types $\mathcal{L}$ such that: $\omega \in \mathcal{L}$ and $\sigma, \tau \in \mathcal{L} \Rightarrow \sigma \wedge \tau \in \mathcal{L}$.

An *intersection type theory* $\mathcal{T}$ is a collection of inequalities among types in a language $\mathcal{L}$, closed under the following axioms and rules:

$$\frac{}{\sigma \leq \sigma} \qquad \frac{}{\sigma \wedge \tau \leq \sigma} \qquad \frac{}{\sigma \wedge \tau \leq \tau} \qquad \frac{}{\sigma \leq \omega} \qquad \frac{\sigma \leq \tau \quad \tau \leq \rho}{\sigma \leq \rho} \qquad \frac{\rho \leq \sigma \quad \rho \leq \tau}{\rho \leq \sigma \wedge \tau}$$

We write $\sigma \leq_{\mathcal{T}} \tau$ for $\sigma \leq \tau \in \mathcal{T}$, or just $\sigma \leq \tau$ when $\mathcal{T}$ is understood. The inequality $\leq$ is a preorder over $\mathcal{L}$ and $\mathcal{L}/_{\leq}$ is an inf-semilattice with top $[\omega]$ (the equivalence class of ω).

Let $\mathcal{F}$ be the set of subsets $F \subseteq \mathcal{L}$ which are non-empty, upward closed and closed under finite intersection. Then one has $\mathcal{F} \simeq \mathsf{Filt}(\mathcal{L}/_{\leq})$, the set of filters over $\mathcal{L}/_{\leq}$, so that the elements of $\mathcal{F}$ are called *filters* as well. A *principal filter* is a set $\uparrow\sigma = \{\tau \in \mathcal{L} \mid \sigma \leq \tau\}$; we write $\mathcal{F}_p$ for the set of principal filters; clearly $\mathcal{F}_p \subseteq \mathcal{F}$.

The poset $(\mathcal{F}, \subseteq)$ is an ω-algebraic lattice, whose set of *compact points* $\mathcal{K}(\mathcal{F})$ is $\mathcal{F}_p$. By this, $\mathcal{F}$ is called a *filter domain*. On the other hand, any ω-algebraic lattice X can be presented as a filter domain, due to the isomorphism $X \simeq \mathsf{Filt}(\mathcal{K}^{op}(X))$ where $\mathcal{K}^{op}(X)$ is $\mathcal{K}(X)$ (the set of compact elements in X) with the inverse ordering. This is an instance of Stone duality in the case of the category **ALG**, which is a particular and simpler case than the general ones studied in [1].

Observing that

$$\sigma \leq \tau \iff \uparrow\tau \subseteq \uparrow\sigma$$

we have that $\mathcal{F}_p \simeq \mathcal{L}/_{\leq^{op}}$ considered as sup-semilattices, where $\uparrow\sigma \sqcup \uparrow\tau = \uparrow(\sigma \wedge \tau)$. Suppose that, given $X \in \mathbf{ALG}$, we have established the isomorphism $\mathcal{L}/_{\leq^{op}} \simeq \mathcal{K}(X)$ (as sup-semilattices) or equivalently $\mathcal{L}/_{\leq} \simeq \mathcal{K}^{op}(X)$ (as inf-semilattices), then we obtain

$$\mathcal{F} \simeq \mathsf{Filt}(\mathcal{L}/_{\leq}) \simeq \mathsf{Filt}(\mathcal{K}^{op}(X)) \simeq X.$$

Lemma 4.2. Let $\mathcal{T}$ be an intersection type theory over the language $\mathcal{L}$ and $(\mathcal{L}, \leq)$ the induced preorder. If X is a domain and $\Theta : \mathcal{L} \to \mathcal{K}(X)$ a surjective map such that

$$\forall \sigma, \tau \in \mathcal{L}\,[\,\sigma \leq \tau \iff \Theta(\tau) \sqsubseteq \Theta(\sigma)\,]$$

then $\mathcal{L}/_{\leq} \simeq \mathcal{K}^{op}(X)$ as inf-semilattices.

Definition 4.3 (Type languages). Fix a domain R of results. For $A \in \{R, C, D\}$, we define the languages of intersection types $\mathcal{L}_A$:

$$\mathcal{L}_R : \rho ::= v \mid \omega \mid \rho \wedge \rho$$
$$\mathcal{L}_C : \kappa ::= \omega \mid \delta \times \kappa \mid \kappa \wedge \kappa$$
$$\mathcal{L}_D : \delta ::= v \mid \omega \mid \kappa \to \rho \mid \delta \wedge \delta$$

where the type constants $v = v_a$ are in one-to-one correspondence with $a \in \mathcal{K}(R)$.

Definition 4.4 (Type theories). The theories $\mathcal{T}_R, \mathcal{T}_C$ and $\mathcal{T}_D$ are the least intersection type theories closed under the following axioms and rules, inducing the preorders $\leq_R, \leq_C$ and $\leq_D$ over $\mathcal{L}_R, \mathcal{L}_C$ and $\mathcal{L}_D$ respectively, where, for $A \in \{R, C, D\}$, $\sigma =_A \tau$ is defined by $\sigma \leq_A \tau \leq_A \sigma$:

$$\overline{v_\perp =_R \omega} \qquad \overline{v_{a \sqcup b} =_R v_a \wedge v_b} \qquad \overline{\omega \leq_C \omega \times \omega} \qquad \overline{\omega \leq_D \omega \to \omega} \qquad \overline{v =_D \omega \to v}$$

$$\overline{(\delta_1 \times \kappa_1) \wedge (\delta_2 \times \kappa_2) \leq_C (\delta_1 \wedge \delta_2) \times (\kappa_1 \wedge \kappa_2)} \qquad \overline{(\kappa \to \delta_1) \wedge (\kappa \to \delta_2) \leq_D \kappa \to (\delta_1 \wedge \delta_2)}$$

$$\frac{\delta_1 \leq_D \delta_2 \quad \kappa_1 \leq_C \kappa_2}{\delta_1 \times \kappa_1 \leq_C \delta_2 \times \kappa_2} \qquad \frac{b \sqsubseteq a \in \mathcal{K}(R)}{v_a \leq_R v_b} \qquad \frac{\kappa_2 \leq_C \kappa_1 \quad \rho_1 \leq_R \rho_2}{\kappa_1 \to \rho_1 \leq_D \kappa_2 \to \rho_2}$$

The filter-domain induced by $\mathcal{T}^A$ is $\mathcal{F}^A$, $\uparrow \rho$ is the principal filter generated by ρ, and $\mathcal{F}_p^A$ is the set of principal filters in $\mathcal{F}^A$. Define $\Theta_R : \mathcal{L}_R \to \mathcal{K}(R)$ by:

$$\Theta_R(v_a) = a, \quad \Theta_R(\omega) = \perp, \quad \Theta_R(\rho_1 \wedge \rho_2) = \Theta_R(\rho_1) \sqcup \Theta_R(\rho_2).$$

Lemma 4.5. $\mathcal{F}^R \simeq R$, and the isomorphism is the continuous extension of the mapping $\uparrow \rho \mapsto \Theta_R(\rho)$ from $\mathcal{F}_p$ to $\mathcal{K}(R)$.

Definition 4.6. We define the following maps:

$$F : \mathcal{F}^D \to [\mathcal{F}^C \to \mathcal{F}^R] \qquad F\,d\,k = \{\rho \in \mathcal{L}_R \mid \exists \kappa \to \rho \in d\,[\kappa \in k]\}$$
$$G : [\mathcal{F}^C \to \mathcal{F}^R] \to \mathcal{F}^D \qquad G\,f = \uparrow\{\bigwedge_{i \in I} \kappa_i \to \rho_i \in \mathcal{L}_D \mid \forall i \in I\,[\rho_i \in f(\uparrow \kappa_i)]\}$$
$$H : \mathcal{F}^C \to (\mathcal{F}^D \times \mathcal{F}^C) \qquad H\,k = \langle \{\delta \in \mathcal{L}_D \mid \delta \times \kappa \in k\}, \{\kappa \in \mathcal{L}_D \mid \delta \times \kappa \in k\}\rangle$$
$$K : (\mathcal{F}^D \times \mathcal{F}^C) \to \mathcal{F}^C \qquad K\langle d, k\rangle = \uparrow\{\delta \times \kappa \in \mathcal{L}_C \mid \delta \in d \ \& \ \kappa \in k\}$$

where $[\_ \to \_]$ is the space of continuous functions.

Theorem 4.7. $\mathcal{F}^D \simeq [\mathcal{F}^C \to \mathcal{F}^R]$ and $\mathcal{F}^C \simeq \mathcal{F}^D \times \mathcal{F}^C$.

Combining Lem. 4.5 with Thm. 4.7, we conclude that the filter-domains $\mathcal{F}^R, \mathcal{F}^D$ and $\mathcal{F}^C$ are solutions of the continuation equations in **ALG**. However, a closer look at their structure exhibits the choice of the type languages and theories in Def. 4.3 and 4.4.

Theorem 4.8 (Solution of Continuation Domain Equations [30]). Let $\mathcal{D}$ be a category of domains, and R a fixed object of $\mathcal{D}$. If C and D are initial/final solutions of the domain equations (1) in the category $\mathcal{D}$ then

$$D \simeq [D \to D] \quad \text{and} \quad D \simeq R_\infty,$$

where R_∞ is the inverse limit with respect to $R_0 = R$ and $R_{n+1} = R_n \to R_n$.

The solution of the continuation domain equations in Thm. 4.8 is obtained by the inverse limit technique, considering:

$$
\begin{aligned}
C_0 &= \{\bot\} \\
D_n &= [C_n \rightarrow R] \\
C_{n+1} &= D_n \times C_n
\end{aligned}
$$

so that in particular $D_0 = [C_0 \rightarrow R] \simeq R$. Following [13], to describe C and D as filter-domains we stratify the languages $\mathcal{L}_A$ according to the rank function:

$$
\begin{aligned}
rk(v) = rk(w) &= 0 \\
rk(\sigma \times \tau) = rk(\sigma \wedge \tau) &= \max\{rk(\sigma), rk(\tau)\} \\
rk(\sigma \rightarrow \tau) &= \max\{rk(\sigma), rk(\tau)\} + 1
\end{aligned}
$$

Let $\mathcal{L}_{A_n} = \{\sigma \in \mathcal{L}_A \mid rk(\sigma) \leq n\}$, then $\leq_{A_n}$ is $\leq_A$ restricted to $\mathcal{L}_{A_n}$ and $\mathcal{F}^{A_n}$ is the set of filters over $\mathcal{L}_{A_n}$.

Recall that if $a \in \mathcal{K}(A)$ and $b \in \mathcal{K}(B)$ then the *step function* $(a \Rightarrow b) : A \rightarrow B$ is defined by $(a \Rightarrow b)(x) = b$ if $a \sqsubseteq x$, and $(a \Rightarrow b)(x) = \bot$ otherwise. The function $(a \Rightarrow b)$ is continuous and compact in the space $[A \rightarrow B]$, whose compact elements are all finite sups of step functions.

Definition 4.9 (Type interpretation). The mappings $\Theta_{C_n} : \mathcal{L}_{C_n} \rightarrow \mathcal{K}(C_n)$ and $\Theta_{D_n} : \mathcal{L}_{D_n} \rightarrow \mathcal{K}(D_n)$ are defined through mutual induction by:

$$
\begin{aligned}
\Theta_{C_0}(\kappa) &= \bot \\
\Theta_{D_n}(v) &= (\bot \Rightarrow \Theta_R(v)) = \lambda k \in C_n.\Theta_R(v) \\
\Theta_{D_n}(\kappa \rightarrow \rho) &= (\Theta_{C_n}(\kappa) \Rightarrow \Theta_R(\rho)) \\
\Theta_{C_{n+1}}(\delta \times \kappa) &= \langle \Theta_{D_n}(\delta), \Theta_{C_n}(\kappa) \rangle
\end{aligned}
$$

Finally, for $A_n = C_n, D_n$:

$$
\begin{aligned}
\Theta_{A_n}(\omega) &= \bot \\
\Theta_{A_n}(\sigma \wedge \tau) &= \Theta_{A_n}(\sigma) \sqcup \Theta_{A_n}(\tau)
\end{aligned}
$$

Proposition 4.10. The filter domains $\mathcal{F}^D \simeq D$ and $\mathcal{F}^C \simeq C$ are the initial/final solutions of the continuation equations in **ALG**. The isomorphisms are given in terms of the mappings Θ_C and Θ_D where $\Theta_C(\kappa) = \Theta_{C_{rk(\kappa)}}(\kappa)$, and similarly for Θ_D.

Remark 4.11. Thm. 4.7 and Prop. 4.10 suggest that $\mathcal{F}^D$ in Thm. 4.7 is a λ-model (see [24]). On the other hand, by Thm. 4.8 it is also isomorphic to a $R_\infty \simeq [R_\infty \rightarrow R_\infty]$ model. To see this from the point of view of the intersection type theory, consider the extension $\mathcal{L}_{D'} = \cdots \mid \delta \rightarrow \delta$ of $\mathcal{L}_D$, adding to $\mathcal{T}_D$ the equation $\delta \times \kappa \rightarrow \rho = \delta \rightarrow \kappa \rightarrow \rho$. In the intersection type theory $\mathcal{T}_{D'}$, the following rules are derivable:

$$
\frac{}{(\delta \rightarrow \delta_1) \wedge (\delta \rightarrow \delta_2) \leq_{D'} \delta \rightarrow (\delta_1 \wedge \delta_2)} \qquad \frac{\delta_1' \leq_{D'} \delta_1 \quad \delta_2 \leq_{D'} \delta_2'}{\delta_1 \rightarrow \delta_2 \leq_{D'} \delta_1' \rightarrow \delta_2'}
$$

By this, $\mathcal{T}_{D'}$ is a *natural equated* intersection type theory in terms of [3], and hence $\mathcal{F}^{D'} \simeq [\mathcal{F}^{D'} \rightarrow \mathcal{F}^{D'}]$ (see [3], Cor. 28(4)). By an argument similar to Prop. 4.10 we can show that $\mathcal{F}^{D'} \simeq R_\infty$, so that by the same Prop. and Thm. 4.8, we have $\mathcal{F}^D \simeq \mathcal{F}^{D'}$.

We end this section by defining the interpretation of types in $\mathcal{L}_A$ as subsets of A.

Definition 4.12 (Semantics). For $A \in \{R, C, D\}$ define $\llbracket \cdot \rrbracket^A : \mathcal{L}_A \to \mathcal{P}(A)$:

$$
\begin{aligned}
\llbracket v_a \rrbracket^R &= \uparrow a = \{r \in R \mid a \sqsubseteq r\} & \llbracket \omega \rrbracket^A &= A \\
\llbracket \delta \times \kappa \rrbracket^C &= \llbracket \delta \rrbracket^D \times \llbracket \kappa \rrbracket^C & \llbracket \tau_1 \wedge \tau_2 \rrbracket^A &= \llbracket \tau_1 \rrbracket^A \cap \llbracket \tau_2 \rrbracket^A \\
\llbracket \kappa \to \rho \rrbracket^D &= \{d \in D \mid \forall k \in \llbracket \kappa \rrbracket^C \, [d(k) \in \llbracket \rho \rrbracket^R]\} \\
\llbracket v_a \rrbracket^D &= \llbracket \omega \to v_a \rrbracket^D = \{d \in D \mid \forall k \in C \, [d(k) \in \llbracket v_a \rrbracket^R]\}
\end{aligned}
$$

Theorem 4.13. For $A \in \{R, C, D\}$: *1.* $\llbracket \sigma \rrbracket^A = \uparrow \Theta_A(\sigma)$.
 2. $\forall \sigma, \tau \in \mathcal{L}_A \, [\sigma \leq_A \tau \Longleftrightarrow \llbracket \sigma \rrbracket^A \subseteq \llbracket \tau \rrbracket^A]$.

5 An Intersection Type Assignment System and Filter Model

We now define a type assignment system for pure $\lambda\mu$, following a construction analogous to that of filter-models for the λ-calculus. The central idea is that types completely determine the meaning of terms; this is achieved by tailoring the type assignment to the term interpretation in the filter description of a model, usually called the *filter*-model. In fact, by instantiating D to $\mathcal{F}^D$ and C to $\mathcal{F}^C$, for closed M the interpretation of a type statement $M : \delta$ (which is in general $\llbracket M \rrbracket^D \in \llbracket \delta \rrbracket^D$) becomes $\delta \in \llbracket M \rrbracket^D$, since $\Theta_D(\delta) = \{d \in \mathcal{F}^D \mid \delta \in d\}$. Representing the environment by a pair of sets of assumptions Γ and Δ, the treatment can be extended to open terms.

A set $\Gamma = \{x_1{:}\delta_1, \ldots, x_n{:}\delta_n\}$ is a *basis* or *variable environment* where $\delta_i \in \mathcal{L}_D$ for all $1 \leq i \leq n$, and the x_i are distinct. Similarly, a set $\Delta = \{\alpha_1{:}\kappa_1, \ldots, \alpha_m{:}\kappa_m\}$ is a *name context*, where $\kappa_j \in \mathcal{L}_C$ for all $1 \leq j \leq m$. A *judgement* is an expression of the form $\Gamma \vdash M : \sigma \mid \Delta$. We write $\Gamma, x{:}\delta$ for the set $\Gamma \cup \{x{:}\delta\}$, and $\alpha{:}\kappa, \Delta$ for $\{\alpha{:}\kappa\} \cup \Gamma$.

Definition 5.1 (Intersection type assignment for $\lambda\mu$).

$$
(Ax): \frac{}{\Gamma, x{:}\delta \vdash x : \delta \mid \Delta} \quad
(\times): \frac{\Gamma \vdash M : \delta \mid \alpha{:}\kappa, \Delta}{\Gamma \vdash [\alpha]M : \delta \times \kappa \mid \alpha{:}\kappa, \Delta} \quad
(\mu): \frac{\Gamma \vdash Q : (\kappa' \to \rho) \times \kappa' \mid \alpha{:}\kappa, \Delta}{\Gamma \vdash \mu\alpha.Q : \kappa \to \rho \mid \Delta}
$$

$$
(\to E): \frac{\Gamma \vdash M : \delta \times \kappa \to \rho \mid \Delta \quad \Gamma \vdash N : \delta \mid \Delta}{\Gamma \vdash MN : \kappa \to \rho \mid \Delta} \quad
(\to I): \frac{\Gamma, x{:}\delta \vdash M : \kappa \to \rho \mid \Delta}{\Gamma \vdash \lambda x.M : \delta \times \kappa \to \rho \mid \Delta}
$$

$$
(\wedge): \frac{\Gamma \vdash M : \sigma \mid \Delta \quad \Gamma \vdash M : \tau \mid \Delta}{\Gamma \vdash M : \sigma \wedge \tau \mid \Delta} \quad
(\omega): \frac{}{\Gamma \vdash M : \omega \mid \Delta} \quad
(\leq): \frac{\Gamma \vdash M : \sigma \mid \Delta}{\Gamma \vdash M : \tau \mid \Delta} \, (\sigma \leq \tau)
$$

where $\delta \in \mathcal{L}_D$, $\kappa, \kappa' \in \mathcal{L}_C$ and $\rho \in \mathcal{L}_R$.

To understand the above rules, we read them backward from the conclusion to the premises. To conclude that MN is a function mapping a continuation with type κ to a value of type ρ, rule $(\to E)$ requires that the continuation fed to M has type $\delta \times \kappa$, where δ is the type of N. This mimics the storage of (the denotation of) N into the continuation fed to MN, which is treated as a stack. Rule $(\to I)$ is just the rule symmetric to $(\to E)$.

Note that the (perhaps unusual) rules $(\to E)$ and $(\to I)$ are instances of the usual ones of the simply typed λ-calculus, observing that $\delta \times \kappa \to \rho \in \mathcal{L}_D$ is equivalent to $\delta \to (\kappa \to \rho) \in \mathcal{L}_{D'}$ so that, admitting types in $\mathcal{L}_{D'}$, the following rules are admissible:

$$(\to E'): \frac{\Gamma \vdash M : \delta \to (\kappa \to \rho) \mid \Delta \quad \Gamma \vdash N : \delta \mid \Delta}{\Gamma \vdash MN : \kappa \to \rho \mid \Delta} \qquad (\to I'): \frac{\Gamma, x : \delta \vdash M : \kappa \to \rho \mid \Delta}{\Gamma \vdash \lambda x.M : \delta \to (\kappa \to \rho) \mid \Delta}$$

In rule $(\times)$, we conclude that the command $[\alpha]M$ has type $\delta \times \kappa$ because $⟦[\alpha]M⟧^C e = \langle ⟦M⟧^D e, e\,\alpha \rangle$, so M must have type δ and the continuation type κ has to be the same as that given to α in the name context.

Rule (μ) expresses (a part of) the concept of swapping the current continuation by stating that $\mu\alpha.Q$ is able to produce a result of type ρ when applied to a continuation of type κ; type ρ is of the result obtained from Q and the (in general different) continuation of type κ'. Nonetheless, κ cannot be ignored, and indeed it contributes to the typing of the command Q since it appears in the name environment; this corresponds to the fact that in the semantics $⟦\mu\alpha.Q⟧^D e\,k = d\,k'$ where $\langle d, k' \rangle = ⟦Q⟧^C e[\alpha := k]$, *i.e.* Q is evaluated in the modified environment $e[\alpha := k]$. We clarify the interaction between $(\times)$ and (μ) in the next example, where this is compared to typing in Parigot's system.

Example 5.2. Consider the following derivations in, respectively, Parigot's system and ours.

$$\frac{\dfrac{\overline{\qquad\qquad\qquad}}{\Gamma \vdash M : A \mid \beta{:}A, \alpha{:}B, \Delta}}{\dfrac{\Gamma \vdash [\beta]M : \perp \mid \beta{:}A, \alpha{:}B, \Delta}{\Gamma \vdash \mu\alpha.[\beta]M : B \mid \beta{:}A, \Delta} \, (\mu)} \, (\perp) \qquad\qquad \frac{\dfrac{\overline{\qquad\qquad\qquad}}{\Gamma \vdash M : \kappa \to \rho \mid \beta{:}\kappa, \alpha{:}\kappa', \Delta}}{\dfrac{\Gamma \vdash [\beta]M : (\kappa \to \rho) \times \kappa \mid \beta{:}\kappa, \alpha{:}\kappa', \Delta}{\Gamma \vdash \mu\alpha.[\beta]M : \kappa' \to \rho \mid \beta{:}\kappa, \Delta} \, (\mu)} \, (\times)$$

Both derivations express that $\mu\alpha.[\beta]M$ constitutes a *context switch*; in our case the computational side of the switch is (very elegantly) made apparent: before the switch, the type $\kappa \to \rho$ expresses that M is a term working with the input-stream κ of terms, and returns ρ; after the context switch, the type $\kappa' \to \rho$ for $\mu\alpha.[\beta]M$ now expresses that the input-stream κ' is taken instead.

The following lemma provides a characterisation of derivability in our system.

Lemma 5.3 (Generation lemma). If $\Gamma \vdash M : \delta \mid \Delta$, then either δ is an intersection, or ω, or:

$$\begin{aligned}
\Gamma, x{:}\delta' \vdash x : \delta \mid \Delta &\iff \delta' \leq \delta \\
\Gamma \vdash \lambda x.M : \delta \times \kappa \to \rho \mid \Delta &\iff \Gamma, x{:}\delta \vdash M : \kappa \to \rho \mid \Delta \\
\Gamma \vdash MN : \kappa \to \rho \mid \Delta &\iff \exists \delta [\Gamma \vdash M : \delta \times \kappa \to \rho \mid \Delta \ \& \ \Gamma \vdash N : \delta \mid \Delta] \\
\Gamma \vdash \mu\alpha.Q : \kappa \to \rho \mid \Delta &\iff \exists \kappa' [\Gamma \vdash Q : (\kappa' \to \rho) \times \kappa' \mid \alpha{:}\kappa, \Delta]. \\
\Gamma \vdash [\alpha]M : \delta \times \kappa \mid \Delta &\iff \alpha{:}\kappa \in \Delta \ \& \ \Gamma \vdash M : \delta \mid \Delta.
\end{aligned}$$

We will now show that our notion of type assignment is closed under conversion. The proofs of the properties of subject expansion and subject reduction are relatively standard; the main theorems follow from the relative substitution lemmas.

Lemma 5.4 (Term substitution lemma). $\Gamma \vdash M[L/x] : \sigma \mid \Delta$ if and only if there exists δ such that $\Gamma, x{:}\delta \vdash M : \sigma \mid \Delta$ and $\Gamma \vdash L : \delta \mid \Delta$.

A similar lemma for structural substitution is not as easily formulated. This is mainly due to the fact that the μ-bound variable gets 're-used' in the result of the substitution, but with a different type, as can be observed from the following example.

Example 5.5. First observe that $(\mu\alpha.[\beta]\mu\gamma.[\alpha]x)N$ reduces to $(\mu\alpha.[\beta]\mu\gamma.[\alpha]xN)$. We can type the first term in our system as follows:

$$x{:}\delta\times\kappa\to\rho \vdash x : \delta\times\kappa\to\rho \mid \alpha{:}\delta\times\kappa, \beta{:}\kappa_1, \gamma{:}\kappa_1$$
$$x{:}\delta\times\kappa\to\rho \vdash [\alpha]x : (\delta\times\kappa\to\rho)\times\delta\times\kappa \mid \alpha{:}\delta\times\kappa, \beta{:}\kappa_1, \gamma{:}\kappa_1$$
$$x{:}\delta\times\kappa\to\rho \vdash \mu\gamma.[\alpha]x : \kappa_1\to\rho \mid \alpha{:}\delta\times\kappa, \beta{:}\kappa_1$$
$$x{:}\delta\times\kappa\to\rho \vdash [\beta]\mu\gamma.[\alpha]x : (\kappa_1\to\rho)\times\kappa_1 \mid \alpha{:}\delta\times\kappa, \beta{:}\kappa_1$$
$$x{:}\delta\times\kappa\to\rho \vdash \mu\alpha.[\beta]\mu\gamma.[\alpha]x : \delta\times\kappa\to\rho \mid \beta{:}\kappa_1 \qquad \boxed{} \quad \Gamma \vdash N{:}\delta \mid \Delta$$
$$\Gamma, x{:}\delta\times\kappa\to\rho \vdash (\mu\alpha.[\beta]\mu\gamma.[\alpha]x)N : \kappa\to\rho \mid \beta{:}\kappa_1, \Delta$$

Using the information of this derivation, the derivation for $(\mu\alpha.[\beta]\mu\gamma.[\alpha]xN)$ implicitly constructed by the reduction is:

$$x{:}\delta\times\kappa\to\rho \vdash x : \delta\times\kappa\to\rho \mid \alpha{:}\kappa, \beta{:}\kappa_1, \gamma{:}\kappa_1 \qquad \boxed{} \quad \Gamma \vdash N{:}\delta \mid \Delta$$
$$x{:}\delta\times\kappa\to\rho \vdash xN : \kappa\to\rho \mid \alpha{:}\kappa, \beta{:}\kappa_1, \gamma{:}\kappa_1$$
$$x{:}\delta\times\kappa\to\rho \vdash [\alpha]xN : (\kappa\to\rho)\times\kappa \mid \alpha{:}\kappa, \beta{:}\kappa_1, \gamma{:}\kappa_1$$
$$x{:}\delta\times\kappa\to\rho \vdash \mu\gamma.[\alpha]xN : \kappa_1\to\rho \mid \alpha{:}\kappa, \beta{:}\kappa_1$$
$$x{:}\delta\times\kappa\to\rho \vdash [\beta]\mu\gamma.[\alpha]xN : (\kappa_1\to\rho)\times\kappa_1 \mid \alpha{:}\kappa, \beta{:}\kappa_1$$
$$x{:}\delta\times\kappa\to\rho \vdash \mu\alpha.[\beta]\mu\gamma.[\alpha]xN : \kappa\to\rho \mid \beta{:}\kappa_1$$
$$\Gamma, x{:}\delta\times\kappa\to\rho \vdash (\mu\alpha.[\beta]\mu\gamma.[\alpha]xN) : \kappa\to\rho \mid \beta{:}\kappa_1, \Delta$$

Notice that here the type for α has changed from $\delta\times\kappa$ to κ.

Lemma 5.6 (Structural substitution lemma). $\Gamma \vdash M[\alpha \Leftarrow L]{:}\sigma \mid \alpha{:}\kappa, \Delta$ if and only if there exists δ such that $\Gamma \vdash L{:}\delta \mid \Delta$, and $\Gamma \vdash M{:}\sigma \mid \alpha{:}\delta\times\kappa, \Delta$.

Using these two substitution results, the following theorems are easy to show.

Theorem 5.7 (Subject expansion). If $M \to N$, and $\Gamma \vdash N{:}\delta \mid \Delta$, then $\Gamma \vdash M{:}\delta \mid \Delta$.

Theorem 5.8 (Subject reduction). If $M \to N$, and $\Gamma \vdash M{:}\delta \mid \Delta$, then $\Gamma \vdash N{:}\delta \mid \Delta$

We define satisfaction in a $\lambda\mu$-model (R, D, C) and validity in the standard way.

Definition 5.9 (Satisfaction and Validity). Let $\mathcal{M} = (R, D, C)$. We define:

$$e \models_{\mathcal{M}} \Gamma, \Delta \quad\Longleftrightarrow\quad \forall x{:}\delta \in \Gamma\,[e(x) \in \llbracket\delta\rrbracket^D]\ \&\ \forall \alpha{:}\kappa \in \Delta\,[e(\alpha) \in \llbracket\kappa\rrbracket^C]$$
$$\Gamma \models_{\mathcal{M}} M : \delta \mid \Delta \Longleftrightarrow \forall e \in \mathsf{Env}\,[e \models_{\mathcal{M}} \Gamma, \Delta \Rightarrow \llbracket M\rrbracket\, e \in \llbracket\delta\rrbracket^D]$$
$$\Gamma \models M{:}\delta \mid \Delta \quad\Longleftrightarrow\quad \forall \mathcal{M}\,[\Gamma \models_{\mathcal{M}} M : \delta \mid \Delta]$$

We can now show soundness and completeness for our notion of intersection type assignment with respect to the filter semantics:

Theorem 5.10 (Soundness). $\Gamma \vdash M : \delta \mid \Delta \Rightarrow \Gamma \models M : \delta \mid \Delta$.

Lemma 5.11 (Filter model). Let $A \in \{D, C\}$. Given an environment $e \in (Var \to \mathcal{F}^D) +$ $(Name \to \mathcal{F}^C)$, we have

$$\llbracket M \rrbracket^{\mathcal{F}^A} e = \{\sigma \in \mathcal{L}_A \mid \exists \Gamma, \Delta \, [e \models \Gamma, \Delta \ \& \ \Gamma \vdash M : \sigma \mid \Delta]\}$$

Theorem 5.12 (Completeness). $\Gamma \models M : \delta \mid \Delta \Rightarrow \Gamma \vdash M : \delta \mid \Delta$.

6 Type Preservation from Parigot's Type System to the Intersection Type System

We conclude this paper by showing that typed $\lambda\mu$-terms have a non-trivial intersection typing in our system, i.e. they can be assigned a type different from ω.

Example 6.1. In Parigot's system for $\lambda\mu$, we can derive $x{:}A \vdash_P \mu\alpha.[\beta]\mu\gamma.[\alpha]x : A \mid \beta{:}C$; We can type the same pure $\lambda\mu$-term in our system as well (where $\Gamma = x{:}\delta \times \kappa \to \rho$):

$$\dfrac{\dfrac{\dfrac{\dfrac{\dfrac{}{x{:}A \vdash x : A \mid \alpha{:}A, \beta{:}C, \gamma{:}C} \, (Ax)}{x{:}A \vdash [\alpha]x : \bot \mid \alpha{:}A, \beta{:}C, \gamma{:}C} \, (\bot)}{x{:}A \vdash \mu\gamma.[\alpha]x : C \mid \alpha{:}A, \beta{:}C} \, (\mu)}{x{:}A \vdash [\beta]\mu\gamma.[\alpha]x : \bot \mid \alpha{:}A, \beta{:}C} \, (\bot)}{x{:}A \vdash \mu\alpha.[\beta]\mu\gamma.[\alpha]x : A \mid \beta{:}C} \, (\mu)$$

$$\dfrac{\dfrac{\dfrac{\dfrac{\dfrac{}{\Gamma \vdash x : \delta \times \kappa \to \rho \mid \alpha{:}\delta \times \kappa, \beta{:}\kappa_1, \gamma{:}\kappa_1} \, (Ax)}{\Gamma \vdash [\alpha]x : (\delta \times \kappa \to \rho) \times \delta \times \kappa \mid \alpha{:}\delta \times \kappa, \beta{:}\kappa_1, \gamma{:}\kappa_1} \, (\times)}{\Gamma \vdash \mu\gamma.[\alpha]x : \kappa_1 \to \rho \mid \alpha{:}\delta \times \kappa, \beta{:}\kappa_1} \, (\mu)}{\Gamma \vdash [\beta]\mu\gamma.[\alpha]x : (\kappa_1 \to \rho) \times \kappa_1 \mid \alpha{:}\delta \times \kappa, \beta{:}\kappa_1} \, (\times)}{\Gamma \vdash \mu\alpha.[\beta]\mu\gamma.[\alpha]x : \delta \times \kappa \to \rho \mid \beta{:}\kappa_1} \, (\mu)$$

We can also type terms that are not typeable in Parigot's system:

$$\dfrac{\dfrac{\dfrac{\dfrac{\dfrac{\dfrac{\dfrac{}{x{:}\kappa \to \rho \vdash x : \kappa \to \rho \mid \beta{:}\kappa', \gamma{:}\kappa \wedge ((\kappa \to \rho) \times \kappa')} \, (Ax)}{x{:}\kappa \to \rho \vdash [\gamma]x : (\kappa \to \rho) \times (\kappa \wedge ((\kappa \to \rho) \times \kappa')) \mid \beta{:}\kappa', \gamma{:}\kappa \wedge ((\kappa \to \rho) \times \kappa')} \, (\times)}{x{:}\kappa \to \rho \vdash [\gamma]x : (\kappa \to \rho) \times \kappa \mid \beta{:}\kappa', \gamma{:}\kappa \wedge ((\kappa \to \rho) \times \kappa')} \, (\leq)}{x{:}\kappa \to \rho \vdash \mu\beta.[\gamma]x : \kappa' \to \rho \mid \gamma{:}\kappa \wedge ((\kappa \to \rho) \times \kappa')} \, (\mu)}{\vdash \lambda x.\mu\beta.[\gamma]x : (\kappa \to \rho) \times \kappa' \to \rho \mid \gamma{:}\kappa \wedge ((\kappa \to \rho) \times \kappa')} \, (\to I)}{\vdash [\gamma](\lambda x.\mu\beta.[\gamma]x) : ((\kappa \to \rho) \times \kappa' \to \rho) \times (\kappa \wedge ((\kappa \to \rho) \times \kappa')) \mid \gamma{:}\kappa \wedge ((\kappa \to \rho) \times \kappa')} \, (\times)}{\vdash [\gamma](\lambda x.\mu\beta.[\gamma]x) : ((\kappa \to \rho) \times \kappa' \to \rho) \times ((\kappa \to \rho) \times \kappa') \mid \gamma{:}\kappa \wedge ((\kappa \to \rho) \times \kappa')} \, (\leq)}{\vdash \mu\gamma.[\gamma](\lambda x.\mu\beta.[\gamma]x) : (\kappa \wedge ((\kappa \to \rho) \times \kappa')) \to \rho \mid} \, (\mu)$$

Notice that $\mu\gamma.[\gamma](\lambda x.\mu\beta.[\gamma]x)$ is not typeable in Parigot's system, since the two occurrences of $[\gamma]$ need to be typed differently, with non-unifiable types.

We can strengthen the first observation and show that we can faithfully embed Parigot's system into ours; to this purpose, we define first an interpretation of Parigot's types into our intersection types.

Definition 6.2. We change the definition of types for Parigot's system slightly by using

$$A, B ::= \varphi \mid \bot_A \mid A \to B \quad (A \neq \bot_C)$$

and change the rule $(\bot)$ into a rule that registers to what type the contradiction was established:

$$(\bot_B) : \frac{\Gamma \vdash M : B \mid \beta{:}B, \Delta}{\Gamma \vdash [\beta]M : \bot_B \mid \beta{:}B, \Delta}$$

Taking an arbitrary $a \neq \bot \in R$, we define $\overline{A}$ and $\underline{A}$ simultaneously through:

$$\overline{\varphi} = (v_a \times \omega) {\rightarrow} v_a \; \text{for all } \varphi \qquad\qquad \underline{\varphi} = v_a \times \omega \qquad \text{for all } \varphi$$

$$\overline{\bot_A} = (\underline{A} {\rightarrow} v_a) \times \underline{A} \qquad\qquad \underline{A {\rightarrow} B} = (\underline{A} {\rightarrow} v_a) \times \underline{B}$$

$$\overline{A {\rightarrow} B} = \overline{A} \times \kappa {\rightarrow} v_a \qquad \text{where } \overline{B} = \kappa {\rightarrow} v_a$$

We also define $\overline{\Gamma} = \{ x{:}\overline{A} \mid x{:}A \in \Gamma \}$ and $\underline{\Delta} = \{ \alpha{:}\underline{A} \mid \alpha{:}A \in \Delta \}$.

We can now show:

Theorem 6.3 (Type preservation). If $\Gamma \vdash_\mathsf{P} M : A \mid \Delta$, then $\overline{\Gamma} \vdash M : \overline{A} \mid \underline{\Delta}$

Example 6.4. To illustrate this result, consider the derivation in $\vdash_\mathsf{P}$ on the left, which gets translated into the one on the right (where $\Gamma = x{:}(v_a \times \omega) {\rightarrow} v_a$):

$$\frac{\dfrac{}{x{:}\varphi \vdash x : \varphi \mid \alpha{:}\varphi, \beta{:}\varphi'}}{\dfrac{x{:}\varphi \vdash [\alpha]x : \bot_\varphi \mid \alpha{:}\varphi, \beta{:}\varphi'}{\dfrac{x{:}\varphi \vdash \mu\beta.[\alpha]x : \varphi' \mid \alpha{:}\varphi}{\vdash \lambda x.\mu\beta.[\alpha]x : \varphi {\rightarrow} \varphi' \mid \alpha{:}\varphi}}}$$

$$\frac{\dfrac{}{\Gamma \vdash x : (v_a \times \omega) {\rightarrow} v_a \mid \alpha{:}v_a \times \omega, \beta{:}v_a \times \omega}}{\dfrac{\Gamma \vdash [\alpha]x : ((v_a \times \omega) {\rightarrow} v_a) \times (v_a \times \omega) \mid \alpha{:}v_a \times \omega, \beta{:}v_a \times \omega}{\dfrac{\Gamma \vdash \mu\beta.[\alpha]x : (v_a \times \omega) {\rightarrow} v_a \mid \alpha{:}v_a \times \omega}{\vdash \lambda x.\mu\beta.[\alpha]x : ((v_a \times \omega) {\rightarrow} v_a) \times (v_a \times \omega) {\rightarrow} v \mid \alpha{:}v_a \times \omega}}}$$

7 Conclusions and Future Work

We have presented a filter model for the $\lambda\mu$-calculus which is an instance of Streicher and Reus's continuation model, and a type assignment system such that the set of types that can be given to a term coincides with its denotation in the model. The type theory and the assignment system can be viewed as the logic for reasoning about the computational meaning of $\lambda\mu$-terms, much as it is the case for λ-calculus. We expect that significant properties of pure $\lambda\mu$ can be characterised via their typing, and we see the characterisation of the strongly normalisable pure terms as the first challenge.

We have also shown that $\lambda\mu$-terms which are typeable in Parigot's first order type system have non-trivial typing in our system in a strong sense. This opens the possibility of a new proof of strong normalisation for typed $\lambda\mu$ and, by using a variant of the system, of de Groote's typeable $\Lambda\mu$-terms.

The investigation of other significant properties of the calculi should also be possible with the same tools, like confluence, standardisation, solvable terms, etc. More significantly, for the $\lambda\mu$-calculus we are interested in the use of the type system for interpreting relevant combinators, like in the case of de Groote's encoding of Felleisen's $\mathcal{C}$ operator, whose types are essentially those of (the η-expansion of) identity, and, in general, for investigating the computational behaviour of combinators representing proofs of non-constructive principles. To this aim a study of the principal typing with respect to the present system would be of great help.

References

1. Abramsky, S.: Domain Theory in Logical Form. APAL 51, 1–77 (1991)
2. Alessi, F., Barbanera, F., Dezani-Ciancaglini, M.: Intersection types and lambda models. TCS 355(2), 108–126 (2006)
3. Alessi, F., Severi, P.: Recursive Domain Equations of Filter Models. In: Geffert, V., Karhumäki, J., Bertoni, A., Preneel, B., Návrat, P., Bieliková, M. (eds.) SOFSEM 2008. LNCS, vol. 4910, pp. 124–135. Springer, Heidelberg (2008)
4. van Bakel, S.: Completeness and Partial Soundness Results for Intersection & Union Typing for $\overline{\lambda}\mu\tilde{\mu}$. APAL 161, 1400–1430 (2010)
5. van Bakel, S.: Sound and Complete Typing for $\lambda\mu$. In: ITRS 2010. EPTCS, vol. 45, pp. 31–44 (2010)
6. Barbanera, F., Dezani-Ciancaglini, M.: de'Liguoro, U. Intersection and Union Types: Syntax and Semantics. I&C 119(2), 202–230 (1995)
7. Barendregt, H.: The Lambda Calculus: its Syntax and Semantics. North-Holland, Amsterdam (1984)
8. Barendregt, H., Coppo, M., Dezani-Ciancaglini, M.: A filter lambda model and the completeness of type assignment. JSL 48(4), 931–940 (1983)
9. Bierman, G.M.: A Computational Interpretation of the $\lambda\mu$-calculus. In: Brim, L., Gruska, J., Zlatuška, J. (eds.) MFCS 1998. LNCS, vol. 1450, pp. 336–345. Springer, Heidelberg (1998)
10. Coppo, M., Dezani-Ciancaglini, M.: An Extension of the Basic Functionality Theory for the λ-Calculus. NDjFL 21(4), 685–693 (1980)
11. Curien, P.-L., Herbelin, H.: The Duality of Computation. In: ICFP 2000, ACM Notes 35.9, pp. 233–243 (2000)
12. David, R., Py, W.: $\lambda\mu$-Calculus and Böhm's Theorem. JSL 66(1), 407–413 (2001)
13. Dezani-Ciancaglini, M., Honsell, F., Alessi, F.: A complete characterization of complete intersection-type preorders. ACM Trans. Comput. Log. 4(1), 120–147 (2003)
14. Dougherty, D., Ghilezan, S., Lescanne, P.: Intersection and Union Types in the $\overline{\lambda}\mu\tilde{\mu}$-calculus. In: ITRS 2004. ENTCS, vol. 136, pp. 228–246 (2004)
15. Dougherty, D., Ghilezan, S., Lescanne, P.: Characterizing strong normalization in the Curien-Herbelin symmetric lambda calculus: extending the Coppo-Dezani heritage. TCS 398 (2008)
16. Gentzen, G.: Investigations into logical deduction. In: Szabo, M.E. (ed.) The Collected Papers of Gerhard Gentzen, p. 68. North Holland, Amsterdam (1969) (1935)
17. Gierz, G., Hofmann, K.H., Keimel, K., Lawson, J.D., Mislove, M., Scott, D.S.: Continuous Lattices And Domains. Cambridge University Press, Cambridge (2003)
18. de Groote, P.: On the relation between the $\lambda\mu$-calculus and the syntactic theory of sequential control. In: Pfenning, F. (ed.) LPAR 1994. LNCS, vol. 822, pp. 31–43. Springer, Heidelberg (1994)
19. Herbelin, H., Saurin, A.: $\lambda\mu$-calculus and $\Lambda\mu$-calculus: a Capital Difference (2010) (manuscript)
20. Hofmann, M., Streicher, T.: Continuation models are universal for $\lambda\mu$-calculus. In: LICS 1997, pp. 387–397 (1997)
21. Hofmann, M., Streicher, T.: Completeness of continuation models for lambda-mu-calculus. I&C 179(2), 332–355 (2002)
22. Krivine, J.-L.: Lambda calculus, types and models. Ellis Horwood, England (1993)
23. Lafont, Y., Reus, B., Streicher, T.: Continuation Semantics or Expressing Implication by Negation. Report 9321, Universität München (1993)
24. Meyer, A.R.: What is a Model of the Lambda Calculus? Inf. Contr. 52(1), 87–122 (1982)
25. Ong, C.-H.L., Stewart, C.A.: A Curry-Howard foundation for functional computation with control. In: POPL 1997, pp. 215–227 (1997)

26. Ong, C.-H.L.: A Semantic View of Classical Proofs: Type-Theoretic, Categorical, and De-notational Characterizations. In: LICS 1996, pp. 230–241 (1996)
27. Parigot, M.: An algorithmic interpretation of classical natural deduction. In: Voronkov, A. (ed.) LPAR 1992. LNCS, vol. 624, pp. 190–201. Springer, Heidelberg (1992)
28. Py, W.: Confluence en $\lambda\mu$-calcul. PhD thesis, Universite de Savoie (1998)
29. Saurin, A.: On the relations between the syntactic theories of $\lambda\mu$-calculi. In: Kaminski, M., Martini, S. (eds.) CSL 2008. LNCS, vol. 5213, pp. 154–168. Springer, Heidelberg (2008)
30. Streicher, T., Reus, B.: Classical logic: Continuation Semantics and Abstract Machines. JFP 11(6), 543–572 (1998)

Approximation Semantics and Expressive Predicate Assignment for Object-Oriented Programming
(Extended Abstract)

Reuben Rowe and Steffen Van Bakel

Department of Computing, Imperial College London,
180 Queen's Gate, London SW7 2BZ, UK
{r.rowe,s.vanbakel}@imperial.ac.uk

Abstract. We consider a semantics for a class-based object-oriented calculus based upon *approximation*; since in the context of LC such a semantics enjoys a strong correspondence with *intersection type assignment systems*, we also define such a system for our calculus and show that it is *sound* and *complete*. We establish the link with between type (we use the terminology *predicate* here) assignment and the approximation semantics by showing an approximation result, which leads to a sufficient condition for head-normalisation and termination.

We show the expressivity of our predicate system by defining an encoding of Combinatory Logic (and so also LC) into our calculus. We show that this encoding preserves predicate-ability and also that our system characterises the normalising and strongly normalising terms for this encoding, demonstrating that the great analytic capabilities of these predicates can be applied to OO.

1   Introduction

Semantics is a well established area of research for both functional and imperative languages; for the functional programming language side, semantics is mainly *denotational*, based on Scott's domain theory [25], whereas for imperative languages it is mainly *operational* [24]. In this paper, we present the first results of our research in the direction of denotational, type-based semantics for object-oriented (OO) calculi, which we aim to extend towards semantics-based systems of abstract interpretation.

Over the years many expressive type systems have been defined and investigated. Amongst those, the *intersection type discipline* (ITD) [14,15,11,2] stands out as a system that is closed under β-equality and gives rise to a filter model; it is defined as an extension of Curry's basic type system for the Lambda Calculus (LC) [10], by allowing term-variables to have many, potentially non-unifiable types. This generalisation leads to a very expressive system: for example, termination (*i.e.* normalisation) of terms can be characterised by assignable types. Furthermore, intersection type-based models and approximation results show that intersection types describe the full semantical behaviour of typeable terms. Intersection type systems have also been employed successfully in analyses for dead code elimination [17], strictness analysis [20], and control-flow analysis [9], proving them a versatile framework for reasoning about programs. Inspired by this expressive power, investigations have taken place of the suitability of intersection

L. Ong (Ed.): TLCA 2011, LNCS 6690, pp. 229–244, 2011.
© Springer-Verlag Berlin Heidelberg 2011

type assignment for other computational models: for example, van Bakel and Fernández have studied [6,7] intersection types in the context of Term Rewriting Systems (TRS) and van Bakel studied them in the context of sequent calculi [4].

Also the *object-oriented* programming paradigm has been the subject of extensive theoretical study over the last two decades. OO languages come in two broad flavours: the *object* (or prototype) based, and the *class* based. A number of formal models has been developed [13,12,21,18,1,19]; for example, the ς-calculus [1] and Featherweight Java (FJ) [19] give elementary models for object based and class-based OO respectively. In an attempt to bring intersection types to the context of OO, in [5] van Bakel and de'Liguoro presented a system for the ς-calculus; it sees assignable types as an *execution predicate*, or *applicability predicate*, rather than as a functional characterisation as is the view in the context of LC and, as a result, recursive calls are typed individually, with different types. This is also the case in our system.

In the current paper we aim to define type-based semantics for class-based OO, so introduce a notion of intersection type assignment for such languages (we will use the terminology *predicates* here, to distinguish our notion of types from the traditional notion of class types). In order to be able to concentrate on the essential difficulties, we focus on Featherweight Java [19], a restriction of Java defined by removing all but the most essential features of the full language; Featherweight Java bears a similar relation to Java as LC does to languages such as ML and Haskell; in fact, we will show it to be Turing complete. We will show that the expected properties of a system based on intersection predicates (*i.e. soundness* and *completeness*) hold, opening up the possibility to define a predicate-based semantics for FJ. In future work, we will look at adding the normal programming features, and investigate which of the main properties we show in this paper are still achievable.

We also define a notion of *approximant* for FJ-programs as a finite, rooted segment – that cannot be reduced – of a [head] normal form; we go on to show an *approximation result* which states that, for every predicate assignable to a term in our system, an approximant of that term exists which can be assigned the same predicate. Interpreting a term by its set of approximants gives an *approximation semantics* and the approximation result then relates the approximation and the predicate-based semantics. This has, as far as we are aware, not previously been shown for a model of OO. The approximation result allows for a predicate-based analysis of termination.

As is also the case for LC and TRS, in our system this result is shown using a notion of computability; since the notion of reduction we consider is *weak*, as in [7] to show the approximation result we need to consider a notion of reduction on predicate derivations. We illustrate the expressive power of our calculus by showing that it is Turing complete through an embedding of Combinatory Logic – and thereby also the embedding of LC. We also recall the notion of Curry type assignment, for which we can easily show a principal predicate property and show a predicate preservation result: types assignable to λ-terms in Curry's system of simple type assignment correspond to predicates in our system that can be assigned to the interpreted λ-terms. This is easily extended to the strict intersection type assignment system for LC [2]; this then implies that the collection of predicate-able OO expressions correspond to the λ-terms that are typeable using intersection types, *i.e.* all semantically meaningful terms.

In [8] we presented a similar system which here has been simplified. In particular, we have removed the *field update* feature (which can be modelled using method calls[1]), which gives a more straightforward presentation of system and proofs. We have decoupled our intersection predicate system from the existing class type system, which shows that the approximation result does not depend on the class type system in any way.

For lack of space, proofs are omitted from this paper; we refer the interested reader to http://www.doc.ic.ac.uk/~rnr07 for a version of this paper with detailed proofs.

2   The Calculus FJ^{ξ}

In this section, we will define our variant of Featherweight Java. It defines *classes*, which represent abstractions encapsulating both data (stored in *fields*) and the operations to be performed on that data (encoded as *methods*). Sharing of behaviour is accomplished through the *inheritance* of fields and methods from parent classes. Computation is mediated by *instances* of these classes (called *objects*), which interact with one another by *calling* (or *invoking*) methods and accessing each other's (or their own) fields. We have removed cast expressions since, as the authors of [19] themselves point out, the presence of *downcasts* is unsound[2]; for this reason we call our calculus FJ^{ξ}. We also leave the constructor method as implicit.

Before defining the calculus itself, we introduce notation to represent and manipulate *sequences* of entities which we will use in this paper.

Definition 1 (Sequence Notation). We use $\bar{n}$ ($n \in \mathbb{N}$) to represent the list $1, \ldots, n$. A sequence $a_1, \ldots, a_n$ is denoted by $\vec{a_n}$; the subscript can be omitted when the exact number of elements in the sequence is not relevant. We write $a \in \vec{a_n}$ whenever there exists some $i \in \{1, \ldots, n\}$ such that $a = a_i$. The empty sequence is denoted by ε, and concatenation on sequences by $\vec{a} \cdot \vec{a'}$.

We use familiar meta-variables in our formulation to range over class names (C and D), field names (f), method names (m) and variables (x). We distinguish the class name Object (which denotes the root of the class inheritance hierarchy in all programs) and the variable this (which is used to refer to the receiver object in method bodies).

Definition 2 (FJ^{ξ} Syntax). FJ^{ξ} programs P consist of a *class table* $\mathcal{CT}$, comprising the *class declarations*, and an *expression* e to be run (corresponding to the body of the main method in a real Java program). They are defined by:

$$
\begin{array}{lll}
\text{e} & ::= & x \mid \text{new } \mathrm{C}(\vec{\text{e}}) \mid \text{e}.f \mid \text{e}.m(\vec{\text{e}}) \\
\text{fd} & ::= & \mathrm{C}f; \\
\text{md} & ::= & \mathrm{D}\ m(\mathrm{C}_1\ x_1,\ \ldots,\ \mathrm{C}_n\ x_n)\ \{\ \text{return } \text{e};\ \} \\
\text{cd} & ::= & \text{class } \mathrm{C} \text{ extends } \mathrm{C}'\ \{\ \overrightarrow{\text{fd}}\ \overrightarrow{\text{md}}\ \} & (\mathrm{C} \neq \text{Object}) \\
\mathcal{CT} & ::= & \overrightarrow{\text{cd}} \\
P & ::= & (\mathcal{CT}, \text{e})
\end{array}
$$

---

[1] We can simulate field update by adding to every class C, for each field f_i belonging to the class, a method C update_$f_i(x)$ { return new C(this.$f_1, \ldots, x, \ldots,$ this.f_n); }.

[2] In the sense that typeable expressions can get stuck at runtime.

From this point, all the concepts defined are program dependent (parametric on the class table); however, since a program is essentially a fixed entity, it will be left as an implicit parameter in the definitions that follow. This is done in the interests of readability, and is a standard simplification in the literature (*e.g.* [19]). Here, we also point out that we only consider programs which conform to some sensible well-formedness criteria: no cycles in the inheritance hierarchy, and fields and methods in any given branch of the inheritance hierarchy are uniquely named. An exception is made to allow the redeclaration of methods, providing that only the *body* of the method differs from the previous declaration (in the parlance of class-based OO, this is called *method override*).

Definition 3 (Lookup Functions).   The following lookup functions are defined to extract the names of fields and bodies of methods belonging to (and inherited by) a class.

1. The function $\mathcal{F}(C)$ returns the list of fields $\overrightarrow{f_n}$ belonging to class C (including those it inherits).
2. The function $\mathcal{M}b(C,m)$ returns a tuple $(\vec{x}, e)$, consisting of a sequence of the method m's (as defined in the class C) formal parameters and its body.

As usual, *substitution* is at the basis of reduction in our calculus: when a method is invoked on an object (the *receiver*) the invocation is replaced by the body of the method that is called, and each of the variables is replaced by a corresponding argument.

Definition 4 (Reduction).   1. A *term substitution* $S = \{x_1 \mapsto e_1, \ldots, x_n \mapsto e_n\}$ is defined in the standard way, as a total function on expressions that systematically replaces all occurrences of the variables x_i by their corresponding expression e_i. We write e^S for $S(e)$.
2. The reduction relation $\rightarrow$ is the smallest relation on expressions satisfying:
   - new C($\overrightarrow{e_n}$).$f_j \rightarrow e_j$, for class name C with $\mathcal{F}(C) = \overrightarrow{f_n}$ and $j \in \overline{n}$.
   - new C($\vec{e}$).$m(\overrightarrow{e'_n}) \rightarrow e^S$, where $S = \{\text{this} \mapsto \text{new } C(\vec{e}),\ x_1 \mapsto e'_1,\ \ldots,\ x_n \mapsto e'_n\}$, for class name C and method m with $\mathcal{M}b(C,m) = (\overrightarrow{x_n}, e)$.

   and the usual congruence rules for allowing reduction in subexpressions.
3. If $e \rightarrow e'$, then e is the *redex* and e' the *contractum*; $\rightarrow^*$ is the reflexive, transitive closure of $\rightarrow$.

This notion of reduction is *confluent*.

3   Approximation Semantics

In this section we define an *approximation semantics* for FJ$^\wp$. The notion of *approximant* was first introduced in [27] for LC. Essentially, an approximant is a partially evaluated expression in which the locations of incomplete evaluation (*i.e.* where reduction *may* still take place) are explicitly marked by the element $\bot$; thus, they *approximate* the result of computations. Intuitively, an approximant can be seen as a 'snapshot' of a computation, where we focus on that part of the resulting program which will no longer change (*i.e.* the observable *output*).

Definition 5 (Approximants).   1. The set of *approximants* FJ$^{\mathscr{C}}$ is defined by the following grammar:

$$
\begin{array}{llll}
\mathsf{a} & ::= & x \mid \bot \mid \mathsf{a}.f \mid \mathsf{a}.m(\overrightarrow{\mathsf{a}_n}) \mid \text{new } C(\overrightarrow{\mathsf{a}_n}) & (n \geq 0) \\
\mathsf{A} & ::= & x \mid \bot \mid \text{new } C(\overrightarrow{\mathsf{A}_n}) & (n \geq 0) \\
& & \mid \mathsf{A}.f \mid \mathsf{A}.m(\overrightarrow{\mathsf{A}}) & (\mathsf{A} \neq \bot, \mathsf{A} \neq \text{new } C(\overrightarrow{\mathsf{A}_n}))
\end{array}
$$

Note that approximate normal forms approximate expressions in (head) normal form. In addition, if we were to extend the notion of reduction so that field accesses and method calls on $\bot$ reduce to $\bot$, then we would find that the approximate normal forms are exactly the normal forms with respect to this extended reduction relation.

The notion of approximation is formalised as follows.

Definition 6 (Approximation Relation).   The *approximation relation* $\sqsubseteq$ is the contextual closure of the smallest preorder on approximants satisfying: $\bot \sqsubseteq \mathsf{a}$, for all a.

The relationship between the approximation relation and reduction is:

Lemma 7. *If* $\mathsf{A} \sqsubseteq \mathsf{e}$ *and* $\mathsf{e} \to^* \mathsf{e}'$, *then* $\mathsf{A} \sqsubseteq \mathsf{e}'$.

Notice that this property expresses that the observable behaviour of a program can only increase (in terms of $\sqsubseteq$) through reduction.

Definition 8 (Approximants).   The set of *approximants* of e is defined as $\mathcal{A}(\mathsf{e}) = \{ \mathsf{A} \mid \exists \mathsf{e}' [\mathsf{e} \to^* \mathsf{e}' \,\&\, \mathsf{A} \sqsubseteq \mathsf{e}'] \}$.

Thus, an approximant (of some expression) is a approximate normal form that approximates some (intermediate) stage of execution. This notion of approximant allows us to define what an approximation model is for FJ$^{\mathscr{C}}$.

Definition 9 (FJ$^{\mathscr{C}}$ Semantics).   An *approximation model* for an FJ$^{\mathscr{C}}$ program is a structure $\langle \wp(\mathbb{A}), [\![\cdot]\!] \rangle$, where the interpretation function $[\![\cdot]\!]$, mapping expressions to elements of the domain, $\wp(\mathbb{A})$, is defined by $[\![\mathsf{e}]\!] = \mathcal{A}(\mathsf{e})$.

As for models of LC, our approximation semantics equates expressions which have the same reduction behaviour, as shown by the following theorem.

Theorem 10. $\mathsf{e} \to^* \mathsf{e}' \Rightarrow \mathcal{A}(\mathsf{e}) = \mathcal{A}(\mathsf{e}')$.

4   Predicate Assignment

We will now define a notion of predicate assignment which is sound and complete with respect to the approximation semantics defined above in the sense that every predicate assignable to an expression is also assignable to an approximant of that expression, and vice versa. Notice that, since in approximants redexes are replaced by $\bot$, this result is not an immediate consequence of subject reduction; we will see that it is the predicate derivation itself which specifies the approximant in question. This relationship is formalised in the next section.

The predicate assignment system defined below uses intersection predicates; it is influenced by the predicate system for the ς-calculus as defined in [5], and can ultimately be seen as based upon the strict intersection type system for LC (see [2] for a survey). Our predicates describe the capabilities of an expression (or rather, the object to which that expression evaluates) in terms of (1) the operations that may be performed on it (*i.e.* accessing a field or invoking a method), and (2) the *outcome* of performing those operations. In this way, our predicates express detailed properties about the contexts in which expressions can be safely used.

More intuitively, our predicates capture the notion of *observational equivalence*: two expressions with the same (non-empty) set of assignable predicates will be observationally indistinguishable. Our predicates thus constitute *semantic predicates*, so for this reason (and also to distinguish them from the already existing Java class types) we do not call them types.

Definition 11 (Predicates). The set of *predicates* (ranged over by ϕ, ψ) and its subset of *strict* predicates (ranged over by σ) are defined by the following grammar (where φ ranges over *predicate variables*, and as for syntax C ranges over class names):

$$\phi, \psi \ ::= \ \omega \mid \sigma \mid \phi \cap \psi$$
$$\sigma \ ::= \ \varphi \mid C \mid \langle f : \sigma \rangle \mid \langle m : (\phi_1, \ldots, \phi_n) \to \sigma \rangle \quad (n \geq 0)$$

It is possible to group information stated for an expression in a collection of predicates into *intersections* from which any specific one can be selected as demanded by the context in which the expression appears. In particular, an intersection may combine different (even non-unifiable) analyses of the *same* field or method.

Our predicates are *strict* in the sense of [2] since they must describe the outcome of performing an operation in terms of a(nother) *single* operation rather than an intersection. We include a predicate constant for each class, which we can use to type objects when a more detailed analysis of the object's fields and methods is not possible[3]. The predicate constant ω is a *top* (maximal) predicate, assignable to all expressions.

Definition 12 (Subpredicate Relation). The subpredicate relation $\unlhd$ is the smallest preorder satisfying the following conditions:

$$\phi \ \unlhd \ \omega \quad \text{for all } \phi \qquad \phi \cap \psi \ \unlhd \ \phi$$
$$\phi \unlhd \psi \ \& \ \phi \unlhd \psi' \ \Rightarrow \ \phi \unlhd \psi \cap \psi' \qquad \phi \cap \psi \ \unlhd \ \psi$$

We write $\sim$ for the equivalence relation generated by $\unlhd$, extended by

$$\sigma \sim \sigma' \ \Rightarrow \qquad \langle f : \sigma \rangle \ \sim \ \langle f : \sigma' \rangle$$
$$\forall i \in \overline{n}[\phi_i' \sim \phi_i'] \ \& \ \sigma \sim \sigma' \ \Rightarrow \ \langle m : (\phi_1, \ldots, \phi_n) \to \sigma \rangle \ \sim \ \langle m : (\phi_1', \ldots, \phi_n') \to \sigma' \rangle$$

We consider predicates modulo $\sim$; in particular, all predicates in an intersection are different and ω does not appear in an intersection. It is easy to show that $\cap$ is associative, so we write $\sigma_1 \cap \ldots \cap \sigma_n$ (where $n \geq 2$) to denote a general intersection.

---

[3] This may be because the object does not contain any fields or methods (as is the case for Object) or more generally because no fields or methods can be safely invoked.

$$(\text{VAR}) : \cfrac{}{\Pi, x{:}\phi \vdash x{:}\sigma}\ (\phi \trianglelefteq \sigma) \qquad (\text{FLD}) : \cfrac{\Pi \vdash e{:}\langle f{:}\sigma \rangle}{\Pi \vdash e.f{:}\sigma} \qquad (\text{JOIN}) : \cfrac{\Pi \vdash e{:}\sigma_1 \ \ldots \ \Pi \vdash e{:}\sigma_n}{\Pi \vdash e{:}\sigma_1 \cap \ldots \cap \sigma_n}\ (n \geq 2)$$

$$(\omega) : \cfrac{}{\Pi \vdash e{:}\omega} \qquad (\text{INVK}) : \cfrac{\Pi \vdash e{:}\langle m{:}(\overrightarrow{\phi_n}) \to \sigma \rangle \quad \Pi \vdash e_1{:}\phi_1 \ \ldots \ \Pi \vdash e_n{:}\phi_n}{\Pi \vdash e.m(\overrightarrow{e_n}){:}\sigma}$$

$$(\text{NEWO}) : \cfrac{\Pi \vdash e_1{:}\phi_1 \quad \ldots \quad \Pi \vdash e_n{:}\phi_n}{\Pi \vdash \text{new}\ C(\overrightarrow{e_n}){:}C}\ (\mathcal{F}(C) = \overrightarrow{f_n})$$

$$(\text{NEWF}) : \cfrac{\Pi \vdash e_1{:}\phi_1 \quad \ldots \quad \Pi \vdash e_n{:}\phi_n}{\Pi \vdash \text{new}\ C(\overrightarrow{e_n}){:}\langle f_i{:}\sigma \rangle}\ (\mathcal{F}(C) = \overrightarrow{f_n}, i \in \overline{n}, \phi_i = \sigma)$$

$$(\text{NEWM}) : \cfrac{\{\text{this}{:}\psi, x_1{:}\phi_1, \ldots, x_n{:}\phi_n\} \vdash e_b{:}\sigma \quad \Pi \vdash \text{new}\ C(\vec{e}){:}\psi}{\Pi\ \vdash\ \text{new}\ C(\vec{e}){:}\langle m{:}(\overrightarrow{\phi_n}) \to \sigma \rangle}\ (\mathcal{M}b(C,m) = (\vec{x_n}, e_b))$$

Fig. 1. Predicate Assignment for FJ$^\mathcal{e}$

Definition 13 (Predicate Environments). 1. A *predicate statement* is of the form
e:ϕ, where e is called the *subject* of the statement.
2. An environment Π is a set of predicate statements with (distinct) variables as subjects; $\Pi, x{:}\phi$ stands for $\Pi \cup \{x{:}\phi\}$ where x does not appear in Π.
3. If $\overrightarrow{\Pi_n}$ is a sequence of environments, then $\bigcap \overrightarrow{\Pi_n}$ is the environment defined as follows: $x{:}\phi_1 \cap \ldots \cap \phi_m \in \bigcap \overrightarrow{\Pi_n}$ if and only if $\{x{:}\phi_1, \ldots, x{:}\phi_m\}$ is the non-empty set of all statements in the union of the environments that have x as the subject.

We will now define our notion of intersection predicate assignment, which is a slight variant of the system defined in [8]:

Definition 14 (Predicate Assignment). Predicate assignment for FJ$^\mathcal{e}$ is defined by the natural deduction system given in Fig. 1. The rules in fact operate on the larger set of approximants, but for clarity we abuse notation slightly and use the meta-variable e for expressions rather than a. Note that there is no special rule for typing $\bot$, meaning that the only predicate which may be assigned to (a subterm containing) $\bot$ is ω.

The rules of our predicate assignment system are fairly straightforward generalisations of the rules of the strict intersection type assignment system for LC to OO: e.g. (FLD) and (INVK) are analogous to ($\to E$); (NEWF) and (NEWM) are a form of ($\to I$); and (OBJ) can be seen as a universal (ω)-like rule for *objects* only. The only non-standard rule from the point of view of similar work for term rewriting and traditional nominal OO type systems is (NEWM), which derives a predicate for an object that presents an analysis of a method. It makes sense however when viewed as an abstraction introduction rule. Like the corresponding LC typing rule ($\to I$), the analysis involves typing the body of the abstraction (*i.e.* the method body), and the assumptions (*i.e.* requirements) on the formal parameters are encoded in the derived predicate (to be checked on invocation). However, a method body may also make requirements on the *receiver*, through the use of the variable this. In our system we check that these hold *at the same time* as typing the method body (so-called *early self typing*). This checking of requirements on the object itself is where the expressive power of our system resides. If a method calls itself recursively, this recursive call must be checked, but – crucially – carries a *different* predicate if a valid derivation is to be found. Thus only recursive calls which terminate at a certain point (*i.e.* which can be assigned ω, and thus ignored) will be permitted by the system.

As is standard for intersection type assignment systems, our system exhibits both subject reduction *and* subject expansion; the proof is standard.

Theorem 15 (Subject reduction and expansion). Let $e \rightarrow e'$; then $\Pi \vdash e':\phi$ if and only if $\Pi \vdash e:\phi$.

5 Linking Predicates with Semantics: The Approximation Result

We will now describe the relationship between the predicate system and the approximation semantics, which is expressed through an *approximation theorem*: this states that for every predicate-able approximant of an expression, the same predicate can be assigned to the expression itself, and vice-versa: $\Pi \vdash e:\phi \Leftrightarrow \exists A \in \mathcal{A}(e)\,[\Pi \vdash A:\phi]$. As for other systems [3,7], this result is a direct consequence of the strong normalisability of derivation reduction: the structure of the normal form of a given derivation exactly corresponds to the structure of the approximant. As we see below, this implies that predicate-ability provides a sufficient condition for the (head) normalisation of *expressions*, *i.e.* a *termination* analysis for FJ^e; it also immediately puts into evidence that predicate assignment is undecidable.

Since reduction on expressions is *weak*, we need to consider derivation reduction, as in [7]. For lack of space, we will skip the details of this reduction; suffice to say that it is essentially a form of cut-elimination on predicate derivations, defined through the following two basic 'cut' rules:

$$\frac{\dfrac{\dfrac{\mathcal{D}_1}{\Pi \vdash e_1:\phi_1} \quad \cdots \quad \dfrac{\mathcal{D}_n}{\Pi \vdash e_n:\phi_n}}{\Pi \vdash \mathsf{new}\ C(\overrightarrow{e_n}):\langle f_i:\sigma\rangle}}{\Pi \vdash \mathsf{new}\ C(\overrightarrow{e_n}).f_i:\sigma} \quad \rightarrow_{\mathcal{D}} \quad \dfrac{\mathcal{D}_i}{\Pi \vdash e_i:\sigma}$$

$$\frac{\dfrac{\dfrac{\mathcal{D}_b}{\mathtt{this}:\psi, x_1:\phi_1,\ldots,x_n:\phi_n \vdash e_b:\sigma} \quad \dfrac{\mathcal{D}_{\mathrm{self}}}{\Pi \vdash \mathsf{new}\ C(\vec{e}'):\psi}}{\Pi \vdash \mathsf{new}\ C(\vec{e}'):\langle m:(\overrightarrow{\phi_n})\rightarrow\sigma\rangle \quad \vdots \quad \dfrac{\mathcal{D}_1}{\Pi \vdash e_1:\phi_1} \cdots \dfrac{\mathcal{D}_n}{\Pi \vdash e_n:\phi_n}}{\Pi \vdash \mathsf{new}\ C(\vec{e}').m(\overrightarrow{e_n}):\sigma}} \quad \rightarrow_{\mathcal{D}} \quad \dfrac{\mathcal{D}_b^S}{\Pi \vdash e_b^S:\sigma}$$

where $\mathcal{D}_b{}^S$ is the derivation obtained from $\mathcal{D}_b$ by replacing all sub-derivations of the form $\langle \mathrm{VAR}\rangle :: \Pi,x_i:\phi_i \vdash x_i:\sigma$ by (a sub-derivation of[4]) $\mathcal{D}_i$, and sub-derivations of the form $\langle \mathrm{VAR}\rangle :: \Pi,\mathtt{this}:\psi \vdash \mathtt{this}:\sigma$ by (a sub-derivation of) $\mathcal{D}_{\mathrm{self}}$. Similarly, $e_b{}^S$ is the expression obtained from e_b by replacing each variable x_i by the expression e_i, and the variable $\mathtt{this}$ by $\mathsf{new}\ C(\vec{e}')$. This reduction creates exactly the derivation for

---

[4] Note that ϕ_i could be an intersection, containing σ.

a contractum as suggested by the proof of the subject reduction, but is explicit in all its details, which gives the expressive power to show the approximation result.

Notice that sub-derivations of the form $\langle \omega \rangle :: \Pi \vdash e : \omega$ do *not* reduce (although e might) - they are already in normal form with respect to derivation reduction. This is crucial for the strong normalisation result, since it decouples the reduction of a derivation from the possibly infinite reduction sequence of the expression which it assigns a predicate to.

This notion of derivation reduction is not only *sound* (*i.e.* produces valid derivations) but, most importantly, we have that it corresponds to reduction on expressions.

Theorem 16 (Soundness of Derivation Reduction). If $\mathcal{D} :: \Pi \vdash e : \phi$ and $\mathcal{D} \rightarrow_{\mathcal{D}} \mathcal{D}'$, then $\mathcal{D}'$ is a well-defined derivation, in that there exists some e' such that $\mathcal{D}' :: \Pi \vdash e' : \phi$, and $e \rightarrow e'$.

The key step in showing the approximation result is proving that this notion of derivation reduction is terminating, *i.e. strongly normalising*. In other words, all derivations have a *normal form* with respect to $\rightarrow_{\mathcal{D}}$. Our proof uses the well-known technique of *computability* [26]; the formal definition of the $Comp(\mathcal{D})$ predicate is, as standard, defined inductively over the structure of predicates:

Definition 17 (Computability). The set of *computable* derivations is defined as the smallest set satisfying the following conditions (where $Comp(\mathcal{D})$ denotes that $\mathcal{D}$ is a member of the set of computable derivations):

1. $Comp(\langle \omega \rangle :: \Pi \vdash e : \omega)$.
2. $Comp(\mathcal{D} :: \Pi \vdash e : \varphi) \Leftrightarrow SN(\mathcal{D} :: \Pi \vdash e : \varphi)$.
3. $Comp(\mathcal{D} :: \Pi \vdash e : C) \Leftrightarrow SN(\mathcal{D} :: \Pi \vdash e : C)$.
4. $Comp(\mathcal{D} :: \Pi \vdash e : \langle f : \sigma \rangle) \Leftrightarrow Comp(\langle \mathcal{D}, \text{FLD} \rangle :: \Pi \vdash e.f : \sigma)$.
5. $Comp(\mathcal{D} :: \Pi \vdash e : \langle m : (\overrightarrow{\phi_n}) \rightarrow \sigma \rangle) \Leftrightarrow$
   $\forall \overrightarrow{\mathcal{D}_n} [\, \forall i \in \overline{n} \,[Comp(\mathcal{D}_i :: \Pi_i \vdash e_i : \phi_i)] \Rightarrow$
   $Comp(\langle \mathcal{D}', \mathcal{D}'_1, \ldots, \mathcal{D}'_n, \text{INVK} \rangle :: \Pi' \vdash e.m(\overrightarrow{e_n}) : \sigma)\,]$

   where $\mathcal{D}' = \mathcal{D}[\Pi' \lessdot \Pi]$ and $\mathcal{D}'_i = \mathcal{D}_i[\Pi' \lessdot \Pi_i]$ for each $i \in \overline{n}$ with $\Pi' = \bigcap \Pi \cdot \overrightarrow{\Pi_n}$, and $\mathcal{D}[\Pi' \lessdot \Pi]$ denotes a derivation of exactly the same shape as $\mathcal{D}$ in which the environment Π is replaced with Π' in each statement of the derivation.
6. $Comp(\langle \mathcal{D}_1, \ldots, \mathcal{D}_n, \text{JOIN} \rangle :: \Pi \vdash e : \sigma_1 \cap \ldots \cap \sigma_n) \Leftrightarrow \forall i \in \overline{n} \,[Comp(\mathcal{D}_i)]$.

As can be expected, we show that computable derivations are strongly normalising, and that all valid derivations are computable.

Theorem 18. *1.* $Comp(\mathcal{D} :: \Pi \vdash e : \phi) \Rightarrow SN(\mathcal{D} :: \Pi \vdash e : \phi)$.
2. $\mathcal{D} :: \Pi \vdash e : \phi \Rightarrow Comp(\mathcal{D} :: \Pi \vdash e : \phi)$

Then the key step to the approximation theorem follows directly.

Theorem 19 (Strong Normalisation). If $\mathcal{D} :: \Pi \vdash e : \phi$ then $SN(\mathcal{D})$.

Finally, the following two properties of approximants and predicate assignment lead to the approximation result itself.

Lemma 20. *1. If $\mathcal{D}::\Pi\vdash a:\phi$ and $a\sqsubseteq a'$ then there exists a derivation $\mathcal{D}'::\Pi\vdash a':\phi$.*
  2. If $\mathcal{D}::\Pi\vdash e:\phi$ and $\mathcal{D}$ is in normal form with respect to $\rightarrow_{\mathcal{D}}$, then there exists A and $\mathcal{D}'$ such that $A\sqsubseteq e$ and $\mathcal{D}'::\Pi\vdash A:\phi$.

The first of these two properties simply states the soundness of predicate assignment with respect to the approximation relation. The second is the more interesting, since it expresses the relationship between the structure of a derivation and the approximant. The derivation $\mathcal{D}'$ is constructed from $\mathcal{D}$ by replacing sub-derivations of the form $\langle\omega\rangle::\Pi\vdash e:\omega$ by $\langle\omega\rangle::\Pi\vdash\bot:\omega$ (thus covering any redexes appearing in e). Since $\mathcal{D}$ is in normal form, there are also no redexes that carry a non-trivial predicate, ensuring that the expression in the conclusion of $\mathcal{D}'$ is a (normal) approximant. The 'only if' part of the approximation result itself then follows easily from the fact that $\rightarrow_{\mathcal{D}}$ corresponds to reduction of expressions, so A is also an *approximant* of e. The 'if' part follows from the first property above and subject expansion.

Theorem 21 (Approximation). $\Pi\vdash e:\phi$ iff there exists $A\in\mathcal{A}(e)$ such that $\Pi\vdash A:\phi$.

In other intersection type systems [3,7], the approximation theorem underpins characterisation results for various forms of termination. Like the LC (and in contrast to the system in [7] for TRS) our predicate system gives a *full characterisation* of normalisability. So predicate-ability gives a guarantee of termination since our normal approximate forms of Definition 5 correspond in structure to expressions in (head) normal form.

Definition 22 ((Head-)Normal Forms). 1. The set of expressions in *head-normal form* (ranged over by H) is defined by:

$$H ::= x\mid\text{new }C(\vec{e})\mid H.f\mid H.m(\vec{e})\quad(H\neq\text{new }C(\vec{e}))$$

2. The set of expressions in *normal form* (ranged over by N) is defined by:

$$N ::= x\mid\text{new }C(\vec{N})\mid N.f\mid N.m(\vec{N})\quad(N\neq\text{new }C(\vec{N}))$$

Notice that the difference between these two notions sits in the second and fourth alternative, where head-normal forms allow arbitrary expressions to be used.

Lemma 23. *1. If $A\neq\bot$ and $A\sqsubseteq e$, then e is a head-normal form.*
  2. If $A\sqsubseteq e$ and A does not contain $\bot$, then e is a normal form.

Thus any predicate, or, more accurately, any predicate derivation other than those of the form $\langle\omega\rangle::\Pi\vdash e:\omega$ (which correspond to the approximant $\bot$) specifies the structure of a (head) normal form via the normal form of its derivation.

Definition 24. 1. A derivation is *strong* if it contains no instances of the rule (ω).
  2. If the only instances of the (ω) rule in a derivation are those typing the arguments to method invocations, then we say it is ω-*safe* .
  3. For a predicate environment Π, if for all $x:\phi\in\Pi$ either $\phi=\omega$ or ϕ does not contain ω at all, then we say Π is ω-safe.

From the approximation result, the following normalisability guarantees are easily achieved.

Theorem 25 (Normalisation).   1. $\Pi \vdash e : \sigma$ if and only if e has a head-normal form.

2. $\mathcal{D} :: \Pi \vdash e : \sigma$ with ω-safe $\mathcal{D}$ and Π only if e has a normal form.

3. $\mathcal{D} :: \Pi \vdash e : \sigma$ with $\mathcal{D}$ strong if and only if e is strongly normalisable.

Notice that we currently do not have an 'if and only if' result for Theorem 25(2), whereas terms with normal forms *can* be completely characterised in LC. This is because derivation expansion does not preserve ω-safety in general. To see why this is the case consider that while an ω-safe derivation may exist for $\Pi \vdash e_i : \sigma$, no ω-safe derivation may exist for $\Pi \vdash$ new $\mathsf{C}(\overrightarrow{e_n}) . f_i : \sigma$ (due to non-termination in the other expressions e_j) even though this expression has the same normal form as e_i.

6   Expressivity

In this section we consider the formal expressivity of our OO calculus and predicate system. We show that $\mathsf{FJ}^\wp$ is Turing complete by considering an encoding of Combinatory Logic (CL). Through the approximation result of the previous section all normal forms of the CL program can be assigned a non-trivial predicate in our system. Thus, we have a predicate-based characterisation of all (terminating) computable functions in OO.

Combinatory Logic is a model of computation defined by H. Curry [16] independently of LC. It defines a higher-order term rewriting system over of the function symbols $\{\mathbf{S}, \mathbf{K}\}$ and the following rewrite rules:

$$\mathbf{K}\,x\,y \;\;\rightarrow\;\; x$$
$$\mathbf{S}\,x\,y\,z \;\;\rightarrow\;\; x\,z\,(y\,z)$$

Our encoding of CL in $\mathsf{FJ}^\wp$ is based on a *curryfied first-order version* of the system above (see [6] for details), where the rules for $\mathbf{S}$ and $\mathbf{K}$ are expanded so that each new rewrite rule has a *single* operand, allowing for the partial application of function symbols. Application, the basic engine of reduction in term rewriting systems, is modelled via the invocation of a method named app belonging to a Combinator *interface*. Since we do not have interfaces proper in $\mathsf{FJ}^\wp$, we have defined a Combinator class but left the body of the app method unspecified to indicate that in a full-blown Java program this would be an interface. The reduction rules of curryfied CL each apply to (or are 'triggered' by) different 'versions' of the $\mathbf{S}$ and $\mathbf{K}$ combinators; in our encoding these rules are implemented by the bodies of five different versions of the app method which are each attached to different subclasses (*i.e.* different versions) of the Combinator class.

Definition 26. The encoding of Combinatory Logic (CL) into the $\mathsf{FJ}^\wp$ program OOCL (Object-Oriented CL) is defined using the class table in Figure 2 and the function $[\![\cdot]\!]$ which translates terms of CL into $\mathsf{FJ}^\wp$ expressions, and is defined as follows:

$$[\![x]\!] \;=\; x \qquad\qquad [\![t_1 t_2]\!] \;=\; [\![t_1]\!] . \mathsf{app}([\![t_2]\!])$$
$$[\![\mathbf{K}]\!] \;=\; \text{new } \mathsf{K}_1() \qquad [\![\mathbf{S}]\!] \;=\; \text{new } \mathsf{S}_1()$$

The reduction behaviour of OOCL mirrors that of CL.

Theorem 27. *For CL terms* $t_1, t_2: t_1 \rightarrow^* t_2$ *if and only if* $[\![t_1]\!] \rightarrow^* [\![t_2]\!]$.

```
class Combinator extends Object {
      Combinator app(Combinator x) { return this; } }
class K₁ extends Combinator {
      Combinator app(Combinator x) { return new K₂(x); } }
class K₂ extends K₁ { Combinator x;
      Combinator app(Combinator y) { return this.x; } }
class S₁ extends Combinator {
      Combinator app(Combinator x) { return new S₂(x); } }
class S₂ extends S₁ { Combinator x;
      Combinator app(Combinator y) { return new S₃(this.x, y); } }
class S₃ extends S₂ { Combinator y;
      Combinator app(Combinator z) {
          return this.x.app(z).app(this.y.app(z)); } }
```

Fig. 2. The class table for Object-Oriented Combinatory Logic (OOCL) programs

Given the Turing completeness of CL, this result shows that our model of class-based OO is also Turing complete. Although this certainly does not come as a surprise, it is a nice formal property for our calculus to have. In addition, our predicate system can perform the same 'functional' analysis as ITD does for LC and CL. This is illustrated by a *type preservation* result. We focus on Curry's type system for CL and show we can give equivalent types to OOCL programs.

Definition 28 (Curry Types). The set of *simple types* is defined by the grammar:

$$\tau ::= \varphi \mid \tau \to \tau$$

Definition 29 (Curry Type Assignment for CL). 1. A *basis* B is a set of statements of the form $x{:}\tau$ in which each of the variables x is distinct.
2. Simple types are assigned to CL-term using the following natural deduction system:

$$(\text{VAR}): \frac{}{\text{B} \vdash_{\text{CL}} x{:}\tau} \ (x{:}\tau \in \text{B}) \qquad (\to\text{E}): \frac{\text{B} \vdash_{\text{CL}} t_1{:}\tau \to \tau' \quad \text{B} \vdash_{\text{CL}} t_2{:}\tau}{\text{B} \vdash_{\text{CL}} t_1 t_2{:}\tau'}$$

$$(\text{K}): \frac{}{\text{B} \vdash_{\text{CL}} \text{K}{:}\tau \to \tau' \to \tau} \qquad (\text{S}): \frac{}{\text{B} \vdash_{\text{CL}} \text{S}{:}(\tau \to \tau' \to \tau'') \to (\tau \to \tau') \to \tau \to \tau''}$$

To show type preservation, we first define what the equivalent of Curry's types are in terms of predicates.

Definition 30 (Type Translation). The function $\llbracket \cdot \rrbracket$, which transforms Curry types into predicates[5], is defined as follows:

$$\begin{aligned} \llbracket \varphi \rrbracket &= \varphi \\ \llbracket \tau \to \tau' \rrbracket &= \langle \text{app} : \llbracket \tau \rrbracket \to \llbracket \tau' \rrbracket \rangle \end{aligned}$$

It is extended to bases by: $\llbracket \text{B} \rrbracket = \{ x{:}\llbracket \tau \rrbracket \mid x{:}\tau \in \text{B} \}$.

---

[5] Note we have *overloaded* the notation $\llbracket \cdot \rrbracket$, which we also use for the translation of CL terms to FJ$^{\wp}$ expressions.

We can now show the following type preservation result.

Theorem 31 (Preservation of Types). *If* $B \vdash_{CL} t{:}\tau$ *then* $\llbracket B \rrbracket \vdash \llbracket t \rrbracket : \llbracket \tau \rrbracket$.

Furthermore, since the well-known encoding of the LC into CL preserves typeability, we also have a type-preserving encoding of LC into $FJ^{\mathcal{L}}$; it is straightforward to extend this preservation result to full-blown strict intersection types. We stress that this result really demonstrates the validity of our approach. Indeed, our predicate system actually has more power than intersection type systems for CL, since there not all normal forms are typeable using strict types, whereas in our system they are.

Lemma 32. *If* e *is a* $\perp$-*free approximate normal form of* OOCL, *then there are* ω-*safe* $\mathcal{D}$ *and* Π *and strict predicate* σ *such that* $\mathcal{D} :: \Pi \vdash e{:}\sigma$.

Since our system has a subject expansion property (and ω-safe typeability is preserved under expansion for the images of CL terms in OOCL), this leads to a complete characterisation of termination for OOCL.

Theorem 33. *Let* e *be an expression such that* $e = \llbracket t \rrbracket$ *for some* CL *term* t; *then* e *has a normal form if and only if there are* ω-*safe* $\mathcal{D}$ *and* Π *and strict predicate* σ *such that* $\mathcal{D} :: \Pi \vdash e{:}\sigma$.

7   Some Observations

In this paper we have shown how the ITD approach can be applied to class-based OO, preserving the main expected properties of intersection type systems. There are however some notable differences between our type system and previous work on LC and TRS upon which our research is based.

Firstly, we point out that when considering the encoding of CL (and via that, LC) in $FJ^{\mathcal{L}}$, our system provides *more* than the traditional analysis of terms as functions: there are untypeable LC and CL terms which have typeable images in OOCL. Let δ be the following CL term: **S (S K K) (S K K)**. Notice that $\delta \delta \rightarrow^* \delta \delta$, *i.e.* it is unsolvable, and thus can only be given the type ω (this is also true for $\llbracket \delta \delta \rrbracket$). Now, consider the term $t = $ **S (K δ) (K δ)**. Notice that it is a normal form ($\llbracket t \rrbracket$ has a normal form also), but that for any term t', **S (K δ) (K δ)** $t' \rightarrow^* \delta \delta$. In a strict system, no functional analysis is possible for t since $\phi \rightarrow \omega$ is not a type and so the only way we can type this term is using ω[6]. In our type system however we may assign several forms of predicate to $\llbracket t \rrbracket$. Most simply we can derive $\emptyset \vdash \llbracket t \rrbracket : S_3$, but even though a 'functional' analysis via the app method is impossible, it is still safe to access the fields of the value resulting from $\llbracket t \rrbracket$ – both $\emptyset \vdash \llbracket t \rrbracket : \langle x{:}K_2 \rangle$ and $\emptyset \vdash \llbracket t \rrbracket : \langle y{:}K_2 \rangle$ are also easily derivable statements. In fact, we can derive even more informative types: the expression $\llbracket K \, \delta \rrbracket$ can be assigned predicates of the form $\sigma_{K\delta} = \langle \mathsf{app} : (\sigma_1) \rightarrow \langle \mathsf{app} : (\sigma_2 \cap \langle \mathsf{app} : (\sigma_2) \rightarrow \sigma_3 \rangle) \rightarrow \sigma_3 \rangle \rangle$, and so we can also assign $\langle x{:}\sigma_{K\delta} \rangle$ and $\langle y{:}\sigma_{K\delta} \rangle$ to $\llbracket t \rrbracket$. Notice that the equivalent λ-term to t is $\lambda y.(\lambda x.xx)(\lambda x.xx)$, which is a *weak* head normal form without a (head) normal form.

---

[6] In other intersection type systems (*e.g.* [11]) $\phi \rightarrow \omega$ is a permissible type, but is equivalent to ω (that is $\omega \leq (\phi \rightarrow \omega) \leq \omega$) and so semantics based on these type systems identify terms of type $\phi \rightarrow \omega$ with unsolvable terms.

The 'functional' view is that such terms are observationally indistinguishable from un-solvable terms. When encoded in $FJ^{\mathfrak{e}}$ however, our type system shows that these terms become meaningful (head-normalisable).

The second observation concerns *principal* types. In the LC, each normal form has a *unique* most-specific type: *i.e.* a type from which all the other assignable types may be generated. This property is important for practical type *inference*. Our intersection type system for $FJ^{\mathfrak{e}}$ does not have such a property. Consider the following program: class C extends Object {C m() {return new C();}}. The expression new C() is a normal form, and so we can assign it a non-trivial predicate, but observe that the set of all predicates which may be assigned to this expression is the *infinite* set $\{C, \langle m : () \rightarrow C \rangle, \langle m : () \rightarrow \langle m : () \rightarrow C \rangle \rangle, \ldots\}$. None of these types may be considered the *most* specific one, since whichever predicate we pick we can always derive a more informative (larger) one. On the one hand, this is exactly what we want: we may make a series of any finite number of calls to the method m and this is expressed by the predicates. On the other hand, this seems to preclude the possibility of practical type inference for our system. Notice however that these predicates are not unrelated to one another: they each approximate the 'infinite' predicate $\langle m : () \rightarrow \langle m : () \rightarrow \ldots \rangle \rangle$, which can be finitely represented by the recursive type $\mu X . \langle m : () \rightarrow X \rangle$. This type concisely captures the reduction behaviour of new C(), showing that when we invoke the method m on it we again obtain our original term. In LC such families of types arise in connection with fixed point operators. This is not a coincidence: the class C was *recursively* defined, and in the face of such self-reference it is not then suprising that this is reflected in our type analysis.

8   Conclusions and Future Work

We have considered an approximation-based denotational semantics for class-based OO programs and related this to a predicate-based semantics defined using an intersection type approach. Our work shows that the techniques and strong results of this approach can be transferred straightforwardly from other programming formalisms (*i.e.* LC and term rewriting systems) to the OO paradigm. Through characterisation results we have shown that our predicate system is powerful enough (at least in principle) to form the basis for expressive analyses of OO programs.

Our work has also highlighted where the OO programming style differs from its functional cousin. In particular we have noted that because of the OO facility for *self-reference*, it is no longer the case that all normal forms have a most-specific (or principal) type. The types assignable to such normal forms do however seem to be representable using recursive definitions. This observation futher motivates and strengthens the case (by no means a new concept in the analysis of OO) for the use of recursive types in this area. Some recent work [22] shows that a restricted but still highly expressive form of recursive types can still characterise strongly normalising terms, and we hope to fuse this approach with our own to come to an equally precise but more concise and practical predicate-based treatment of OO.

We would also like to reintroduce more features of full Java back into our calculus, to see if our system can accommodate them whilst maintaining the strong theoretical properties that we have shown for the core calculus. For example, similar to $\lambda\mu$ [23], it seems natural to extend our simply typed system to analyse the exception handling features of Java.

References

1. Abadi, M., Cardelli, L.: A Theory of Objects. Springer, Heidelberg (1996)
2. van Bakel, S.: Intersection Type Assignment Systems. TCS 151(2), 385–435 (1995)
3. van Bakel, S.: Cut-Elimination in the Strict Intersection Type Assignment System is Strongly Normalising. NDJFL 45(1), 35–63 (2004)
4. van Bakel, S.: Completeness and Partial Soundness Results for Intersection & Union Typing for $\lambda\mu\tilde{\mu}$. APAL 161, 1400–1430 (2010)
5. van Bakel, S., de'Liguoro, U.: Logical equivalence for subtyping object and recursive types. ToCS 42(3), 306–348 (2008)
6. van Bakel, S., Fernández, M.: Normalisation Results for Typeable Rewrite Systems. IaC 2(133), 73–116 (1997)
7. van Bakel, S., Fernández, M.: Normalisation, Approximation, and Semantics for Combinator Systems. TCS 290, 975–1019 (2003)
8. van Bakel, S., Rowe, R.: Semantic Predicate Types for Class-based Object Oriented Programming. In: FTfJP 2009 (2009)
9. Banerjee, A., Jensen, T.P.: Modular Control-Flow Analysis with Rank 2 Intersection Types. MSCS 13(1), 87–124 (2003)
10. Barendregt, H.: The Lambda Calculus: its Syntax and Semantics. North-Holland, Amsterdam (1984)
11. Barendregt, H., Coppo, M., Dezani-Ciancaglini, M.: A filter lambda model and the completeness of type assignment. JSL 48(4), 931–940 (1983)
12. Cardelli, L., Mitchell, J.C.: Operations on Records. MSCS 1(1), 3–48 (1991)
13. Cardelli, L.: A Semantics of Multiple Inheritance. IaC 76(2/3), 138–164 (1988)
14. Coppo, M., Dezani-Ciancaglini, M.: An Extension of the Basic Functionality Theory for the λ-Calculus. NDJFL 21(4), 685–693 (1980)
15. Coppo, M., Dezani-Ciancaglini, M., Venneri, B.: Functional characters of solvable terms. Zeitschrift für Mathematische Logik und Grundlagen der Mathematik 27, 45–58 (1981)
16. Curry, H.B.: Grundlagen der Kombinatorischen Logik. AJM 52, 509–536, 789–834 (1930)
17. Damiani, F., Prost, F.: Detecting and Removing Dead-Code using Rank 2 Intersection. In: Giménez, E. (ed.) TYPES 1996. LNCS, vol. 1512, pp. 66–87. Springer, Heidelberg (1998)
18. Fisher, K., Honsell, F., Mitchell, J.C.: A lambda Calculus of Objects and Method Specialization. NJ 1(1), 3–37 (1994)
19. Igarashi, A., Pierce, B.C., Wadler, P.: Featherweight Java: a minimal core calculus for Java and GJ. ACM Trans. Program. Lang. Syst. 23(3), 396–450 (2001)
20. Jensen, T.P.: Types in Program Analysis. In: Mogensen, T.Æ., Schmidt, D.A., Sudborough, I.H. (eds.) The Essence of Computation. LNCS, vol. 2566, pp. 204–222. Springer, Heidelberg (2002)
21. Mitchell, J.C.: Type Systems for Programming Languages. In: Handbook of TCS, vol. B, ch. 8, pp. 415–431 (1990)
22. Nakano, H.: A Modality for Recursion. In: LICS, pp. 255–266 (2000)
23. Parigot, M.: An algorithmic interpretation of classical natural deduction. In: Voronkov, A. (ed.) LPAR 1992. LNCS, vol. 624, pp. 190–201. Springer, Heidelberg (1992)
24. Plotkin, G.D.: The origins of structural operational semantics. JLAP 60-61, 3–15 (2004)

25. Scott, D.: Domains for Denotational Semantics. In: Nielsen, M., Schmidt, E.M. (eds.) ICALP 1982. LNCS, vol. 140, pp. 577–613. Springer, Heidelberg (1982)
26. Tait, W.W.: Intensional interpretation of functionals of finite type I. JSL 32(2), 198–223 (1967)
27. Wadsworth, C.P.: The relation between computational and denotational properties for Scott's D_∞-models of the lambda-calculus. SIAM J. Comput. 5, 488–521 (1976)

Author Index